MODELING AND SIMULATION TOOLS FOR EMERGING TELECOMMUNICATION NETWORKS

MODELING AND SIMULATION TOOLS FOR EMERGING TELECOMMUNICATION NETWORKS

Needs, Trends, Challenges and Solutions

Edited by

A. Nejat Ince
Istanbul Technical University

Ercan Topuz
Istanbul Technical University

 Springer

A. Nejat Ince
Istanbul Technical University

Ercan Topuz
Istanbul Technical University

ISBN-10: 1-4614-9804-X
ISBN-13: 978-1-4614-9804-9
Printed on acid-free paper.

ISBN-10: 0-387-34167-6
ISBN-13: 978-0387-34167-6

9 8 7 6 5 4 3 2 1

Spin: 11611622

springer.com

ACKNOWLEDGEMENTS

As Chairman of COST Action 285 and co-editor of this book I wish to express my sincere thanks to all the members of the Management Committee for their full and active participation in the studies embraced by the Action including the decision to sponsor this symposium in which they presented the results of their individual research in different aspects of modeling and simulation of communication networks. I would particularly like to mention here Prof Dr Axel Lehmann and Prof Dr Ercan Topuz who, as Deputy Chair and Technical Secretary respectively of the Action 285, made unique contributions to the organisation and success of the Symposium.

The symposium was very much enriched and gained much breadth and depth by the participation of many experts in the field from outside the Action Group, from the United States of America, and Europe who willingly accepted our invitation to attend and contribute to our deliberations. It would be invidious to single out names but I would like to mention Dr Arnold Bragg who played a very important role inside the Committee as well as in the preparation and conduct of the symposium. I owe them all many thanks and much gratitude.

Last but by no means least I would like to express my appreciation to the COST Office Scientific Secretariat for the administrative and financial support given to the Action and to Prof Dr Ulf Schmerl for making the facilities and staff of his Faculty of Informatics at the University of the German Federal Armed Forces in Munich, available for the symposium. Finally It gives me pleasure to acknowledge the support I received from Mr Zerhan Ener and Mr Semih Ener as well as from the staff of Springer Publishers in the production of this book.

Nejat Ince

PREFACE

The papers which appear in this book were written by their authors based on their presentations made at a symposium hosted by The Fakultaet für Informatik of Universitaet der Bunderswehr München on 8-9 September 2005. The symposium was organised under the eagis of COST Telecommunications Action 285 entitled :

Modeling and Simulation Tools for Research in Emerging Multiservice Telecommunications
Needs, Trends, Challenges, and Solutions

COST- the acronym for European **CO**operation in the field of **S**cientific and **T**echnical research is the oldest and widest European intergovernmental network for cooperation in research. Established by the Ministerial Conference in November 1971, COST is presently used by the scientific communities of 35 European countries to cooperate in common research projects supported by national funds.

The funds provided by COST- less than 1% of the total value of the projects- supported the COST cooperation networks (COST Actions) through which, with only around 20 million Euro per year, more than 30,000 European scientists are involved in research having a total value which exceeds 2 billion Euro per year. This is the financial worth of the European added value which COST achieves.

A "bottom up approach" (the initiative of launching a COST Action comes from the European scientists themselves), "a la carte participation" (only countries interested in the Action participate), "equality of access" (participation is open also to the scientific communities of countries not belonging to the European Union) and "flexible structure" (easy implementation and light management of the research initiatives) are the main characteristics of COST.

As precurser of advanced multidisciplinary research COST plays a very important role in the realisation of the European Research Area (ERA) anticipating and complementing the activities of the Framework

Programmes, constituting a "bridge" towards the scientific communities of emerging countries , increasing the mobility of researchers across Europe and fostering the establishment of "Network of Excellence" in many key scientific domains such as : Physics, Chemistry, Telecommunications and `information Science, Nanotechnologies, Meteorology, Environment, Medicine and Health, Forests, Agriculture and Social Sciences. It covers basic and more applied research and also addresses issues of pre-normative nature or societal importance.

Currently there are some twenty actions in the Telecommunications and Information Science and Technology area one of which is COST Action 285. The main objective of this action is to enhance existing tools and develop new modeling and simulation tools for research in emerging multiservice telecommunications networks in the areas of :

- Model Performance Improvements,
- Multilayer Traffic Modeling,

and
- The important issue of evaluation and validation of the new modeling tools.

The studies related to the above activities are carried out by members of the Action Group , with inputs from invited experts/scientists from academia and industry when deemed necessary, and are coordinated at the Management Committee Meetings (MCM)held two or three times a year. Members participate in other related projects and activities nationally and internationally (e.g. COST, IST, ITU, ETSI, ATM Forum) provide opportunities for formal/informal contacts and for dissemination of results.

The Management Committee for COST Action 285 consists of:

Chairman	: Prof Dr Nejat Ince (TR)
Deputy Chairman	: Prof Dr Axel Lehmann (D)
Technical Secretary	: Prof Dr Erçan Topuz (TR)
Other Members	: There are up to two representatives from Bulgaria, Denmark,France, Germany, Hungary, Ireland, Italy, Macedonia, Malta, Norway, Slovenia, Spain, Turkey,Switzerland, and The United Kingdom.

The Management Committee decided early in the year 2004 to invite external experts/scientists, specialising on the subjects of interest to Action 285, from other COST Actions, software houses, telecommunications companies, universities and government research institutions of not only the COST Countries but also of other continents. A letter of invitation was sent out to known experts and institutions to participate in a symposium with the major aim of harnessing ideas and proposals for improved and new languages and tools to enable network designers, developers and operators to model and simulate networks and services of emerging and future telecommunications systems.

From the papers submitted for presentation at the symposium the text of twenty four of them were selected for inclusion in this book. The symposium presentations were made in four sessions as follows;

Session 1 : Multilayer Traffic and Multimedia Behaviour,
Session 2 : Quality of Simulations,
Session 3 : Accelerated Simulation Methods,
Session 4 : Verification, Validation and Credibility of Simulations.

The contributors and their coordinates are given in the list herewith attached.

The symposium covered a wide spectrum of subjects dealing coherently with nearly all the important aspects of simulation modeling and tools for the design and performance evaluation techniques and systems particularly the emerging ones.

It is hoped and expected that this book , which is the proceedings of the symposium, will be found useful as a reference work for practicing engineers and academic researchers.

Nejat Ince,
Ankara.

TABLE OF CONTENTS

ALPHABETICAL LIST OF CONTRIBUTORS

Prof. Osman Balci
Department of Computer Science
660 McBryde Hall, MC 0106
Virginia Tech Blacksburg,
Virginia 24061, USA
http://manta.cs.vt.edu/balci/
balci@vt.edu

Ilia Baldine
Center for Advanced Network
Research RTI International, Inc.
Box 12194 Research Triangle
Park, NC 27709 USA

Dr. Markus Becker
University of Bremen
Communication Networks, Otto-
Hahn-Allee - NW1 28359
Bremen, Germany
mab@comnets.uni-bremen.de

Dr. Dirk Brade
Kungl Tekniska Högskolan
Department for Electronics and
Computer Science 10044
Stockholm
dirk.brade@eads.com

Dr. Arnold Bragg
Center for Advanced Network
Research RTI International, Inc.
Box 12194 Research Triangle
Park, NC 27709 USA
abragg@rti.org

Dr. Cristian Bungarzeanu
EPFL STI-TCOM, Station 11,
CH-1015 Lausanne, Switzerland
cristian.bungarzeanu@epfl.ch

Prof. Miguel Calvo
Universidad Politécnica de
Madrid. ETSI Telecomunicación.
C. Universitaria s/n. 28040
Madrid, Spain
miguel.calvo@gr.ssr.upm.es

Dr. Giovanna Carofiglio
Dipartimento di Elettronica
Politecnico di Torino, Italy
carofiglio@mail.tlc.polito.it

Prof. Leandro de Haro
Universidad Politécnica de
Madrid. ETSI Telecomunicación.
C. Universitaria s/n. 28040
Madrid, Spain
leandro@gr.ssr.upm.es

Mike Devetsikiotis
Electrical & Computer
Engineering, North Carolina
State University Raleigh, NC
27695 USA

Prof. Richard Fujimoto
College of Computing, Georgia
Institute of Technology Atlanta,
Georgia 30332-0280
fujimoto@cc.gatech.edu

Dr. Ivan Ganchev
MIEEE Deputy Director,
Telecom Research Centre ECE
Department, University of
Limerick, Limerick, IRELAND
Ivan.ganchev@ul.ie

F. Javier García
Universidad Politécnica de
Madrid. ETSI Telecomunicación.
C. Universitaria s/n. 28040
Madrid, Spain

Laura García
Universidad Politécnica de
Madrid. ETSI Telecomunicación.
C. Universitaria s/n. 28040
Madrid, Spain
lgg@gr.ssr.upm.es

Dr. Michele Garetto
Dipartimento di Elettronica
Politecnico di Torino, Italy
garetto@mail.tlc.polito.it

Dr. Paola Giaccone
Dipartimento di Elettronica
Politecnico di Torino, Italy
giaccone@mail.tlc.polito.it

Prof. Carmelita Görg
Communication Networks,
University of Bremen, Otto-
Hahn-Alle - NW1, 28359
Bremen, Germany
cg@comnets.uni-bremen.de

Randall Guensler
School of Civil and
Environmental Engineering,
Georgia Institute of Technology
Atlanta, Georgia 30332-0355

Dr. Dan He
Centre for Communication
System Research, University of
Surrey Guildford, Surrey
United Kingdom
D.He@surrey.ac.uk

Michael Hunter
School of Civil and
Environmental Engineering,
Georgia Institute of Technology
Atlanta, Georgia 30332-0355

Prof. Nejat Ince
FIEEE, Member of the
International Academy of
Astronautics, Paris, Member of
the Russian Academy of
Technological Sciences, Member
of the New York Academy of
Sciences, Istanbul Technical
University Istanbul, Turkey
nejatince@ttnet.net.tr

Prof. Villy. B. Iversen
COM.DTU, Technical
University of Denmark, DK-
2800 Kgs. Lyngby, Denmark
vbi@com.dtu.dk

Dr. Rachid El Abdouni Khayari
Universitaet der Bundeswehr München, Deutschland Institut für Technische Informatik, WernerHeisenberg Weg 39, 85577 Neubiberg, Deutschland
rachid@informatik.unibw-muenchen.de

Dr. Gerta Köster
Siemens AG-Corporate Technology, Otto-Hahn-Ring 6 D-81730 München
gerta.koester@siemens.com

Nino Kubinidze
ECE Department, University of Limerick, Limerick, IRELAND

Dr. Marko Lackovic
Ericsson Research & Development Center, Krapinska 45, HR-10000 Zagreb, Croatia
marko.lackovic@ericsson.com

Prof. Axel Lehmann
Universitaet der Bundeswehr München, Deutschland Institut für Technische Informatik, WernerHeisenberg Weg 39, 85577 Neubiberg, Deutschland
lehmann@informatik.unibw-muenchen.de

Prof. Emilio Leonardi
Dipartimento di Elettronica Politecnico di Torino, Italy
leonardi@mail.tlc.polito.it

Prof. Marko Ajmone Marsan
Dipartimento di Elettronica Politecnico di Torino, Italy
ajmone@mail.tlc.polito.it

Ramón Martínez
Universidad Politécnica de Madrid. ETSI Telecomunicación. C. Universitaria s/n. 28040 Madrid, Spain

Alberto Martínez
Universidad Politécnica de Madrid. ETSI Telecomunicación. C. Universitaria s/n. 28040 Madrid, Spain

Dr. Máirtín O'Droma
FIEE, SMIEEE Director, Telecoms Research Centre, ECE Department, University of Limerick, Limerick, Ireland
martin.odroma@ul.ie

William F. Ormsby
Naval Surface Warfare Center Dahlgren Division, Code T12 17320 Dahlgren Road Dahlgren, Virginia 22448, USA
william.f.ormsby@navy.mil

Ms. Selin Parlar
Istanbul Technical University Maslak, Istanbul, Turkey
selin@cscrs.itu.edu.tr

Harry Perros
Computer Science, North Carolina State University Raleigh, NC 27695 USA

Dr. Siegfried Pohl
Institute for Technology of
Intelligent Systems (ITIS e.V.)
Werner-Heisenberg-Weg 39 D-
85577 Neubiberg, Germany
Siegfrie.pohl@unibw.de

Dr. Stoyan A. Poryazov
Institute of Mathematics and
Informatics, Acad. G. Bonchev
Str. Block 8 1113 Sofia, Bulgarie
stoyan@cc.bas.bg

Dr. Francesco Potortì
ISTI-CNR, Pisa (IT)
Potorti@isti.cnr.it

Dr. Matevz Pustišek
University of Ljubljana (SI)
matevz.pustisek@fe.uni-lj.si

S.V. Raghavan
Department of Computer Science
and Engineering, Indian Institute
of Technology, Madras,
Chennai, India

Ignacio Álvarez Salcidos
Dpt. Señales, Sistemas y
Radiocomunicaciones.
Universidad Politécnica de
Madrid

Dr. Augustine Samba
Department of Computer Science
Kent State University, 44242,
Kent OH USA
asamba@cs.kent.edu

Manuel García-Sánchez
Dpt. Teoría de la Señal y
Comunicaciones. Universidad de
Vigo.

Ass. Prof. Emiliya T. Saranova
College of Telecommunications
and Post, Studentski grad.1 Acad
Stefan Mladenov Str 1700 Sofia,
Bulgaria.

Dragan Savić
University of Ljubljana (SI)
dragan.savic@ltfe.org

Dr. Enrico Schiattarella
Dipartimento di Elettronica
Politecnico di Torino, Italy
schiattarella@mail.tlc.polito.it

Prachee Sharma
Department of Computer Science
and Engineering, University at
Buffalo, SUNY, USA

Prof. Marcus Siegle
Universitaet der Bundeswehr
München, Deutschland Institut
für Technische Informatik
WernerHeisenberg Weg 39,
85577 Neubiberg, Deutschland
siegle@informatik.unibw-
muenchen.de

Prof. S.N. Stepanov
Sistema Telecom 125047,
Moscow, 1st Tverskay-
Yamskaya 5 Russia
stepanov@sistel.ru

Dan Stevenson
Center for Advanced Network
Research RTI International, Inc.
Box 12194 Research Triangle
Park, NC 27709 USA

Dr. Zhili Sun
Centre for Communication
System Research, School of
Electronics and physical
Sciences, University of Surrey
Guildford, Surrey, GU2 7XH
United Kingdom
Z.Sun@surrey.ac.uk

Dr. Alberto Tarello
Dipartimento di Elettronica
Politecnico di Torino, Italy
tarello@mail.tlc.polito.it

Prof. Ercan Topuz
Istanbul Technical University
Maslak, Istanbul, Turkey
topuz@ehb.itu.edu.tr

Prof. Satish K. Tripathi
Provost and Executive Vice
President for Academic Affairs,
Department of Computer Science
and Engineering University at
Buffalo,
SUNY, USA
tripathi@buffalo.edu

Bernd-Ludwig Wenning
Communication Networks,
University of Bremen, Otto-
Hahn-Allee - NW1, 28359
Bremen, Germany
wenn@comnets.uni-bremen.de

Hao Wu
College of Computing Georgia
Institute of Technology Atlanta,
Georgia 30332-0280
wh@cc.gatech.edu

Mr. Bo Zhou
Centre for Communication
System Research University of
Surrey Guildford, Surrey
United Kingdom
B.Zhou@surrey.ac.uk

CHAPTER 1

European Concerted Research Action COST 285 Modeling and Simulation Tools for Research in Emerging Multiservice Telecommunications

A. Nejat Ince

Istanbul Technical University

Abstract. *This paper contains the keynote address given at the Symposium by the Chairmain of the COST Action 285. It outlines The studies undertaken by the members of the Action with the objective to enhance existing modeling and simulation tools and to develop new ones for research in emerging multiservice telecommunication networks.*

The paper shows how the scope of COST Action 285 has been enriched by the contributions made at the symposium.

1. INTRODUCTION

As a background and introduction to the symposium we shall give here a review of the COST Action 285. The aim is to show how much the Action has been enriched in breadth and depth by the contributions made by the attendants, and particularly by the invited experts.

The main objective of the Action is to enhance existing and develop new modeling and simulation tools for research in emerging multiservice telecommunication networks. To achieve these aims several studies are undertaken in the following three areas identified to be of major importance for the objectives of the Action:

i)   Model Performance Improvement
- Research in the area of new modeling tools based on analytic and hybrid cores for quick solutions of network design and performance evaluation problems.
- Research in multimedia traffic.
- Research on how to improve the quality of the statistical analysis tools imbedded in commercial simulation products.

ii) Multilayer Traffic Models
- Research in the area of multilayer traffic modeling

iii) The important issue of evaluation and validation of new modeling tools.

1.1 Model Performance Improvement

1.1.1 Analytic and Hybrid Cores for Reducing Simulation Time

One of the research objectives of this Action is to investigate the strength and limitations of modeling tools which are based on analytic cores (e.g., Analytic Engines Inc. NetRule) and use steady-state queuing theory formulas and mathematical modeling techniques to produce quick solutions. These tools, which do not capture nuances of protocol behaviors can, however, provide good first-order approximations and may be the only reasonable approach for very large networks.

Another approach to reduce simulation time is the hybrid simulation in which one may focus the simulation on a portion of the network of special interest and model the remainder of the network using the analytic method. It will be interesting to investigate the strength and limitations of hybrids in terms of gain in speed and loss in accuracy.

1.1.2   Statistical Analysis Tools

Experience with available simulation products has shown that the quality and depth of statistical analysis tools imbedded in them are rather disappointing and often require the simulation output captured to be analysed in a statistical package like SAS. A more elegant solution would be to have the simulation tool itself "package" the output. This needs therefore to be investigated.

1.1.3   Multimedia Traffic

Multimedia traffic studies carried out in COST 256 [1] and elsewhere give evidence that different protocols behave very differently at different time scales. A study needs therefore to be undertaken of the network behaviour at different time scales, e.g., 1,10,100 second intervals, including the method of validation of simulation results.

1.2 Multilayer Traffic Modeling

1.2.1 Assessment of the State-of-the Art Development

To prepare an exhaustive annotated compilation of teletraffic models which will comment on any applicability to multilayer traffic modeling and analysis and also on how commercial packages (e.g., OPNET) address multilayer modeling.

Some commercial packages handle aggregate "background" traffic quite well via analytical representations and flow models. If and when enhancements such as abstracting lower-layer protocols, perhaps via analytical representations, appear in the literature they may be used as tools for research on multilayer traffic models.

1.2.2   Multilayer Traffic Models

Some recently proposed models for TCP and their potential in addressing multilayer traffic modeling needs to be investigated. Some of these models (e.g., those proposed by Chadi Barakat, Inria France) focus on window sizing and network issues separately but not independently, and if they can be coupled with a source level-model the outcome might be a valuable step toward multilayer modeling. It would be interesting to try combining good source and lower-layer models into two-layer or three-layer traffic models and to begin studying the convolutional and confounding effects.

1.3   Model Verification, Validation and Credibility (VVC)

1.3.1   Model Verification

The fundamental building blocks of a simulation are the real-world problem entity being simulated, a conceptual model representation of that entity and the computer model implementation of the conceptual model.

Conceptual model validity, software verification, and operational validity along with data validity are the technical processes that must be addressed to show that a model is credible. The technical experts of the Action, as well as experts from other institutions may be consulted to review the conceptual model to judge if it is valid. This activity – called face validity analysis – is performed to support the technical process of conceptual model validity.

Verification is the process of determining that a model operates as intended. Verification process is also referred to as "debugging". The aim is to find and remove unintentional errors in the model's logic and in its implementation.

Verification involves performing test runs in which a comprehensive set of tests are devised to generate ideally all errors and that all error generated during test runs are recognized.

Test cases that explore non-typical operating conditions are most likely to expose modeling errors (increase arrival rate and/or reduce service rate; increase the rate of infrequent events, etc.). Verification of the model should start at the module level. It is paramount that the programming language used presents facilities such as an interactive debugger and the capability of performing a localised "trace" of the flow of events within and between modules.

1.3.2   *Model Validation*

Validation is the process of assessing an acceptable level of confidence that the inferences drawn from the model correctly represent the real-world system modeled. The aim is to determine whether or not the simplifications and omissions of detail knowingly made in the model introduce unacceptably large errors in the results.

Validation focuses on three questions [1] :

   i)   Conceptual Validity : Does the model adequately represent the real-world system?

   ii)   Operational Validity : Are the model-generated behavioural data characteristic of the actual system?

   iii) Credibility : Does the user have confidence in the model's results?

Three variations of the real-world system have to be distinguished :

- The system exists and its outputs can be compared with the simulated results
- The system exists but is not available for direct experimentation.
- The system is yet to be designed; validation here is in the sense of convincing others and ourselves that observed model behaviour represents the proposed reference system if it is implemented.

The methods used for model validation are testing for reasonableness, for model structure and data and model behaviour. Model's validation will be based on comparison of the results obtained from different analytical and simulation models and, where they exist, measurements in real telecommunication systems.

2.   COST 285 INTERIM STUDY RESULTS

Study of the problems and issues outlined above is being performed by the members of the Action Group. They are expected to complete their programme of work in the time frame assigned to the Action (2003-2007) by utilising their own potential which may be augmented, where necessary, by external contributions from symposia organised by the Group as well as from other sources. In this regard it should be mentioned that a formal association has been established with the Center for Advanced Network Research of RTI International Inc., USA, for cooperative research in the areas of mutual interest.

We give below a summary of the studies so far achieved with appropriate pointers to the papers in the present book.

3. MODEL PERFORMANCE IMPROVEMENT

3.1 Existing Design Tools

Simulation and modeling for research and design accounts for a substantial fraction of engineering computation. With valid and credible models, simulation is often dramatically more cost-effective than are real experiments, which can be expensive, dangerous, or, in fact, impossible because a new system may not yet be available.

Scientists and engineers deal with the complexity of the real world in terms of simplified models. There are however techniques that may be used for deciding on an appropriate level of model detail, for validating a simulation model, and for developing a credible model.

One of the basic questions is "Can network design tools help one to find the combination (complicated mix of application, protocols, device and link technologies, traffic flows and routing algorithms) that is right for the problem in hand?". The most challenging decision is to determine how accuracy can be traded off with speed considering the user's role (network designers, sales/marketing, network managers) in the design process.

From a preliminary review and evaluation made of this topic it is clear that the network design tools available on the market are either general-purpose or special purpose tools. The first and most important decision to make for the user is to select the right tool to use. This implies a challenging decision to determine how much accuracy to sacrifice for simulation speed which, depending on the application, can span four orders of magnitude, from tens of seconds to days.

A check list was developed [2] which addresses the needs of the following four categories of users (in terms of user features such as user interface, modeling paradigm, network topology, network types, supported traffic models, level of customisation, etc.,) :

i. <u>Researchers and developers</u> want to reduce development costs and risks by testing the effects of new or modified protocols, devices, architectures, components design, and traffic models in the lab or on the workbench. They need complete control of simulated behavior at the programming language level, and want the language to provide a rich set of special-purpose modeling functions. They usually simulate discrete events (packets transiting a router, protocol retransmissions, etc.), and must often simulate billions of events in order to mimic actual behavior. They want accuracy over run-time speed.

However, they also want tools that support rapid and reliable simulations. Simulations in industry are expected to mimic the behavior of a specific product that may still be in the design phase, and whose features depend on ever-changing market forecasts. At the same time, reliability is a must, since business decisions are based on simulation results. (See Chapter 9 for a discussion of the challenges of simulation tool development in the telecommunication industry.)

ii. <u>Network designers</u> specify and build new networks, or overhaul existing ones. They want to reduce design time and improve design accuracy, ensure that designs meet performance requirements without overbuilding, and identify potential bottlenecks and overloads. They need an extensive library of link technologies, devices, architectures, and protocols to build or upgrade the network, and tools to accurately predict its performance.

Low rate high latency data services will co-exist with high rate low latency real-time multimedia applications in next generation networks, generating time-varying demands on the quality of service (QoS) and network resources. (See Chapter 2 for a summary of challenges in the design of next-generation networks having dynamic resource reservation schemes and extreme variability in demand patterns and for a potential solution using a learning, prediction and correction (LPC) architecture.)

iii. <u>Network managers and network engineers</u> operate networks, troubleshoot and solve performance problems, and make sure that service-level agreements are met. If the existing network is large, they need to import topology and traffic data from other tools. The set of alternatives is usually enormous, so they want to evaluate scenarios in tens of minutes rather than hours. (See Chapter 8 for an in-depth discussion of network management issues and requirements for emerging telecommunications networks, and for modeling tools required to move from today's centralized monitoring and management systems to emerging systems that require highly coordinated real-time control capabilities, automated decision making, and an integrated view of end-to-end managed network entities.)

iv. <u>Sales and field staff</u> want to show customers reasonably accurate representations of how a product, service or technology will improve the customer's network and support customer's business case. They want an intuitive tool that runs on a laptop computer and can be mastered in 1-2 days. They need fast execution speed (tens of seconds per scenario rather than tens of minutes), and extensive presentation features

3.2   Shortcomings of Existing Tools

Even though many of the features indicated in the list above are common in many of the existing commercial tools used or known by the Action members, most if not all of these tools suffer from the following shortcomings:

i. Adequate information is not given generally about the mathematical models underlying the programming objects (models) and algorithms used in their implementation, nor about the validation and credibility issues of the software.

ii. The quality and depth of statistical analysis tools imbedded in the commercial simulation products are rather disappointing and often require the simulation output to be captured in a separate statistical package [see Section 3.4 below]

iii. Little or no support is provided for open source operating systems. It would be very desirable to have tools to run under Linux, FreeBSD, etc. In the experience of some users Microsoft Windows variants, even XP, lack the resilience and robustness required for long simulation runs.

iv.   Little or no run-time debugging capability is provided. Tools that allow viewing (or at least logging and viewing later) the event list, task scheduler, and other dynamic kernel components at run time would be very desirable.

v.   Some commercial packages are prohibitively expensive for universities and non-profit R&D institutions. This is not a technical shortcoming per se, but it severely limits the scope of tools available to a large population of modeling and simulation practitioners.

vi.   Some open source simulation kernels are unstable, perhaps due to the collegial way in which they are modified, debugged, and supported; two different kernel releases can provide inconsistent results with the same model suite and parameter set. This is considered to be a major limitation of open source simulation systems. (See Chapter 12 for a discussion of compatibility issues and shortcomings, and a summary of potential enhancements for a widely-used open source network simulator.)

3.3 Analytic and Hybrid Cores for Reducing Simulation Time

3.3.1 Hybrid Techniques

Hybrid techniques are being investigated for modeling two types of telecommunications systems that are far too complex for commercial analytical or discrete event simulation tools. These studies are very relevant to the Action because they use a combination of methods to reduce simulation time (DES, analytical models, close-form expressions, emulation, etc.) and because they appear to be a feasible way of modeling the very large and/or very complex infrastructures deployed in emerging multi-service telecommunication networks:

i.   The first is ultra high capacity optical networks, which are widely deployed in the core of emerging multi-service telecommunication networks. These networks have aggregate traffic volumes 4-5 orders of magnitude larger than what today's discrete event and hybrid simulators are able to handle. Modeling them in (near) real time requires a combination of techniques: emulators to mimic the network's control plane; an inference engine to deduce data plane behavior and performance from traffic observed in the control plane; analytics to model core network behaviors; a fast hybrid simulator to inject traffic and network impairments; and a supervisory kernel to interconnect and manage the components [3].

ii.   The second is computing Grids, which some believe will be the next major computing paradigm. Grids are a major backbone technology in multi-service research testbeds, and are focus areas in at least three COST Actions (282, 283, 291) [4]. Grids are enormously complex dynamic systems. They require new modeling and simulation methods and tools to understand the behavior of components at various protocol layers and time scales, and to decipher the complex functional, spatial, and temporal interactions among the components. A "grid-in-lab" tool is under development, which is scalable modular, configurable, and plug-and-play, for modeling grids and analysing the performance of applications and middleware services offered on Grids (see Chapter 11).

Hybrid techniques can also be used for transmission network design – e.g., using the object oriented paradigms of Cosmos and Nyx tools. Cosmos is used as the support for system and component development, behavior description, and simulation and analytic calculation development. Nyx is used to describe optimization procedures using general heuristic search techniques. The combination of these two development environments and their component approach enables reusability and shortening of the new application development time (see Chapter 17).

3.3.2   Analytic Solution

On the analytic solution front, we have a number of important new results. Markov chains have been used successfully to model burst errors on channels and at the access level for UMTS/DVB-T systems. New channel measurements on indoor environments are being conducted to improve Markov chain methods, and to validate them (see Chapter 16).

This activity is being extended to models on Power Line Communication (PLC) systems which are expected to be used for communicating over power lines [5]. (The EU is leading the world in the development of PLC-based communications.)

Verification of simulation using a computer and simulation/emulation using DSPs/FPGAs has been studied. Comparison between MATLAB/C simulation programs and real time implementation on DSPs/FPGAs has been performed. The implemented systems are UMTS modems and smart antennas for UMTS. Moreover, the antenna pattern produced by the smart antenna (which includes a UMTS modem and RF transmitters/receivers on it) has been measured, and the mutual coupling effect between RF chains and antennas has been corrected to obtain results close to simulation (see Chapter 21).

A Monte Carlo approach for modeling and simulation of an ultra high performance optical switching network technology – burst switching – has also been proposed. These networks typically present enormously difficult challenges because they lack sufficiently accurate models, and may require excessively large computational resources. New analytical modeling techniques are required to produce results with reasonable run times (see Chapter 15.)

Another novel analytical approach has been proposed to evaluate the end-to-end performance of optical networks that employ slotted (fixed length) optical packets. For a given topology and traffic matrix, one can estimate the end-to-end loss ratio analytically; the approach can also be used to develop a network dimensioning (sizing) method (see Chapter 18). This is impossible with discrete-event simulators because of the very large number of events required to generate a single rare-event loss.

Fluid models of IP networks have been proposed to break the scalability barrier of traditional performance evaluation approaches, both simulative (e.g., ns-2) and analytical (e.g., queues and Markov chains). Fluid models adopt a deterministic description of the average source and network dynamics through a set of (coupled) ordinary differential equations that are solved numerically, obtaining estimates of the time-dependent behavior of the IP network. These models are scalable, i.e., their complexity is independent of the number of TCP flows and of link capacities (see Chapter 6).

An analytical solution for optimal dimensioning of multi-service lines is also proposed. The solution is based on a method to convert recursions for global state probabilities of multi-service models into stable form. The approach is numerically stable because it deals with normalised values of global state probabilities used for estimating main stationary performance measures (see Chapter 7).

An analytical model is proposed that is sufficiently general for a broad family of virtual circuit switching systems. Virtual circuits are widely used in many telecommunications and networking architectures: wireline and wireless telephony; packet-switched networks; MPLS and ATM networks; optical networks; and in emerging multi-service networks. The model is used to derive analytical expressions for traffic intensity, blocking probability, and quality of service (QoS) dimensioning of resources (see Chapter 24).

3.4   Statistical Analysis Tools

Our discussions on the topic of Statistical Analysis Tools show that no "elegant solution" appears to have been implemented; however, XML may be a solution for data interfacing between network simulation tools (like OPNET, NS) and statistical analysis tools (like SAS, MATLAB) as it is supported by a number of systems/vendors. However XML adds a large amount of overhead to data; HDF5 (Hierarchical Data Format 5) is a far more efficient method which, however, does not appear to support discrete-event simulators and statistical tools (e.g. OPNET and SAS) at this time. Action members have agreed that perhaps a framework/strategy/approach/methodology would be a reasonable deliverable, and that support for two (or more solutions), targeting small- and large-scale simulations, are likely recommendations.

There are several solutions to the problem, each with tradeoffs reflecting the problem's dimensions (performance, size, deployment, etc.); perhaps one should (1) identify the dimensions and the range of solutions in each, and (2) systematically lay out an approach for presenting the problem at large. The Action concluded therefore that "Structured Formats for Simulation Results" outlined in [6] might be used to build upon for the deliverable. Sometimes an enumeration of alternatives and where each fits (or does not fit) is a more valuable contribution than a so-called 'point solution'. Helping practitioners understand the issues and build a strategy is perhaps all one can do; there is no "one size fits all" solution.

A software package has been designed and is being implemented that can:

i.   Import simulation or measurement results, in the form of trace files, into a common data structure starting with ns2 and tcpdump.

ii.   Launch post-processing (i.e. filtering and calculations) applications on the imported data, then complementing the raw data with the results of post-processing, storing results of post-processing separately and then feeding them to other applications (e.g. for presentation).

iii. Export data into various output formats (ASCII, XML, etc.).

Typical input structures are being identified to assess which commercial statistical tools support HDF 5.

3.5 Multimedia Traffic

The invariants in network traffic has been investigated [7]. It is concluded that different time scales need to be used at different protocol layers, mathematical formulas are available to describe traffic at different protocol layers, and that one has to be careful to understand and to identify the parameters or the characteristics of the traffic to be used for modeling and simulation purposes. We also have results that show that system performance strongly depends on the incoming workload. Some characteristics, like heavy tail distributions, temporal locality, and frequency of references may depend on certain types of multimedia content (see Chapter 4).

Several multiplicative SARIMA (s, p, d, q) models based on 30 frame-per-second MPEG-4 multimedia traces from TU-Berlin have been developed and their behaviour at different time scales is being examined. Preliminary results suggest that the dominant factor at time scales up to 1 sec. is the seasonal effect induced by the MPEG-4 "group of pictures (GOP)" structure. So-called "scene length" is the dominant factor at longer time scales (see Chapter 3).

IP Network traffic measurement and modeling at the packet level as well as network performance with self-similar traffic has been carried out and discussed [8, 9].

Auto Regressive Integrated Moving Average (ARIMA) models, combined with Generalized Auto Regressive Conditional Heteroscedasticity (GARCH) models, have been used to try to capture both short-and long-range dependence characteristics of the underlying data, especially for highly bursty data (see Chapter 5).

A parameter estimation procedure and an adaptive prediction scheme for the ARIMA/GARCH model has been developed which is capable of capturing the non-stationary characteristics of Internet data [10].

4.   MULTILAYER TRAFFIC MODELS

It has been shown that the topics of Multilayer Traffic Models and Statistical Analysis Tools as well as Model Verification /Validation and Credibility are rather intractable problems which do not have elegant universal solutions. Consequently one has to be content with "framework solutions", and with solutions that fit a specific requirement (e.g., HDF 5 for managing large-scale simulation output).

These problems have a large number of dimensions (N-dimensional space) and the so-called solutions we see published represent cases in which someone has fixed most dimensions and solved for only a single point, line, or plane. A particular research result may be important, but it usually does little more than firmly establish one point/line/plane in N-space. Most neither provide insight about the N-dimensional performance response surface, nor attempt to explain how sensitive one parameter might be to other parameters. Another problem is that researchers make limiting assumptions so as to arrive at closed-form expressions. These assumptions ignore the potential convolutional, confounding, and interaction effects among the dimensions, and some of them are of little value because the assumptions are too limiting to be useful.

To demonstrate what can be achieved with multilayer traffic models, Padhye et al's [12] TCP Reno analytical model (layer 4), and plus layer 2 and 3 extensions have been used to investigate whether analytical multi-layer traffic models might provide credible results in (near) real time.

Preliminary results suggest that such a model can provide a reasonable, steady-state, first-order approximation to TCP over a wireless infrastructure; TCP over an optical infrastructure; TCP over a source-routed IP network infrastructure; and TCP over a diffserv-like IP network infrastructure (see Chapter 3). The issues yet to be addressed are interactions among various layers, and "steady state" as TCP technically never reaches steady state in operational networks.

5. VERIFICATION, VALIDATION AND CREDIBILITY

The goals in this area are defined as model quality improvement and credibility increase of existing tools for telecommunication modeling and simulation, and introduction of new credible modeling and simulation tools for telecommunication analysis (and especially for multilayer traffic).

The Action's main contribution to this field would be in the area of introducing a framework/methodological approach for credibility assessment. Credibility depends on the perceived correctness, and on the perceived suitability of modeling and simulation. Analysis of correctness – named verification – refers to consistency and completeness testing. Analysis of suitability – mostly named validation – refers to tests of capabilities, fidelity, and of accuracy. It is expected that a report entitled "A Generalised Process for the Verification and Validation of Model and Simulation Results" will soon be made available by Prof A. Lehmann.

Chapter 20 describes an approach using fuzzy value tree analysis to support the verification, validation, and accreditation of models and simulations. The process of accrediting a model can be divided into gathering and evaluating results of conducted verification and validation (V&V) activities, and aggregating these (raw) results into a total value: the accreditation decision. Classical decision analysis is one possibility to support the accreditation decision by using a structured, scientifically justified approach. Fuzzy value tree analysis can be used to model subject matter experts statements, quantify both value and knowledge about the model under study, and distinguish between compensational and non-compensational attributes in the value tree.

Chapter 23 explains why it is mandatory to use more than one multimedia traffic trace for model validation, and what descriptors of the traffic features have to be chosen. Traces should also represent different areas, and reflect various user behaviour. The authors present a tool kit to help modelers in analyzing measured traces and describes how to use the tools to validate their traffic models.

6.   NETWORK EMULATION AND NETWORK TOMOGRAPHY

Two new subjects have been identified for inclusion in our program of work; Network Emulation and Network Tomography for which it is considered there is adequate knowledge and expertise in some members of the MC who are willing to pursue these subjects in the timeframe foreseen for the Action [11].

Network emulation is a technique that is usually applied to testing experimental network protocols, especially in networks that use IP as the network layer. It is nearly impossible to thoroughly test an experimental protocol in an operational network because the test team cannot control the characteristics of the network, and cannot isolate the network from unforeseen disruptions or other effects caused by the experimental protocol. Emulators run unmodified protocol stacks on a small number of "instances" – commodity single-board PCs, or processes executing on multi-tasking hosts, or both – in near real time. The instances are physically interconnected over a simulated network core. The core realistically routes traffic, simulates network bandwidth and other resources, and models congestion, latency, loss, and error. Current implementations emulate about 100,000 packets per second per core node instance.

Network tomography is a discipline that borrows techniques from signal processing and medical imaging to solve network monitoring and inference problems in which some aspects of the network are not directly observable. For example, one can use tomographic techniques to estimate network performance (e.g., packet delay and loss) based on traffic measurements taken at a small set of nodes or links, or between a small set of sources and destinations. Other tomographic techniques can be used to infer a network's topology with a very small number of direct measurements.

7. NEW APPLICATIONS

Several new applications were identified at the Symposium. The first is an integrated simulation of communication networks and logistical networks based on a simulator for logistic processes. The combined simulation can be used to measure the influence that an improved communication has on logistical transport processes (see Chapter 13).

The second is an evaluation of vehicular networks (with vehicle-to-vehicle and vehicle-to-infrastructure communication) using analysis, simulation, and field experiments. System evaluation methodologies for networked in-vehicle computing systems are essential to understand system behaviors as well as to assess alternate approaches toward realizing a variety of computing and information services for these emerging networks (see Chapter 14).

The third is an assessment of a layered architecture for modeling and simulation for network centric military systems. This poses significant technical challenges, as the network-centric system must interconnect computers, databases, mobile devices, people (users), processes, satellites, sensors, and weapons into a globally networked distributed complex system (see Chapter 19).

Chapter 22 explains the need for credible modeling and simulation in the context of network based defense. Military platforms become participants in "the network", share information, and provide services that allow them to sense, decide, and act beyond their individual capabilities in a cooperative and distributed manner. The information exchange occurs through an information infrastructure, the so called "infostructure", which constitutes a communication network spanned by the participating nodes. Modeling and simulation are considered to be the mandatory enablers to optimize its design, maintenance, use, and improvement.

REFERENCES

[1]   Ince, Nejat, editor ; "Modeling and Simulation Environment for Satellite and Terrestrial Communication Networks", Kluwer Academic Publishers, Boston, USA, 2002.

[2]   Bragg, A., "Features to Consider When Evaluating Network Design Tools", (TD/285/05/07)*.

[3]   Bragg, A., "Modeling and Analysis of Ultrahigh Capacity Optical Networks", (TD/285/05/07).*

[4]   "2005, About COST", COST Office Publication, March 2005.

[5]   Henry, Paul, "Interference Characteristics of Broadband Power Line Communications Systems Using serial Medium Voltage Wires", IEEE Com. Mag., April 2005.

[6]   Pustisek,M., "Structured Formats for Simulation Results", (TD/285/04/10).*

[7]   Sun, Z., "Invariants in Network Traffic", (TD/285/04/02).*

[8]   Liang, L. and Sun, Z., "IP Traffic Measurement and Modeling at Packet Level", (TD/285/03/23).*

[9]   Zhou, B. and Sun, Z., "Simulation Study of self-Similar Traffic and Network Performance", (TD/285/03/24).*

[10] He, D. and Sun, Z., "Internet traffic Prediction using ARIMA/GARCH Model",(TD/285/04/17).*

[11] Ganchev, I., "Network Tomography Measurement in Relation to Network Simulation and Emulation", (TD/285/03/32).*

[12] Padhye, J.,et al., "Modeling TCP Reno Performance: A Sample Model and Its Empirical Validation", Proc.IEEE/ACM Trans.Networking, Vol.8,No.2, April 2000.

---

*   The above references with their document numbers shown may be accessed at the Action's website www.cost285.itu.edu.tr

CHAPTER 2

Challenges in Design of Next Generation Networks

Satish K.Tripathi [1], Prachee Sharma [1], S.V. Raghavan [2]

[1] Department of Computer Science and Engineering
University at Buffalo, USA

[2] Department of Computer Science and Engineering
Indian Institute of Technology Madras, Chennai, India

Abstract. *Low rate high latency data services will co-exist with high rate low latency real-time multimedia applications in the next generation networks. Increasing volume of multimedia flows in an environment with heterogeneity in bandwidth, propagation medium and statistical characteristics of traffic can be expected to generate time-varying demands on the quality of service (QoS) and network resources. In such non-stationary environment, dynamic resource reservation schemes operating in harmony with the variability in demand patterns may provide efficient mechanisms for resource utilization and guarantee QoS compliance. In this work we identify the challenges that need to be addressed in designing a three level core, distribution and edge (CDE) hierarchical network using time-varying resource allocation mechanisms. Learning, prediction and correction (LPC) architecture based upon integration of operational CDE network with online simulation proposed as a design alternative to contemporary networks.*

1. INTRODUCTION

The support for heterogeneity in services, participating network devices and transportation medium is driving the evolution of the next generation communication networks (NGN) [1]. The traffic in NGN is required to be supported on relatively unconstrained desktop computers as well as over personal devices that may be constrained by power, processing capability and bandwidth availability. The propagation medium is increasingly non-uniform varying between wired and wireless environments. The traffic is increasing in volume and exhibiting non-stationary statistical behavior. In the midst of increasing heterogeneity, expectations from the network regarding the quality of the delivered services are becoming more demanding than ever. The real time traffic such as voice, video or multimedia is driven by different quality of service (QoS) requirements. Due to variability in contribution of each traffic class to the aggregate traffic pattern, the QoS constraints can be expected to vary with time. Current QoS sensitive resource allocation schemes are based upon network flow classification and priority based queuing/admission control mechanisms. Resources allocated along the transmission path may be fixed for a user or for a type of data-flow. For example, resource assignment in digital subscriber line (DSL) is fixed at the time an access is provisioned. Instead, an architecture that provides mechanisms that allow rates to be selected or changed more often, potentially on-the-fly may be preferable [2]. QoS control based upon resource reservation schemes [3]-[6] guarantees availability of "tunnels" of predetermined bandwidth for the desired traffic types identified by "labels". Though the reservation may adapt dynamically to changing number of multicast users and routes as suggested in [25], it may not respond to fluctuations in user requirements and remain fixed for the lifetime of the flow. Once the resource allocation is made, network conditions may change considerably over time resulting in the availability of alternative, more efficient resource allocation strategies. Consequently, static allocation schemes may result in inefficient resource utilization and scale poorly in presence of increasing volumes of multimedia traffic. To support QoS compliance in next generation networks, some fundamental changes in the core, distribution and edge (CDE) network technology may be required. This paper identifies the factors that may trigger changes in the current CDE fabric to support the QoS compliant services in NGN. Online simulation supported by demand prediction and learning models may prove to be viable tools for design, deployment and refinement of the current CDE technologies. An alternative network design that continuously predicts, corrects and learns the network configuration parameters is proposed. The proposed architecture is based upon integrating the operational CDE network with online simulations. The research challenges arising from these changes are discussed and a possible future direction of network evolution is outlined.

Section 2 outlines the architecture and the design problem considered in this work. Section 3 describes the issues in the present environment that may make the design changes in the CDE network necessary. Section 4 discusses present QoS implementation techniques and Section 5 describes the proposed enhancements to the present technology. Section 6 concludes the paper.

2.   QoS COMPLIANT NETWORK

2.1   Architecture

Contemporary networks are based upon the concept of a hierarchical three tier architecture comprising of Core, Distribution and Edge (CDE) networks. The core network is typically supported over wired or fiber-optic connections. Distribution and edge networks are a heterogeneous combination of wired, wireless and/or free-space optical links connecting end-users to the core network. The input traffic from all end-users is collected by the edge network and then forwarded to the distribution network. Several such distribution networks feed the traffic to the network core. In the reverse direction, the distribution network takes the input from the core and distributes the traffic to the intended users connected to the edge network. Such a hierarchical architecture is pervasive in today's networks and can exist at several levels. At global level, an optical core can pervade across several continents and nations. At national level, a core network may connect several cities. At an organizational level, for example, in a university, the core network can feed several functional units such as campuses or colleges. The end users for the core network in each case may vary depending upon the level of the core and distribution network of interest. At the organizational level, end user pool may comprise of individual users. For a national or global level core network, the end-users may be the Internet service providers (ISPs) or organizations with inter-continental/international presence.

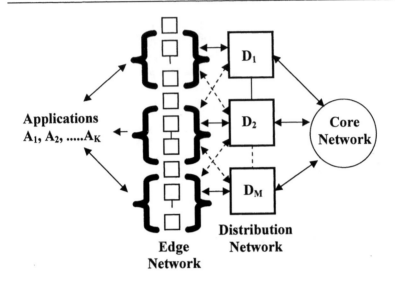

Fig. 1 Core, Distribution, Edge (CDE) Network Architecture

Fig. 1 shows the three-tier CDE network architecture. The core is connected to several nodes D_i, i=1,2,... in the distribution network. Each node D_i acts as a collection and distribution center for the traffic originating from and destined to the nodes E_i, i=1,2,... in the edge network. The edge network may be connected to one or more distribution nodes through a primary path shown by solid lines and several alternate links shown by dashed lines. Multiple connections between edge and distribution network nodes may share the network load to ensure adequate load-balancing and provide redundant connectivity for improved reliability. The input traffic at each edge node comprises of an inhomogeneous mix of *delay sensitive* real-time traffic like video and voice in conjunction with *delay-insensitive* services such as e-mails. The multiplexed traffic generated by applications A_i, i=1,2,... serves as an input to node E_1 located in the edge network and may be destined towards another user connected to node E_2. If both the edge nodes E_1 and E_2 are connected to the same distribution node D_i, the traffic may follow a path $E_1 \rightarrow D_i \rightarrow E_2$. If the nodes E_1 and E_2 are connected to different distribution networks at nodes D_i and D_j, $i \neq j$ respectively, one of the possible paths the traffic could be sent is $E_1 \rightarrow D_i \rightarrow C \rightarrow D_j \rightarrow E_2$ through the network core C.

Packets transmitted by users are received at each of the intermediate nodes along the communication path and stored until the next link is free, and then forwarded towards the intended destination. Such communications incur delays, jitter in the output stream and may yield an unacceptable traffic quality at the destination. Unacceptable quality may refer to high transmission delays and error rate that are in excess of the delays and   errors introduced by the transmission medium [26]. The primary reason for these delays is unavailability of sufficient network resources such as bandwidth, buffer-space and processing power/capability at one or several nodes/links in the CDE network hierarchy. Increased real-time traffic in future networks may be expected to drastically change the Internet traffic profile and impose more stringent delay and jitter requirements. One way to deal with traffic inhomogeneity is to provide resource reservation for each flow along the communication path. This approach works well only when traffic patterns and quality of service requirements are known, for example as time-averaged statistical metric. Due to variability in the   number of network users and generated traffic patterns, reservation based approaches are likely to result in inefficient resource utilization. To   support real-time services, a need for extending the present network   architecture may be necessary.

2.2 Design Problem

Let the QoS requirements for each application A_i be defined by a vector Q_i = $[Q_i^1, Q_i^2,..., Q_i^N]$ comprising of N possible QoS descriptors. The descriptor Q_i^k describes the QoS requirement of the kth traffic stream generated by the ith application. For example, a network download application involving download of a data-file and real-time video may be characterized by different quality guarantees represented by Q_i^1 and Q_i^2 respectively. In general, the QoS descriptor $Q_i^k = f(b_i^k, p_i^k, D_i^k, d_i^k)$ may be represented as some function $f(.)$ of bandwidth b_k, acceptable probability of error p_k, delay D_k and jitter d_k. All QoS parameters can be combined into effective bandwidth B_i^j representation [9] resulting in $Q_i = [B_i^1, B_i^2,..., B_i^N]$. Depending upon resource allocation policies, effective bandwidth can be a function of space related parameters such as buffer size/occupancy and probability of buffer-overflow, and time related parameters such as delay and jitter. Effective bandwidth can be computed as the bandwidth shared at a node by multiplexed sources and is asymptotically bounded by a desired queue length and delay [9,10]. Using exponential bounds for queue length and delays, several approximations for effective bandwidth can be derived for traffic characterized as periodic sources, fluid sources, Gaussian sources and general on-off processes [9]. These approximations may be adequate only for small buffer-overflow probability and large buffer-size [10].

When an application is invoked, several traffic streams may be active either simultaneously or intermittently. An application such as an Internet browser may initiate downloads of real-time audio, video, e-mail as a function of time, each requiring different QoS guarantees. As a result, the QoS requirement of an application will vary temporally and can be described by a vector $Q_i(t) \subseteq Q_i$, where $\subseteq$ denotes a subset. The contemporary approaches to guarantee QoS compliance are based upon a-priori resource reservation schemes [3,4]. Prior to initiating communication, a network path is assigned such that the effective bandwidth requirements for the application are met. Consequently for A_i, the assigned bandwidth at the jth intermediate node along the communication path follows

$$\sum_{k=1}^{N_i(t)} B_i^k \leq B_i^k(t), j = 1,2... \tag{1}$$

where, $B^{Ej}(t)$ on the right hand side denotes the bandwidth assigned at a node E_j in the edge network at time t for the number of active traffic flows $N_i(t)$. The cardinality of the vector representing QoS requirements of A_i is assumed to be $|Q_i(t)| = N_i(t)$, $N_i(t) \leq N$ and as a function of the generated traffic pattern that changes with time. In general, the bandwidth assignments in Eq.(1) should hold for all the intermediate nodes in the core, distribution and edge networks for all time instants $t:[0:\infty]$.

When a node supports several applications, the resource allocation over these applications is constrained by the capacity of the node. The demand on node E_j at time t can be specified in terms of the number of applications $M^{Ej}(t)$ supported at the node E_j at an instant t as

$$D^{Ej}(t) = \sum_{i=1}^{M^{Ej}(t)} B_i^{Ej}(t). \tag{2}$$

$M^{Ej}(t)$ is assumed to be less than or equal to the total number of applications M that can possibly be supported on the network. For all time instants $t:[0:\infty]$,

$$D^{Ej}(t) \leq C^{Ej}, j = 1,2.... \tag{3}$$

where, C^{Ej} denotes the capacity the edge network. The relationship expressed in Eq.(3) must hold for resource assignment at core and distribution networks as well. When the traffic exceeds the node capacity or exhibits any deviation in QoS requirements from the contracted QoS, traffic shapers [7] can be used to guarantee Eq.(3).

The resource allocation problem in Eq.(1) can be cast in terms of several optimization problems subjected to capacity constraint in Eq.(3). For example, resources can be allocated to maximize the number of multiplexed traffic streams $N_i(t)$ for an application; maximize number of applications $M^{Ej}(t)$ supported at a node with a fixed/known $N_i(t)$ and/or minimize resource consumption collectively at all nodes (or at each node) for a given number applications and traffic patterns. The optimization is presumably a highly complex operation due to large number of applications and potential routing paths available.

Communication networks are inherently evolutionary in nature and non-stationary in time. Considering long lifetime of communication networks, the maximum possible applications and traffic streams M and N respectively may also show variability, albeit at a very large time-scale. Assuming M, N as constant, the number of applications $M^{Ej}(t)$ at a node and/or the associated QoS descriptors $N_i(t)$ for the ith application may change over short intervals of time. Co-existence of multimedia and non-multimedia flows can potentially result in a very large range of variability in QoS demands. In response to variability in $M^{Ej}(t)$, $N_i(t)$, the bandwidth allocation $B_i^{Ej}(t)$ may require adaptation throughout the CDE network to maintain the QoS and the integrity of relationships in Eq.(1), (3). In response to changes in demand at time t, the resource adaptation may be feasible only after an interval Δt due to processing and propagation delays inherent in the network. To support real-time applications, one of the design goals could be to *make the interval Δt as close to zero as possible*. The resource allocation for an application must be addressed in a broader context of efficiency of overall resource allocation to maximize of simultaneously supported applications and QoS classes. Consider the residual capacity of node E_j at time t to be

$$R^{Ej}(t) = C^{Ej} - D^{Ej}(t). \tag{4}$$

Variability in traffic patterns and network usage may result in the demand $D^{Ej}(t+\Delta t) \gg R^{Ej}(t)$ exceeding the residual capacity of the node at time $t+\Delta t$. This may occur due to increase in the number of the existing applications rendering $M^{Ej}(t+\Delta t) > M^{Ej}(t)$, or, the change in quality commitment required by the network due to changes in the number of quality descriptors $N_i(t+\Delta t)$ characterizing the existing applications. Excess demand generated by new applications can be curtailed by filtering all the input flows through admission control algorithms [8] and excess demand generated during the life-time of existing applications can be clipped by traffic shaping algorithms [7]. Such admission control and/or traffic shaping approaches may result in inefficient resource utilization when several nodes collectively are able to satisfy the new demand yielding

$$D^{Ej}(t+\Delta t) \le \sum_j C^{Ej}. \tag{5}$$

To accommodate changes in demand and efficiently utilize the excess capacity distributed across several nodes will require redistributing new and existing applications between alternate available nodes. We envisage that NGN will have the flexibility to "context switch" the applications/traffic across network nodes similar to modern distributed operating systems that are capable of context switching between several processing units. Current technology may be enhanced using reactive or proactive approaches for dynamic resource allocation. A reactive method computes resource allocation at time $t+\Delta t$, $\Delta t > 0$ *after* the change in demand is experienced at time t. In a proactive solution, resources may be reserved at time $t-\Delta t$ in anticipation of changes in the demand patterns *before* the actual demand arises at time t. A reactive approach may satisfy QoS requirements for delay insensitive applications and proactive solution may be a promising alternative for delay sensitive real-time applications. The expected benefits from such a scheme are improved resource utilization and support for larger number of users/applications.

3.   CHALLENGES IN QoS BASED DESIGN

Development of an efficient resource allocation plan requires scalability of resource allocation policy with network variability. Temporal variability in network conditions can be primarily attributed to traffic non-stationarity, bandwidth asymmetry and heterogeneity in QoS requirements for users. Each aspect is discussed in this section.

3.1   Traffic Non-Stationarity

The current communication environment is marked by an unprecedented growth in the number of users and traffic volume. To estimate the future growth patterns, one way is by using the growth trends in the past asindicators. ARPAnet between 1971-75 [21], number of Internet hosts between 1981-2000 [32], and the world-wide-web (WWW) traffic [12,20,22] has shown exponential growth patterns. Different protocols contributing to aggregate traffic show strong variation with time [12] and sites [13,14]. Traffic in the future is projected to be an aggregation of telephonic, multimedia, and Internet traffic with the contribution of each traffic class being a function of the growth in the respective domain [23]. Data traffic is projected to exceed voice traffic volume over the next few years [24]. The increased traffic volume may be attributed to improved penetration of communication technology for business, educational and entertainment purposes. The growth in the number of users and variation in their usage profile is promoting spatial and temporal diversity in the traffic volume and aggregate traffic patterns. The key to success in bracing these changes lies in understanding the viability, efficiency and flexibility of the present technology in adapting to heterogeneity in traffic volume and profile. Due to complexity of performance analysis tasks involved and possibly large margins of errors in projections of future trends, simulation based analysis aided by traffic modeling may be a promising approach to allow telecommunication technology to evolve with the growth in demand.

Poisson queuing models pioneered by A.K.Erlang [35] are traditionally used to describe the network traffic. As communication networks are maturing towards providing high-rate multimedia services, the traffic patterns are evolving from simple mutually independent Poisson arrivals to self-similar traffic patterns. Long-tailed inter-arrival time distributions and self-similarity in the traffic over, theoretically infinite range of time-scales, were Paxson and Floyd's key findings [11]. Analysis of traffic flows generated by user interaction with distributed systems [36-39] and measurements for wide and local area networks [40-43] has revealed existence of random, bursty traffic patterns, clusters of packets and hyper-exponentially distributed times of user dialog. Temporal dependence in traffic resulting from self-similarity has resulted in a need to re-evaluate network designs that evolved around Poisson traffic models. These traffic patterns may not readily map to pure stochastic model framework. Complex large-dimensional stochastic models or approximations leading to simpler models may be required to understand and characterize the traffic in NGN.

3.2 Bandwidth Asymmetry

Bandwidth asymmetry results due to higher bandwidth in the forward direction from a server/router to a host as compared to the backward direction from a host to a server/router. Due to asymmetrical bandwidth, latencies in upstream and downstream directions may differ drastically resulting in severe performance bottlenecks [15]. Asymmetrical routes may result when different routes are pursued by upstream and downstream traffic, consequently aggravating the propagation delays [16]. Heterogeneity may exist due to interactions among different protocols, different levels of load and congestion in different parts of the network, topological differences and variation in traffic patterns generated at a site and across different sites [45]. Network asymmetry exists in the capabilities of the end-user that may range from resource-starved personal digital assistants (PDAs) to powerful desktop machines. Network asymmetry poses a formidable challenge in reaching QoS targets [15-19]. TCP, that is the most widely used protocol in today's networks, is adversely affected by these asymmetries due to its inherent feedback mechanism based upon transmission and reception of acknowledgement packets (ACKs). Poor TCP performance can be expected when ACKs are lost or congest the network due to bandwidth asymmetry or incur disproportionately high cost in one direction. Several techniques such as header compression or rescheduling of packet transmissions have been proposed to mitigate effects of bandwidth and medium asymmetry [15] [19].

In addition to performance of underlying protocols, the quality of delivered data is highly dependent upon the quality of wireless links in heterogeneous wired/wireless networks. In current deployments, the service level agreement with a mobile internet user may be difficult to satisfy due to limited resource availability in parts of the network the mobile user may be moving into. Wireless links transition from favorable channel conditions where communication may be as good as the wired link, to unfavorable conditions where errors become highly probable. There is a need to identify new methods to guarantee QoS in such hostile propagation environments. Cross-layer design approaches combining functions of several network layers and adapting parameters based upon multi-layer feedbacks may be one of the solutions [60].

3.3   Heterogeneous QoS Requirements

Real-time applications like video and voice exhibit low tolerance to latencies and jitter compared to traffic such as e-mail and data-exchange. Due to variable bit rate behavior, real-time traffic flows exert time varying demands on the network resources. For example, the QoS demand generated by i^{th} application may comprise of video and voice traffic and can be represented as a vector $Q_i = [B_i^1, B_i^2]$ where, B_i^j represents the effective-bandwidth requirement for the jth traffic type. Bandwidth $B_i^{Ek}(t)$ required to route the traffic through node E_k may change with time as voice or video or both are transmitted. One way to deal with demand variability is to allocate $B_i^{Ek}(t)$ equal to the maximum demand $B_i^{Ek}(t) = Max[\Sigma_j B_i^j]$ or average demand $Avg[\Sigma_j B_i^j]$. When the variability in the demand is very high, such an allocation scheme may result in under-utilization of network resources. Alternatively, the bandwidth allocation may be computed as a function of time to be approximately equal to the demand $D_i^{Ek}(t)$. Amongst contending traffic, time varying resource allocation may be done with preferential or deferential unfairness. Real-time applications, such as voice or interactive video, can be given priority or preferential services avoiding severely penalizing data applications. Undesirable traffic in the network such as denial-of-service, worm-generated traffic or web surfing to destinations exists in the network. Though, these flows may contain real-time video, voice traffic traditionally classified as services requiring high QoS levels, may not merit preferential treatment.

Different users supported by the network may have different perspective of service requirements from the network. From the Internet service providers' viewpoint, designing network functionality based upon cost measures such as maximum bandwidth utilization may be desirable. Such cost-centric approach towards network design may not always result in   desirable QoS levels from the end-users perspective. To obtain the appropriate levels of service for individual users through efficient the bandwidth allocation may be done in accordance with the demand patterns. The number of supported users may be maximized by the utilization of the excess capacity.

4. GUARANTEEING QoS in CONTEMPORARY NETWORKS

The current approaches adopted in contemporary networks to guarantee QoS are outlined in this section.

4.1 Methodology

Achieving QoS through resource reservation is described in [3]. Classification is the first QoS function to occur, often repeatedly at various stages of policy enforcement. The incoming packets are classified at the router into flows. All packets belonging to the same flow obey a predefined rule, have similar quality requirements and receive identical service by the router. In the next step, based upon packet markings received during the classification phase, packets might be discarded by a policer or a congestion-avoidance mechanism. Packets that are not discarded are subject to queuing to prioritize and protect various traffic types when congestion happens to the transmission link. These packets are scheduled for transmission on the egress link, where shaping might occur to control bursts and ensure that outgoing traffic conforms to any service level agreement (SLA) that is in effect on the ingress of the next hop. Congestion-avoidance mechanisms and queuing algorithms may be deployed during this phase to ensure quality compliance. Queuing/scheduling algorithms manage the front of a queue, whereas congestion-avoidance mechanisms manage the tail of a queue.

Despite use of congestion-avoidance mechanisms, network congestion is a common occurrence. Whenever packets enter a device faster than they can exit it, the potential for congestion exists. Congestion-management such as packet compression or fragmentation apply to interfaces that may experience congestion. This could be the case in either a WAN or a LAN environment because of speed mismatches between ingress and egress links. Large data packets may require fragmentation and interleaving to minimize delay incurred at the network interfaces waiting for resources to become available. Smaller packet sizes typically allow usage of small, intermittent availability of resources at network interfaces and consequently reduce the delay/jitter.

4.2 Technologies for QoS Implementations

Mid-90s witnessed an evolution of best effort services into integrated services model. Integrated services requires all the participating network elements such as links and IP routers residing in the propagation path to support mechanisms to control the quality of service delivered to the transmitted packets. QoS requirements are communicated by the application to the network either through controlled-load services [26] or guaranteed services [27]. For controlled load services, clients requesting services provide an estimate of the traffic volume generated, in return benefiting from the reserved network resources provided by the network. The client may be denied service by the network in absence of adequate resources. Guaranteed service support is designed for continuous fixed rate network flows and ensures bounded maximum queuing delay and bandwidth. Resource reservation setup protocol (RSVP) [3] is responsible for allocation and maintenance of network resources along the transmission path during the entire life-time of the transmission.

Differentiated services model was introduced in the late 1990s to address deficiencies of integrated services. As against integrated service that is based upon flow-based resource reservation, DiffServ uses packet markings to classify and treat each packet independently. Features such as expedited forwarding [28,29] to provide a strict-priority based service; assured forwarding [30] with markdown and dropping schemes for excess traffic; class selectors [31] to provide code points that can be used for backward compatibility with IP precedence models are incorporated in DiffServ model. Although the DiffServ model scales well to the internet, it offers no specific bandwidth guarantees to the packets.

In the post DiffServ era, more sophisticated techniques, such as Multiprotocol Label Switching (MPLS) and Virtual Private Networks (VPNs) were proposed. The first major deployment of RSVP technology came with Multiprotocol Label Switching (MPLS) traffic engineering (TE). MPLS traffic engineering automatically establishes and maintains label-switched paths (LSPs) across the backbone by using RSVP to establish and guarantee "tunnels" of predictable bandwidth. RSVP operates at each LSP hop and is used to signal and maintain LSPs based on the MPLS TE calculated path [3,4].

In the recent years, QoS paradigms for wired networks are evolving to support QoS on wireless networks. The QoS control for wireless networks includes management of medium access and physical layer functionality.

IEEE 802.11e standard [46,47] specifies enhancements to the medium access protocol for IEEE 802.11 a, b and g standards to enable QoS support for different traffic classes in local area networks (LAN). IEEE 802.11e provides traffic classification, improved channel access and packet scheduling functions and QoS signaling that assists the resource reservation for the specifies traffic stream.

IEEE 802.16 broadband wireless access standard aims to provide fixed wireless access between the subscriber station (residential or business customers) and the Internet Service Provider (ISP) through the BS. The medium access control (MAC) is built to accommodate a point to multipoint topology. The MAC addresses the high-speed QoS requirements with a flexible design of uplink channels from mobile station (MS) to base-station (BS) and downlink channels from BS to MS. The BS has full control on the bandwidth allocation on both channels. Access allocation is provided by the BS via a request-grant mechanism where the MS explicitly requests access. The principal mechanism of IEEE 802.16 standard for providing QoS support is to associate a packet with a unidirectional flow of packets that provides a particular QoS. QoS provisioning is provided for real-time applications with constant bit rate (CBR) such as voice over IP; real-time applications with variable bit rate (VBR) such as streaming video and audio; non-real-time applications which require better service than best effort, such as high-bandwidth FTP.

The IEEE 802.15 group focuses on standards that will cover Wireless Personal Area Networks (WPANs). So far IEEE 802.15 has introduced one standard, referred to as IEEE 802.15.1, which standardized parts of Bluetooth. In addition, the group works on additional standards that will include a high data rate WPAN, referred to as IEEE 802.15.3, and a low data rate WPAN, referred to as IEEE 802.15.4. The group is also developing recommended practices to facilitate the coexistence of IEEE 802.15 and IEEE 802.11, referred to as IEEE 802.15.2. The QoS paradigms considered in these standards include channel access scheme combined with flow classification to enable a certain level of QoS support. There are a number of QoS mechanisms (i.e., admission control, packet scheduling) that are still undefined and the network designer may implement per-flow QoS solutions for which the standard defines some necessary QoS mechanisms such as per-flow classification. Some of these standards also provide optional priority that may be used to implement differential services.

4.3   Simulation and Measurement Based Analysis

Simulations provide means to explore a synthesized, abstract representation of the network. Present simulations are primarily system-centric, used for understanding intractable systems, experimenting with new technologies and addressing design issues in the present network [44,45]. With the integration of QoS paradigms into the network fabric at the individual user level, simulation paradigms may also require a shift towards user-centric experimentation.

Simulation of core or distribution network may be used to address issues such as: How many VPNs can the core/distribution network support; when is it required to upgrade link/router capacities; is the router/link overloaded; how to shape the incoming/cross traffic at a node for quality compliance; When is it sufficient to enforce admission control and when is enhancement of network resources required. When interactive multimedia flows are supported over the internet, simulations may aid in assigning bandwidth to individual users and determining the number of such applications/users that can be supported. Present technology enables QoS compliant flow through resource reservation. In the near future, traffic engineering based upon MPLS technology may ensure QoS compliance. To effectively implement these technologies to provide adequate QoS, parameter such as bandwidth, link and buffer capacities, processing power required for packet classification/forwarding need to be selected. These parameters are specific to needs of the local network and typically exhibit temporal and spatial variability across the network. Simulations may provide a tractable means of tuning these parameters to demand patterns. At the end user level, simulation can aid in determining desired system behavior in response to changes in input traffic patterns and QoS requirements. At the system level simulation may aid in fine-tuning the network configuration parameters and verify compliance to SLA. These simulations may be carried out for C, D or E networks at different stages of the network life-cycle. Simulations may be used as a tool to facilitate capacity planning during network conception and design phase. After network deployment, simulations may aid in providing bottle-neck analysis to allow capacity enhancements and network flow reconfigurations. During the operational phase, simulations may also be used to provide guidelines for system tuning resulting in performance improvements.

5. GUARANTEEING QoS IN NEXT GENERATION NETWORKS

In section 2, following goals were identified for design of QoS compliant services in next generation networks: Eq. (1) Minimization of network response time yielding a small time interval between initiation of demand and allocation of resources Eq. (2) Dynamic load-balancing to absorb QoS fluctuations in the flow and utilize excess capacity distributed across different network nodes. The first goal facilitates effective implementation real-time applications and the second goal ensures scalability of the resource allocation scheme in presence of variability in the network conditions. It was pointed out that the non-stationarity of traffic pattern, asymmetrical bandwidth and heterogeneous QoS requirements preclude the use of static design approaches for next generation networks and an adaptive approach towards resource allocation may be necessary. In this section learning, prediction and correction model is proposed that may be integrated with CDE network to address dynamic resource allocation issues.

5.1 Methodology

For real-time applications, proactive design involving a-priori allocation of network resources based upon predicted demand may be a promising approach. This solution has two parts - demand prediction and resource allocation. There are several methods that can potentially be used for prediction. For example, hidden Markov models may be utilized for demand prediction using the first order dependency [55]. Linear or non-linear time-series models may be used [58,59] if higher order memory is a more appropriate representation of the system. For example, using a pth order linear time series, the demand $D_i^{Ek}(t+\Delta t)$ can be estimated to be a function of demand in the past weighted by parameters α_k as

$$\hat{D}(t + \Delta t) = \sum_{k=0}^{p} \alpha_k D(t - k\Delta t) + \beta(t + \Delta t) \qquad (6)$$

where, $\beta(t+\Delta t)$ is the noise in the model and can be modeled as a zero mean Gaussian distributed process. The superscript E_k is dropped in the above equation for brevity. The model order p and the coefficients α_k may be evaluated by fitting the model to an observed/measured time-series and can be assumed constant for a stationary stochastic process. However, the network traffic seldom exhibits stationary statistical behavior across geographical locations and time, as pointed out earlier. Therefore, the choice of a single model and a fixed set of parameters cannot be universally assumed to represent the network characteristics accurately. Due to evolutionary nature of the network traffic, the models and its parameters may become obsolete requiring real-time selection of appropriate models and parameters. For example, a more appropriate model for network demand pattern may allow p and α in Eq.(6) to be functions of time. In addition to a mechanism for selection of an appropriate model, a "learning mechanism" that allows adaptation of the model with traffic may be required. Learning techniques such as hidden Markov models, artificial neural networks, genetic algorithms, or some regression methods could possibly be employed to learn the behavior [53-57]. The choice of genetic algorithms [53,54] is driven by its capability to simultaneously learn and predict the workload unlike other techniques such as artificial neural networks and hidden Markov models [55] that require separate training and operational phases.

5.2 Proposed Solution

Dynamic selection of model and an appropriate set of associated parameters require consideration of continuous variability in traffic and environment. We propose integration of network operations with online simulations as one of the promising methods for implementation of dynamic control. Fig. 2 shows the proposed architecture comprising of Workload Learning, Prediction and Correction (LPC) models integrated with online real-time simulations, both interacting with the operational CDE network. Learning, prediction, and correction model interacts with online simulation to arrive at an optimal parameter set using dynamic workload as an input. The adaptive workload model that describes demands placed by the input-requests on the various system resources need to learn the workload behavior of the system continuously in order to predict the future behavior. Based upon the expected behavior of the workload, simulation model may simulate the impact of several alternative configuration choices and an "optimal" parameter set may be chosen to configure the operational CDE network. The effect of changing the network parameters is collected in form of a feedback that is used as an input towards learning in LPC model. In the learning phase of the LPC model, the network feedback is utilized and prediction error and effectiveness of selected configuration parameters is evaluated. This information may be utilized in the simulation model to improve the simulation parameters/approach and strategy of parameter selection. In prediction model, the feedback from the learning model could be utilized in selecting alternative prediction model or model parameters in successive simulation cycles.

The prediction, simulation and learning models are assumed implemented at each node in the CDE network. To be effective, these models should have the visibility of at least the immediate hierarchical neighbor. This would imply that the dynamic resource allocation strategy implemented in the core network should consider the impact of resulting delay, jitter, buffer occupancy and losses on the distribution network nodes. The selection of configuration parameters at the distribution network nodes should similarly consider the impact on the edge and core networks. At the edge network nodes, it may be sufficient to consider the impact of alternate resource allocation plan on the users and the distribution network nodes. Such an integrated approach can be used during capacity planning to determine the design parameters and expected network performance by replacing the actual real-time workload models by anticipated workload descriptions. During the network operational phase, simulations can assist in performing bottleneck analysis and network performance tuning using the real-time measurements.

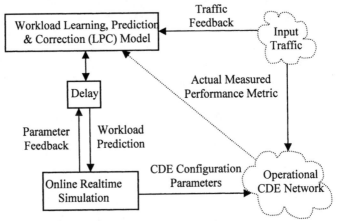

Fig. 2   Integrated Simulation and Operational Architecture for Next Generation Networks

5.3 Implementation Issues

Workload characterization and prediction has been of interest since 70s [49,50]. For typical batch or interactive systems used in 70s, static workload models were adequate to represent the user behavior. Dynamic workload models were introduced in 80s to represent variability in user behavior followed by generative workload models in the recent years to bridge the gap between the user's application oriented view of the load and the actual load (physical, resource oriented requests) submitted to the system. Calzarossa *et.al.* [51, 52] provides a survey of workload characterization issues in standalone batch and interactive computer systems, database systems, communication networks and parallel distributed systems. For communication networks, layered approach of workload characterization is prevalent. An increasingly complex model for the network can be derived from combining workload descriptions of the participating processing nodes that in turn are described by combining the behavior of individual user inputs. The motivation behind workload models so far is to create an abstract, parametric representation of network load for artificial regeneration of traffic in simulation environments. Workload models facilitate *offline* performance studies without requiring actual measurements. The integrated LPC model proposed in this work requires a paradigm shift in simulation approaches that require *online* parameter estimation and learning. Implementation of the proposed architecture that integrates LPC workload model with CDE networks requires combining the workload measurement and modeling techniques with real-time services. The network traffic has exhibited variability in statistical behavior during different times of the day, with changes in constituent multiplexing flows and even at the granularity levels such as packets, blocks and user arrival patterns and traffic bursts. A mechanism for selection of appropriate workload model on-the-fly for each of these scenarios may be the next step. In the CDE hierarchy, bursty behavior of the traffic pattern may be expected to diminish in the network closest to the core network. Differences in traffic behavior across the CDE hierarchy warrants a careful selection of learning models that can best represent the evolving network features. Contemporary simulation approaches are primarily off-line and use static inputs. C, D and E networks are simulated separately with minimal or no mutual interaction. Steps in integrating these simulation approaches may be required to reinforce mutual learning for estimation and adaptation of optimal parameter selection.

6. CONCLUSIONS

Several challenges exist in supporting QoS compliant bursty real-time traffic in next generation networks. The primary challenge is provisioning resources within reasonable delay constraints. The resource reservation and allocation schemes are required that are scalable with non-stationarity in traffic patterns, inhomogenity in network conditions such as bandwidth, propagation environment and conflicts in QoS requirements. Workload learning, prediction and correction (LPC) model that takes the traffic as an input and continuously interacts with online simulation may be integrated with operational CDE networks. A-priori resource reservation using workload prediction is expected to address the timing constraints in the real-time traffic. Demand forecasts provided by predictive workload models are input to online simulation. Online simulation is envisaged to weigh several configuration choices and provide an optimal parameter set to accommodate the anticipated load. The scalability of reservation schemes is addressed through learning and correction approaches that utilize feedback from online simulation and network to update the simulation methods and prediction models.

REFERENCES

[1] Technical Report DSL Forum TR-094, "Multi-Service Delivery Framework for Home Networks", August 2004Produced by: The Architecture and Transport Working Group andDSL Home Technical Working Group, Editor: Mark Dowker, Bell Canada,http://www.dslforum.org/aboutdsl/tr_table.html

[2] Technical Report DSL Forum TR-059, "DSL Evolution - Architecture Requirements for the Support of QoS-Enabled IP Services", September 2003, Produced by: Architecture and Transport Working Group, Editor: Tom Anschutz, BellSouth Telecommunications, http://www.dslforum.org/aboutdsl/tr_table.html

[3] R. Braden, L. Zhang, S. Berson, S. Herzog, S. Jamin, " Resource ReSerVation Protocol (RSVP) -- Version 1 Functional Specification", September 1997

[4] D. Awduche, L. Berger, D. Gan, T. Li, V. Srinivasan, G. Swallow "RSVP-TE: Extensions to RSVP for LSP Tunnels", RFC 3209, December 2001

[5] S. Blake, D. Black, M. Carlson, E. Davies, Z. Wang and W. Weiss, "An Architecture for Differentiated Services", RFC 2475, December 1998

[6] D. Awduche, J. Malcolm, J. Agogbua, M. O'Dell, J. McManus, "Requirements for Traffic Engineering Over MPLS", RFC 2702, September 1999

[7]   O. Bonaventure, S. De Cnodder, "A Rate Adaptive Shaper for Differentiated Services", RFC 2963, October 2000

[8]   H. G. Perros, K. M. Elsayed. "Call Admission Control Schemes: A review", *IEEE Communications Magazine*, 1996, vol.34 no.11, pp. 82-91

[9]   F.Kelly, "Notes on Effective Bandwidths", Stochastic Networks: Theory and applications, vol. 4 of Royal Statistical Society Lecture Notes Series, Oxford University Press (1996), pp. 141-168

[10] M. Schwartz, "Broadband Integrated Networks", Prentice Hall 1996, chap. 6

[11] V. Paxson and S. Floyd, "Wide-area traffic: The failure of Poisson modeling", *IEEE/ACM Transactions on Networking*, vol.3, no.3, pages 226-244 (1996).

[12] V.Paxson, "Growth Trends in Wide-Area TCP Connections", *IEEE Network*, July-Aug. 1994, pp.8-17

[13] V.Paxson, "Empirically Derived Analytic Models of Wide-Area TCP Connections", *IEEE/ACM Transactions on Networking*, vol.2, no.4, Aug. 1994

[14] P.Danzing, , "An Empirical Workload Models for Deriving Wide-Area TCP/IP Network Simulations", Internetworking: Research and Experience, vol.3, no.1, pp. 1-26, 1992

[15] H. Balakrishnan, V.N. Padmanabhan, "How Network Asymmetry Affects TCP", *IEEE Communications Magazine*, April 2001

[16] Y. He, M. Faloutsos, S. Krishnamurthy, "Quantifying Routing Asymmetry in the Internet", *IEEE GLOBECOM 2005*, St. Louis

[17] S. Savage, A. Collins, E. Hoffman, J. Snell and T. Anderson, "The End-to-end Effects of Internet Path Selection", *ACM SIGCOMM*, pages 289-299, September 1999.

[18] C. Labovitz, J. Malan, F. Jahanian, "Internet Routing Instability", *ACM SIGCOMM*, August 1997

[19] H. Balakrishnan, V. N. Padmanabhan, G. Fairhurst, and M. Sooriyabandara, "TCP Performance Implications of Network Asymmetry", *Internet Draft, IETF PILC Working Group*, March 2001

[20] M.J.Riezenman, "Optical Nets Brace for Even Higher Traffic", *IEEE Spectrum*, Jan 2001, pages44-46

[21] L.Kleinrock, "Queuing Systems Volume II: Computer Applications", John Wiely & Sons, 1976

[22] L.G.Roberts, "Beyond Moore's Law: Internet Growth Trends", *Computer*, Jan. 2000, pp.117-118

[23] L.G.Kazovsky, G.Khoe, M. Oskar van Deventer, "Future Telecommunications Networks: Major Trend Projections", *IEEE Communications Magazine*, Nov. 1998, pages.122-127

[24] M.N. Sadiku, "Next Generation Networks", *IEEE Potentials*, April/May 2002, pp.6-8

[25] M. Pullen, M. Myjak, C. Bouwens, "Limitations of Internet Protocol Suite for Distributed Simulation in the Large Multicast Environment", RFC 2502, February 1999

[26] J. Wroclawski, "Specification of the Controlled-Load Network Element Service", RFC 2211, September 1997

[27] S. Shenker, C. Partridge, R. Guerin, "Specification of Guaranteed Quality of Service", RFC 2212, September 1997

[28] B. Davie, A. Charny, J.C.R. Bennett, K. Benson, J.Y. Le Boudec, W. Courtney, S. Davari, V. Firoiu, D. Stiliadis, "An Expedited Forwarding PHB (Per-Hop Behavior)", RFC 3246, March 2002

[29] V. Jacobson, K. Nichols, K. Poduri, "An Expedited Forwarding PHB", RFC 2598, June 1999

[30] J. Heinanen, F. Baker, W. Weiss, J. Wroclawski, "Assured Forwarding PHB Group", June 1999

[31] K. Nichols, S. Blake, F. Baker, D. Black, "Definition of the Differentiated Services Field (DS Field) in the IPv4 and IPv6 Headers", RFC 2474, December 1998

[32] M.Lottor, "Internet Growth (1981-1991)", RFC 1296, Jan. 1992

[33] B. Rajagopalan, J. Luciani, D. Awduche, "IP over Optical Networks: A Framework", March 2004

[34] M. Crawford, "Transmission of IPv6 Packets over FDDI Networks", RFC 2467, December 1998

[35] A.K.Erlang, "The theory of probabilities and telephone conversations", *Nyt Tidsskrift Matematik* B 20:33 (1909), English Translation E. Brockmeyer, H.L.Halstrom and A.Jensen (1948), The life and works of A.K.Erlang, The Copenhagen Telephone Company, Copenhagen.

[36] E. G. Coffman and R. C. Wood, "Interarrival statistics for time sharing systems", *ACM Communications,* vol.9, pages 5000-5003 (1966).

[37] P. E. Jackson and C. D. Stubbs, "A study of multi-access computer communications", *AFIPS Conference Proceedings*, 1969, vol.34, pp.491-504

[38] E. Fuchs and P. E. Jackson, "Estimates of distributions of random variables for certain computer communications traffic models", *ACM Communications* vol.13, pages 752-757 (1970).

[39] P. F. Pawlita, "Traffic measurements in data networks, recent measurement results, and some implications", *IEEE Trans. Communication* vol.29, no.4, pages 525-535 (1981).

[40] R. Jain and S. A. Routhier, "Packet trains- measurements and a new model for computer network traffic", *IEEE Journal Selected Areas Communication* vol. 4, no.6, pages 986-995 (1986).

[41] J. F. Schoch and J. A. Hupp, "Performance of the Ethernet local network", *ACM Communications* vol.23, no.12, pages 711-721 (1980).

[42] D. N. Murray and P. H. Enslow, "An experimental study of the performance of a local area network", *IEEE Communication Magazine* 22: 48-53 (1984).

[43] F. A. Tobagi, "Modeling and measurement techniques in packet communication networks", *Proceedings of IEEE* 66: 1423-1447 (1978).

[44] V.Paxson, S.Floyd, "Why We Don't Know How to Simulate the Internet", *Proceedings of the 1997 Winter Simulation Conference*

[45] S.Floyd, V.Paxson, "Difficulties in Simulating the Internet", *IEEE/ACM Transactions on Networking*, vol.9, no.4, Aug. 2001

[46] "Draft Supplement to International Standard for Information echnology, Telecommunications and Information Exchange between systems LAN/MAN - Specific requirements", IEEE 802.11e/D2.0, Nov 2001.

[47] P.Garg, R. Doshi, R. Greene, M. Baker, M. Malek, X. Cheng, "Using IEEE 802.11e MAC for QoS over Wireless", 22nd *IEEE International Performance Computing and Communications Conference*, IPCCC 2003, Phoenix, Arizona

[48] J. Yang, J. Ye, S. Papavassiliou, N. Ansari, "A Flexible and Distributed Architecture for Adaptive End-to-End QoS Provisioning in Next-Generation Networks", *IEEE Journal on Selected Areas in Communications*, vol. 23, no. 2, February 2005

[49] D. Ferrari, "Workload Characterization and Selection in Computer Performance Measurement", *Computer*, 1972, vol.5, no.4, pp.18-24

[50] A.K. Agrawala, J.M. Mohr, R.M. Bryant, "An Approach to the Workload Characterization Problem", *Computer*, 1976, pages 18-32

[51] M. Calzarossa, L. Massari, D. Tessera, "Workload Characterization Issues and Methodologies", *Lecture Notes In Computer Science*; vol. 1769, Performance Evaluation: Origins and Directions, 2000, Springer-Verlag London, UK, pages 459-481

[52] M. Calzarossa and G. Serazzi, "Workload characterization: A Survey", *Prceedings of the IEEE*, vol.81, no.8, pp.1136-1150, August 1993, http://citeseer.ist.psu.edu/calzarossa93workload.html

[53] S. V. Raghavan, N. Swaminathan, J. Srinivasan, "Predicting Behavior Patterns Using Adaptive Workload Models", *Proceedings of the 7th International Symposium on Modeling, Analysis and Simulation of Computer and Telecommunication Systems* (MASCOTS), 1999

[54] N. Swaminathan, J. Srinivasan, S.V.Raghavan, "Bandwidth-Demand Prediction in Virtual Path in ATM Network Using Genetic Algorithms", *Computer Communications*, July 1999, vol.22, no. 12, pp.1127-1135

[55] S.V. Raghavan, Guenter Haring, V. Srinivasan V, N. Vishnu Priya, "Learning Generators for Workloads", *Proc. of Workshop on Workload Characterization in High Performance Computing Environments*, MASCOTS'98, July 1998, Montreal, Canada, pages 1-81 to 1-88

[56] G. Haring, "On State-Dependent Workload Characterization of Software Resources", Proc. *ACM Sigmetrics Conf*erence, pages 51-57, 1982.

[57] M. Calzarossa and G. Serazzi, "A Characterization of the Variation in Time of Workload Arrival Patterns", *IEEE Trans. on Computers*, 1985, vol.C-34, no.2, pp.156-162

[58] Fan Zhang, J.L.Hellerstein, "An Approach to On-line Predictive Detection", *Proceedings of 8th International Symposium on Modeling, Analysis and Simulation of Computer and Telecommunication Systems*, 2000, 29 Aug.-1 Sept. 2000, pp.549 - 556

[59] J. Ilow, "Forecasting network traffic using FARIMA models with heavy tailed innovations", Proceedings of *IEEE International Conference on Acoustics, Speech, and Signal Processing*, 2000. ICASSP '00, 5-9 June 2000, vol. 6, pages 3814-3817

[60] M. van Der Schaar, N Sai Shankar, "Cross-layer Wireless Multimedia Transmission: Challenges, Principles, and New Paradigms", *IEEE Wireless Communications*, Volume 12, Issue 4, Aug. 2005, page: 50 - 58

CHAPTER 3

An Empirical Approach For Multilayer Traffic Modeling And Multimedia Traffic Modeling At Different Time Scales

Arnold Bragg

Center for Advanced Network Research
RTI International, Inc. Box 12194
Research Triangle Park, NC 27709 USA

Abstract. *We describe empirical approaches for multilayer traffic modeling – i.e., models that span several protocol layers – and for modeling multimedia traffic at various time scales.*

Multilayer traffic modeling is challenging, as one must deal with disparate traffic sources; control loops; the effects of network elements; cross-layer protocols; asymmetries in bandwidth, session lengths, and application behaviors; and an enormous number of potential confounding effects among the various factors.

We summarize experiments that combine an analytical transport layer model (layer 4) with layer 1/2/3 components to investigate whether analytical multilayer traffic models might provide credible outcomes in (near) real time. Preliminary results suggest that such models can provide reasonable, steady-state, first-order approximations of behaviors that span several protocol layers.

Multimedia traffic modeling is also challenging, as many types of multimedia traffic have characteristic statistical signatures induced by their encoders. Traffic analysts have proposed a number of feasible models for multimedia traffic, but it is not clear which is best.

We summarize experiments using multiplicative SARIMA(s,p,d,q) models of MPEG-4 multimedia traces at various time scales. Preliminary results suggest that the seasonal effect induced by MPEG's 'group of pictures' encoding is the dominant factor at time scales up to a few tens of seconds, while scene length predominates at longer time scales.

1. INTRODUCTION

Multilayer traffic modeling is challenging, as one must deal with disparate traffic sources (greedy, bursty, periodic, non-stationary, autocorrelated, short- and long-range dependent, stochastic, streaming, elastic, multiplexed, etc.); control loops (TCP, traffic shapers, etc.); the effects of network elements (blocking in switches, active queue management in routers, etc.); cross-layer protocols (mobile and wireless protocol stacks, λ routing and IP routing in optical networks, etc.); asymmetries in bandwidth, session lengths, and application behaviors; and an enormous number of potential convolutional, confounding, and interaction effects among these factors.

Multimedia modeling is also challenging, as many types of multimedia traffic have characteristic statistical signatures induced by their encoders; e.g., MPEG-4 segments typically have an IBB PBB PBB PBB "group of pictures" (GOP) structure with I/P/B framing. Traffic analysts have proposed a number of feasible stochastic processes and models; e.g., non-linear threshold AR models, fractional ARIMAs and seasonal ARIMAs, DARs, GBARs, nested ARs, various LRD models, MMRPs, $M/G/\infty$ models, fluid flow models, TES, ARTA, ARCH, GARCH, etc. It is not clear which (if any) is best.

We investigate empirical approaches for multilayer traffic modeling, and for modeling multimedia traffic at different time scales, and describe preliminary results herein.

1.1   Relevance to Cost Action 285

Multilayer traffic modeling aligns with COST Action 285's Task 2.2(See Chapter 1); viz., *"some recently proposed models for TCP and their potential in addressing multilayer traffic modeling will be investigated. Some of these models ... focus on window sizing and network issues separately but not independently ... It would be interesting to try combining good source and lower-layer models into two-layer or three-layer traffic models and to begin studying the convolutional and confounding effects."*

Modeling multimedia traffic at different time scales aligns with COST Action 285's Task 1.3; viz., *"multimedia traffic studies ... give evidence that different protocols behave very differently at different time scales. A study will therefore be undertaken of the network behavior in different time scales, e.g., 1, 10, 100 second intervals, including the method of validation of simulation results."*

1.2   Approach For Multilayer Traffic Modeling

We have added components that mimic layer 1/2/3 effects to Padhye's TCP Reno analytical model (layer 4) in order to investigate whether analytical multilayer traffic models might provide credible results in (near) real time [21, 19, 20]. We use the analytical model for the transport layer, and we vary specific parameters to represent the effects of components at layers 1/2/3. E.g., wireless networks are lossy but have short end-to-end delays. We use the model to investigate the impact of various TCP parameters at relatively high loss rates and with very short round trip times to simulate TCP over a wireless (layer 1/2) infrastructure having different protocol time-out intervals and congestion window sizes. (See Appendix I for a brief description of Padhye's analytical model.)

Padhye's model uses mathematical formulas to estimate the TCP source sending rate and TCP end-to-end throughput at steady state. Preliminary results suggest that such models can provide reasonable first-order approximations for TCP over a wireless infrastructure; TCP over an optical infrastructure; and TCP over a source-routed IP network infrastructure. Such multilayer models obviously capture neither the dynamics of TCP nor nuances of the protocol's behavior; however, they may be the only practical approach for simulating TCP flows over large and complex networks.

The issues yet to be addressed are the lack of "steady state", as TCP technically never reaches steady state in operational networks; and protocol variations, as there are tens of operational TCP implementations. We summarize experiments and results to date in Section 2.

1.3   Approach For Modeling Multimedia Traffic At Different Time Scales

We have developed several multiplicative SARIMA(s,p,d,q) models based on empirical 25-30 frame per second MPEG-4 multimedia traces, and have examined the behavior of these models with different sample sizes and at different time scales (with aggregation). Preliminary results suggest that the seasonal effect induced by the MPEG-4 GOP encoding is the predominant factor at time scales up to a few tens of seconds, while scene length is the predominant factor at longer time scales. (See Appendix II for a brief description of the ARIMA family of stochastic processes and models.)

The issue yet to be addressed is how one might automate the modeling step, as determining the best SARIMA model is labor-intensive. We summarize experiments and results to date in Section 3.

2.   MULTILAYER TRAFFIC MODELING

As noted, we use Padhye's analytical TCP Reno model [21, 19, 20]. (The model and parameters are described in Appendix I.) The model estimates TCP throughput $T(p)$ based on the packet loss rate p, the round trip time RTT, the TCP timeout interval T_0, and the maximum TCP congestion window size W_m. We note that RTT, T_0, W_m, and p are strongly coupled with layers 1/2/3.

2.1   An Analytical TCP Throughput Model

Figure 1 shows two plots of TCP Reno throughput $T(p)$ in packets received per unit time over a range of packet loss rates p and round trip times RTT. Logarithmic transformations of $T(p)$, p, and RTT are plotted in the rightmost figure for comparison. The TCP timeout interval T_0 and the maximum TCP congestion window size W_m are fixed at 0.243 seconds and

6 respectively for these two plots. TCP throughput is abysmal at packet loss rates above 5% in this scenario.

A common mistake made when modeling layer 4 TCP traffic is failing to understand nuances in throughput behavior induced by different network infrastructures and protocols – i.e., factors at protocol layers 1/2/3 that are coupled in some way with RTT, T_0, W_m, and p.

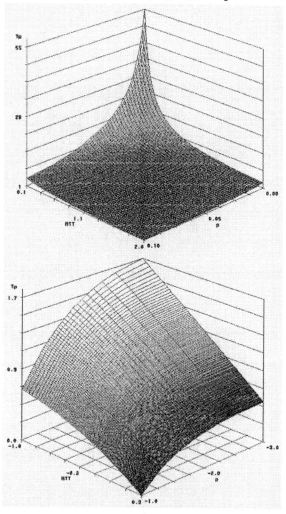

Fig. 1 TCP Throughput (Tp) vs. Round Trip Delay (RTT) and loss rate (p), actual (top) and log scales.

Figure 2 shows two plots of TCP throughput $T(p)$ at packet loss rates p of 0.01 (left) and 0.05 (right) with a round trip delay RTT of 0.500 seconds, while varying both the TCP timeout interval T_0 and the maximum TCP congestion window size W_m. The implication is that one can drastically improve TCP throughput up to a point by simply increasing the maximum congestion window size W_m. Note that one could even achieve a lower TCP throughput at a packet loss rate of 1% (vs. 5%) if W_m is chosen unwisely, which may occur when moving from one layer 1/2/3 infrastructure to another.

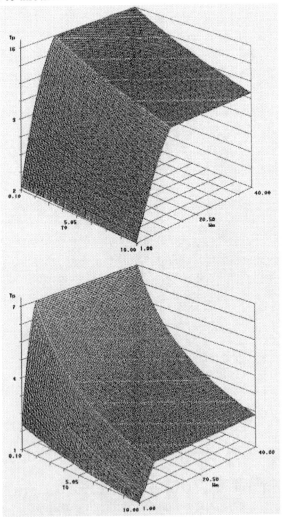

Fig. 2 TCP Throughput (Tp) vs. Timeout interval (T0) and window size (Wm) at loss rates of 1% and 5%.

As noted, Padhye's model assumes that a steady state is achieved, which never occurs in operational networks. Operational TCP variants also have significantly different behaviors. E.g., Figure 3 (from [16]) shows the throughput characteristics of TCP Reno, TCP New Reno, and TCP SACK over time. TCP SACK achieves a consistently higher throughput in this scenario, but all three variants show variability in throughput over time.

2.2 TCP Over Optical Networks

We simulated TCP over a wide-area optical core network infrastructure by focusing on the very low packet loss rates[1] and relatively long round trip times experienced in these networks. Figure 4 (left) shows a plot of TCP throughput $T(p)$ at packet loss rates p between 10^{-4} and 10^{-5} and round trip delays RTT between 1 and 4 seconds. Note that TCP throughput is nearly constant at these low packet loss rates.

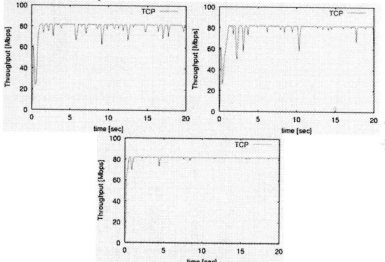

Fig. 3 Throughput vs. time for TCP Reno, New Reno, and SACK (From [16]).

Figure 4 (right) shows a plot of TCP throughput $T(p)$ at a packet loss rate p of 10^{-5}, round trip delays RTT between 1 and 4 seconds, and a maximum TCP congestion window size W_m between 1 and 100. Again, we see that W_m is extremely influential even in an optical network with a

---

[1] As optical networks have bit error rates below 10^{-10} and almost no core congestion that results in packet loss.

completely different layer 1/2/3 infrastructure. Note also that *T(p)* does not increase linearly with decreasing RTT at higher values of W_m.

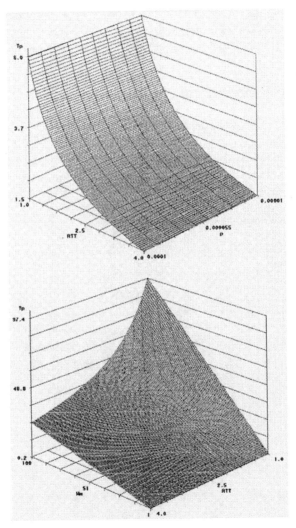

Fig. 4 TCP throughput (Tp) vs. RTT and P (top), and RTT and Wm (bottom) at very low packet loss rates.

As noted, Padhye's model assumes that steady state is achieved. E.g., Figure 5 (from [10]) shows the empirical (and highly variable) throughput characteristics of TCP over different buffer sizes in the ESNet[2] (which is somewhat representative of the behavior of the maximum TCP congestion window size W_m). Compare Figure 5 with the gross performance shown on the rightmost plot of Figure 4 (the Wm axis).

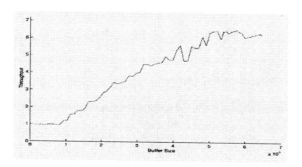

Fig. 5 Throughput vs. Buffer Size (From [10]).

2.3 TCP Over Wireless Networks

Next, we simulated TCP over a wireless network infrastructure by focusing on the relatively high packet loss rates and relatively short round trip times encountered in these networks. Figure 6 (left) shows a plot of TCP throughput $T(p)$ at a packet loss rate p of 0.05, round trip delays RTT between 0.1 and 1 second, T_0 =2.495, and a maximum TCP congestion window size W_m between 1 and 10. Note the ridge near $W_m = 6$. Figure 6 (right) shows a plot of TCP throughput $T(p)$ at a packet loss rate p of 0.05, round trip delays RTT between 0.1 and 1 second, and a TCP timeout interval T_0 between 1 and 7 seconds. Reducing the timeout interval has an impact on throughput, particularly at small RTTs, and is often overlooked when investigating TCP performance using discrete event simulation of wireless networks. So W_m and T_0, affected in part by factors at lower protocol layers, can be influential to TCP throughput in lossy wireless networks.

---

[2]   ESNet is an OC-3 network connecting various US Department of Energy Research Laboratories. The plot shows TCP throughput between Argonne and Lawrence Berkeley National Laboratories over ESnet [ES01].

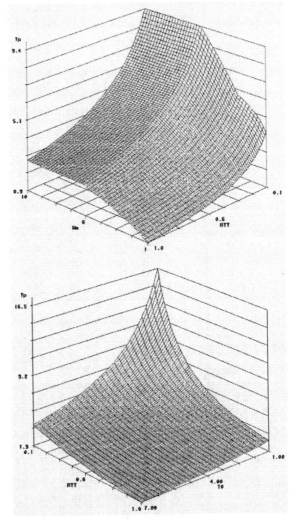

Fig. 6 TCP throughput (Tp) vs. RTT and Wm (top), and RTT and T0 (bottom) at high packet loss rates.

2.4 TCP Over Source-Routed Networks

Next, we simulated TCP over a source-routed network by focusing on the ability of source routing to achieve relatively tight RTT bounds under balanced loads. Figure 7 (left) shows a plot of TCP throughput $T(p)$ at a packet loss rate p of 0.01, round trip delays RTT between 2.0 and 2.2

seconds, T_0 = 2.495, and a maximum TCP congestion window size W_m between 10 and 15. Note again the ridge near W_m = 12. Figure 7 (right) shows a plot of TCP throughput at a packet loss rate of 0.01, W_m = 6, round trip delay between 2.0 and 2.2 seconds, and a timeout interval between 0.1 and 5.0 seconds.

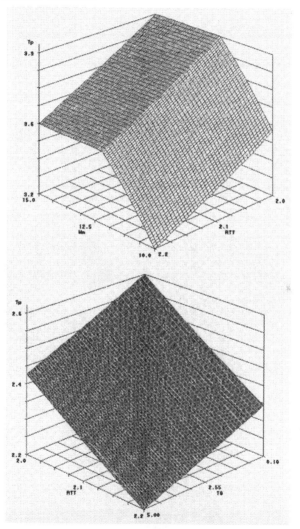

Fig. 7 TCP throughput (Tp) vs. RTT and Wm (top), and RTT and T0 (bottom) with source routing.

Note that increasing W_m produces better throughput only up to a certain point, while T_0 has no similar asymptotic bound (at least in the range used in this scenario). Note also the differences between Figures 6 and 7, which confirm that factors controllable at layers 1/2/3 can dramatically affect TCP throughput under different loss regimes. T_0 and W_m (to a point) are clearly influential to TCP throughput in IP networks using layer 3 source routing.

We concluded the experiment by fixing RTT at 2 seconds and varying both the maximum TCP congestion window size W_m and the timeout interval T_0 at three slightly different packet loss rates over a layer 3 source-routed network scenario. Figure 8 (left) shows a plot of TCP throughput $T(p)$ at a packet loss rate p of 0.010 and a round trip delay RTT of 2.0 seconds. Note again the ridge near $W_m = 12$. Figure 8 (center) and Figure 8 (right) are identical except that the packet loss rates are 0.013 and 0.014 respectively. Slightly increasing the loss rate p shifts the W_m-induced ridge to the right, which suggests a very interesting interrelationship between $T(p)$, p, W_m, and T_0. Again, factors controllable at layers 1/2/3 can significantly affect TCP throughput under slightly different loss regimes.

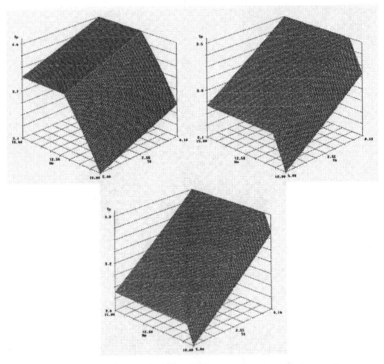

**Fig. 8 TCP throughput (Tp) vs. Wm and T0 with layer 3 source
routing at slightly different packet loss rates.**

2.5 Preliminary Conclusions

Combining an analytical transport layer model (layer 4) with layer 1/2/3 components confirms that analytical multilayer traffic models can provide reasonable first-order steady-state approximations in (near) real time. Using an analytical model for the transport layer, and varying its parameters to mimic the effects of components at other layers has disclosed some interesting effects that we have confirmed with discrete event simulation. This approach to multilayer modeling cannot capture the dynamics of TCP, stochastic nuances of the protocol's behavior, or variations in TCP implementations. It is also not clear that this approach can capture less influential behaviors arising from convolutional, confounding, and interaction effects among various factors.

3. MULTIMEDIA TRAFFIC MODELING AT DIFFERENT TIME SCALES

Modeling multimedia traffic is challenging for a number of reasons. Traffic patterns depend on the data's structure (MPEG's group of pictures [GOP] construct, the use of I/P/B/PB macroblocks, etc.); the compression algorithm used (MPEG-4, H.261/H.263, etc.); the mode (low, intermediate, or high for MPEG-4); the video format; the spatial-resolution encoding method used; the codec's frame rate; and implementation-specific features of various codecs [11]. Traffic streams also depend on content – i.e., the amount of inter-frame redundancy (which is high with static subjects and low with moving subjects), scene changes (e.g., a soliloquy followed by swordplay), and the time scale.

We illustrate several of these components using 60-minute MPEG-4 and H.263 video traffic traces from the Telecommunication Networks (TKN) Group, Department of Electrical Engineering, Technical University Berlin (http://www.tkn.tu-berlin.de/research/trace/trace.html) [24][3].

---

[3] The trace library contains MPEG-4 and H.263 traces prepared by capturing uncompressed YUV information at 25 frames per second in QCIF format. YUV sequences were encoded using both MPEG-4 and H.263. MPEG-4 is encoded at three different quality levels: low, intermediate, and high; H.263 is encoded at four target bit rates: 16 Kbit/sec, 64 Kbit/sec, 256 Kbit/sec, and variable bit rate. YUV is the color space used in PAL television broadcasting; Y is the luminance component, and U and V are the chrominance components. See http://www.tkn. tu-berlin.de/research/trace/trace.html for more information.

3.1   Data Structure Influence

Some multimedia traffic is structured as it leaves the source; e.g., Figure 9 shows a one second MPEG-4 segment with an IBB PBB PBB PBB frame structure consisting of 5 GOPs and 60 frames. Unless noted, frame sizes herein are in bytes and time is the inverse frame rate (where the rate is typically 24-30 frames per second)[4]:

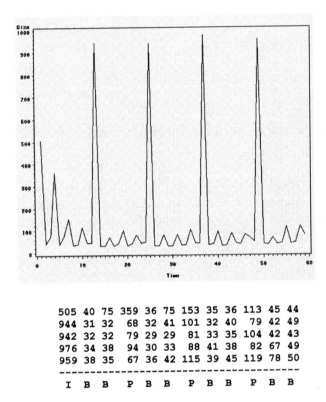

```
505 40 75 359 36 75 153 35 36 113 45 44
944 31 32  68 32 41 101 32 40  79 42 49
942 32 32  79 29 29  81 33 35 104 42 43
976 34 38  94 30 33  88 41 38  82 67 49
959 38 35  67 36 42 115 39 45 119 78 50
-----------------------------------------
 I  B  B   P  B  B   P  B  B   P  B  B
```

Fig. 9 Frame size vs. Time for a 60 Frame MPEG-4 Segment showing the GOP structure.

---

[4]   Data from Star Wars IV , MPEG 4 , low quality frame trace, . Terse_StarWarsIV_10_14_18.dat [TU05].

3.2 Compression Algorithm Influence

Traffic structure depends on the compression algorithm used. Figure 10 shows MPEG-4 (left) and H.263 (right) frame sizes vs. time for the same one minute video segment. Note the characteristic I, P, and B MPEG-4 structure. Time is the inverse frame rate – i.e., 1/25 seconds for MPEG-4, and milliseconds for H.263.

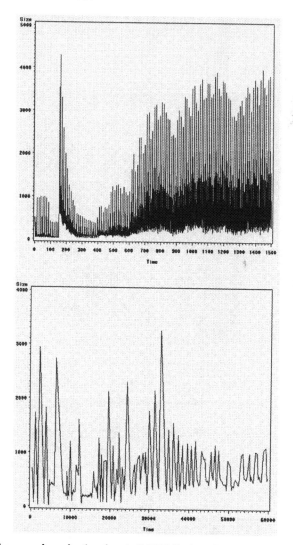

Fig. 10 Frame sizes in 1 minute MPEG-4 and H.263 traces for the same video segment.

3.3 Mode Influence

Traffic structure also depends on the mode (for MPEG-4) or target bit rate (for H.263), which is commonly referred to as the trace quality or resolution. Figure 11 shows MPEG-4 frame sizes vs. time for the same one minute video segment at low, intermediate, and high MPEG-4 quality. Note that the I, P, and B distinctions tend to diminish with increasing quality. Time in Figure 11 is the inverse frame rate – i.e., 1/25 seconds.

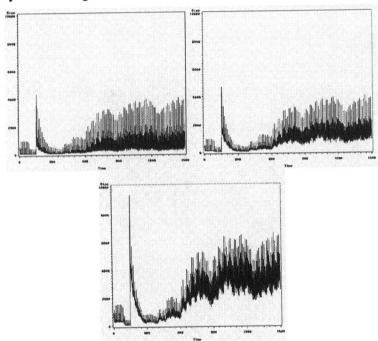

Fig. 11 Frame sizes in 1 minute MPEG-4 and H.263 traces for the same video segment.

Figures 12 and 13 show the empirical frame size distributions for MPEG-4 and H.263 traces, respectively. MPEG-4 traces represent low (left), medium, and high quality modes; H.263 traces represent target bit rates of 16 (left), 64, and 256 Kbits/second. All are based on the same source file[5]. (Both Figures are from [24].) Note the trimodal nature of the low quality MPEG-4 trace (which reflects the I/P/B frame types), and the bimodal nature of all three H.263 traces (which reflects the P and PB

---

[5]   Star Wars IV trace, 60 minute segment.

frame types). Note also that MPEG-4 tends toward unimodality at higher quality.

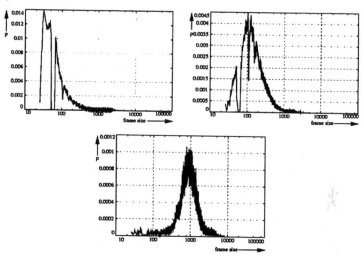

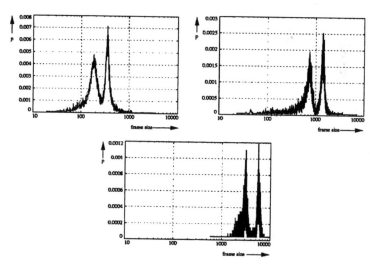

Fig. 12 Frame size distributions from MPEG-4 low, medium, and high quality traces (from [24]).

Fig. 13 Frame size distributions from H.263 16, 64, and 256 Kbit/S traces (from [24]).

3.4 Implementation-Specific Encoding Influence

Traffic structure also depends on various implementation-specific features of codecs. Figure 14 shows MPEG-4 frame sizes vs. time for the same 10 second video segment encoded by two different codecs (with all otherparameters identical). Note that the second (right) codec has more variability in frame size. Time in Figure 14 is the inverse frame rate – i.e., 1/30 seconds for these codecs.

3.5 Inter-Frame Redundancy Influence

Traffic structure also depends on the amount of inter-frame redundancy (which is higher with stationary subjects and lower with moving subjects). Figure 15 shows MPEG-4 frame sizes vs. time for traces from three locales – a parking facility, an office, and a lecture hall. The traces are of different subjects and cannot be compared per se. However, one might expect that a parking facility would tend to have a higher sustained volume of movement and a correspondingly larger frame size than the other two locales (assuming all other factors are equivalent). Time in Figure 15 is the inverse frame rate – i.e., 1/25 seconds.

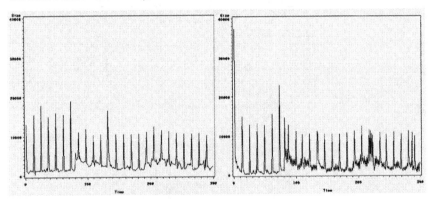

Fig. 14 Frame sizes from two different MPEG-4 codecs for the same 10 second segment.

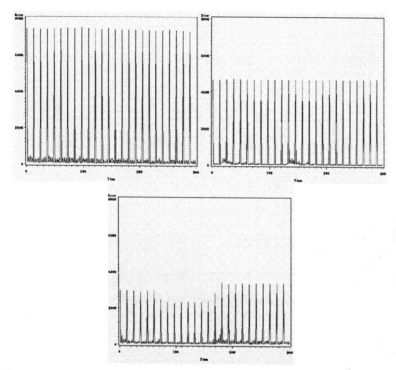

Fig. 15 Frame sizes from 10 second low quality MPEG-4 traces for parking facility, office, and lecture hall locales.

Figure 16 shows the empirical frame size distributions for low quality MPEG-4 traces for the three locales – parking facility, office, and lecture hall respectively. (These Figures are from [24].) Note that while the largest (I frame) sizes are consistent with the traces in Figure 15, the three frame size distributions are quite different in other respects. E.g., the office camera (middle) records a very large number of quite small frames, perhaps reflecting times when the office is not occupied. In contrast, one might imagine that the parking facility's camera reflects periods of no automobile traffic, periods with 1-2 vehicles in motion, and periods with many vehicles in motion.

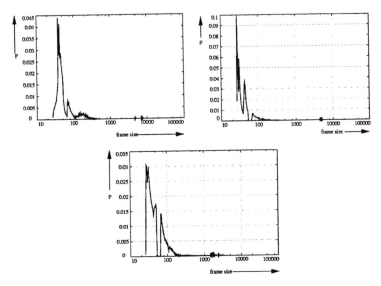

Fig. 16 Frame size distributions from low quality MPEG-4 traces for parking lot, office, and lecture hall scenarios (from [24]).

3.6   Scene Lifetime Influence

Figure 17 shows the effects of scenes within a contiguous video segment. The Figure shows a 40- second segment recorded at low MPEG-4 quality, but there appear to be four distinct 'scenes':

- Frames 1-150 show a regular pattern with little non-GOP variability
- Frames 151-390 show exponential decay in the I-frame sizes, and variability in P and B sizes
- Frames 391-610 show no trend, but somewhat more variability relative to P and B sizes
- Frames 611-1000 show far more variability compared to other segments, and perhaps evidence of a 'seasonal' or periodic component.

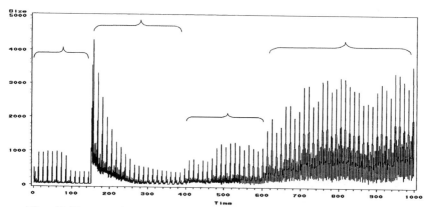

Fig. 17 Frame sizes varying by scene in a 40 second low quality MPEG-4 trace.

3.7   Modeling Traffic at Its Characteristic Rate

Traffic experts have proposed a number of model families to take advantage of the characteristic statistical signatures induced by multimedia encoders (e.g., MPEG-4 GOPs). There is a large body of research in this area dating back to the mid-1980s[6] – e.g., non-linear threshold AR models, fractional ARIMAs, DARs, GBARs, nested ARs, various LRD models, MMRPs, $M/G/\infty$ models, fluid flow models, TES, ARTA, ARCH, GARCH, techniques based on decomposition and recombination, etc. Many of these models and techniques are valid. It is not clear if one family is best, or whether one family might be best suited for building realistic traffic generators.

E.g., the trace in Figure 17 changes rather dramatically over time. Various sub-segments are non-stationary in both the mean and the variance, so transformations (e.g., differencing, logarithmic, Box-Cox) will likely be required in order to analyze them. They also have unmistakable periodic components induced by the encoder which will likely require additional transformations (e.g., seasonal differencing).

---

[6]   The resource requirements and demand characteristics of video traffic were documented during the ATM era; e.g., [9, 13, 14, 15, 17, 18, 22, 23, 25] and modeled in [4, 5/6, 7/8].

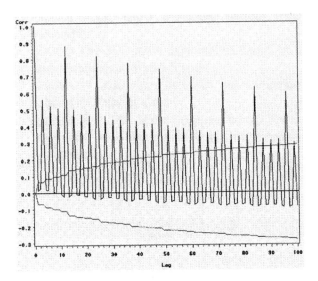

Fig. 18 Autocorrelation Function.

If we plot the autocorrelation function of a 1,000 frame MPEG-4 sub-segment to 100 lags, we see strong seasonal components at multiples of lags 3 and 12 (Figure 18). This is typical of a multiplicative seasonal process. E.g., a multiplicative seasonal autoregressive integrated moving average process (ARIMA × S-ARIMA) is one candidate. However, if we consider low quality MPEG-4 segments of different lengths (e.g., 2, 4, 6, 8, 10, 33 67, and 333 seconds) and plot their frame sizes vs. frame numbers from these traces (Figures 19-21), we see different behaviors in each. So, when modeling traffic traces at characteristic rates (i.e., all frames without aggregation), it is likely that segments of different sizes have significantly different ARIMA(s,p,d,q) representations (Table 1.)

Table 1. ARIMA Family Models for Traces in Figures 19-21.

Sample Size in Frames (Figure)	S(t)	Series Variance	Seasonal Differencing at Lag(s)	Moving Average at Lag(s)	Model (See note[7])	Model Variance Estimate
60 (Fig. 19 left)	Frame size	Stationary	12	3	$(1 - B^{12}) \times S(t) = (1 - 0.70\ B^3) \times e(t)$ (See note [8])	4,849
120 (Fig. 19 right)	Frame size	Stationary	12	3	$(1 - B^{12}) \times S(t) = (1 - 0.21 B^3) \times e(t)$	4,667
180 (Fig. 20 left)	Frame size	Borderline stationary	1, 12	1, 3, 12	$(1 - B^1)(1 - B^{12}) \times S(t) = (1 - 0.76\ B^1)(1 + 0.33\ B^3)(1 - 0.75\ B^{12}) \times e(t)$	154,336
240 (Fig. 20 right)	Log of frame size	Non-stationary	1, 12	1, 3, 12	$(1 - B^1)(1 - B^{12}) \times S(t) = (1 - 0.23\ B^1)(1 - 0.11\ B^3)(1 - 0.54\ B^{12}) \times e(t)$	0.1766
300 (Fig. 21)	Log of frame size	Non-stationary	1, 12	1, 3, 12	$(1 - B^1)(1 - B^{12}) \times S(t) = (1 - 0.27\ B^1)(1 - 0.10\ B^3)(1 - 0.55\ B^{12}) \times e(t)$	0.1546

[7] Method is conditional least squares. The $e(i) \sim$ i.i.d.$(0, \sigma(e))$, where $\sigma(e)$ is the variance of the set $\{e(i)\}$: $e(i) =$ actual $S(i) -$ predicted $S(i)$ $\forall$ i in the sample. Mean terms are not significant in any of these models.

[8] Model: $(1 - B^{12}) \times S(t) = (1 - 0.70\ B^3) \times e(t)$ is a standard shorthand notation for $S(t) - S(t-12) = e(t) - 0.70\ e(t-3)$. The backshift operator B operates on the $S(t)$ and $e(t)$ terms.

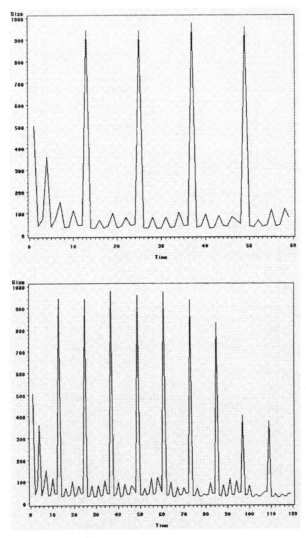

Fig. 19 Low quality MPEG-4 frame size vs. frame number from 2 and 4 second traces.

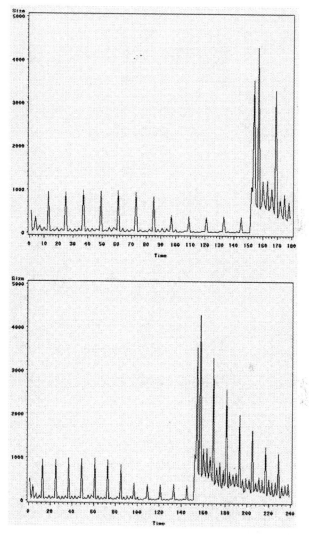

Fig. 20 Low quality MPEG-4 frame size vs. frame number from 6 and 8 second traces.

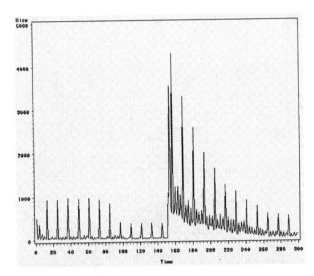

Fig. 21 Low quality MPEG-4 frame size vs. frame number from a 10 second trace.

Some of the models in Table 1 are excellent fits. E.g., the first two models capture the seasonal effect of the large I frames repeating at intervals of 12, and their moving average terms capture a secondary seasonal effect of the P frames' structure in the IBB PBB PBB PBB pattern (Figure 19). The pattern is very regular with little variability.

Other models in Table 1 are less successful. E.g., the third model's series is borderline stationary due to the introduction of a second "scene" near frame 150 and a rather dramatic exponential decay pattern in the I frame sizes. However, the fourth and fifth models have sufficient data to capture the effects of the two distinct processes representing both scenes, and they do so once the series' non-stationary variance is corrected via a logarithmic transformation on the frame size. (Note that a more realistic analysis of this data would model individual scenes or scene segments rather than using a rigid segment size.)

Figures 22 and 23 depict longer segments of the trace (and their models are not included herein.) Note in Figure 22 (left) that there appears to be a longer range seasonal component that is present in frames 611-1000 and continues beyond frame 1000 (Figure 22, right). This could be an indication of long-range dependence in frame sizes, which is characteristic of a self-similar stochastic process. The component is less pronounced in a much longer 333 second trace (Figure 23).

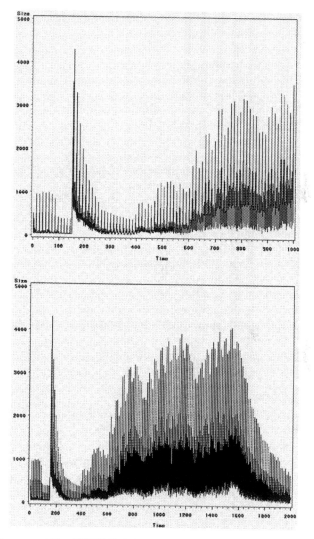

Fig. 22 Low quality MPEG-4 frame size vs. frame number from 33 and 67 second traces.

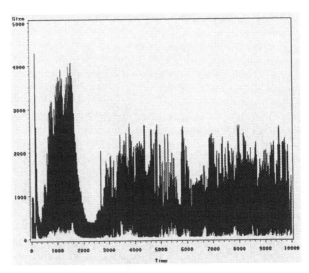

Fig. 23 Low quality MPEG-4 frame size vs. frame number from a 333 second trace.

3.8   Modeling Traffic Aggregates at Different Time Scales

There are several ways to model traffic at different time scales (e.g., at 1, 10, 100 second intervals). One way is to include every frame (as in Section 3.7) when developing the model. Another way is to aggregate frames and model the aggregate. However, if we aggregate every 12 frames then we almost certainly lose the seasonal effects induced by the MPEG-4 GOP structure. To illustrate, we aggregate an 80 second (2,400 frame) segment in 12-frame GOP units (resulting in 200 aggregates) and plot the 12:1 aggregate results in Figure 24. Note that the time axis in Figure 24 is in units of 12/30 = 0.4 second increments, reflecting 12 frame-times.

First-order differencing is required to fit an ARIMA model as the mean is non-stationary The 'best' ARIMA model is a third-order moving average on the *difference* D(t), where[9]:

$$D(t) = S(t) - S(t-1) \text{ for frame size } S(i), \text{ and}$$
$$(1 - B^1) \times D(t) = (1 + 0.33\ B^1)(1 - 0.19\ B^2)(1 - 0.18\ B^3) \times e(t).$$

---

[9]   Method is conditional least squares, and the model's mean term is not significant.

As expected, aggregation has caused us to lose the seasonal component with period 12.

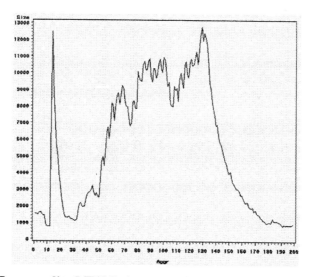

Fig. 24 Low quality MPEG-4 aggregate size vs. number from an 80 second trace with 12:1 aggregation.

Figures 25-27 illustrate how behaviors change when aggregated over different timescales. Figure 25 (left) is a 50 minute trace with 12:1 (1 GOP) aggregation; Figure 25 (right) is the same trace with 24:1 (2 GOP) aggregation. The 12:1 aggregation corresponds to an aggregate time of 0.400 seconds, and the 24:1 aggregation corresponds to an aggregate time of 0.800 seconds. Note that the two plots are grossly similar in appearance.

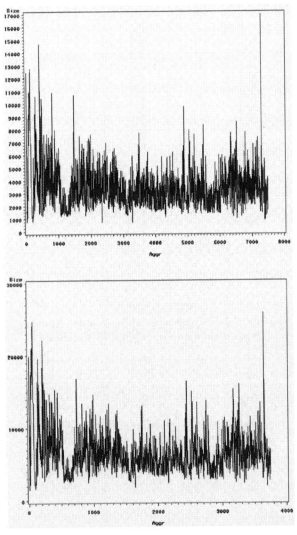

Fig. 25 Low quality MPEG-4 aggregate size vs. number, 60 minute trace, 12:1 and 24:1 aggregation.

Figure 26 (left) is a 50 minute trace with 240:1 (20 GOP) aggregation; Figure 26 (right) is the same trace with 2400:1 (200 GOP) aggregation. The 240:1 aggregation corresponds to an aggregate time of 8.000 seconds, and the 2400:1 aggregation corresponds to an aggregate time of 80.000 seconds. Note that the two plots are not similar in appearance, although the leftmost somewhat resembles the plots in Figure 25.

Figure 27 is a 50 minute trace with 24000:1 (2000 GOP) aggregation. The 24000:1 aggregation corresponds to an aggregate time of 800.000 seconds. Note that the plot does not resemble the plots in Figures 24 or 25.

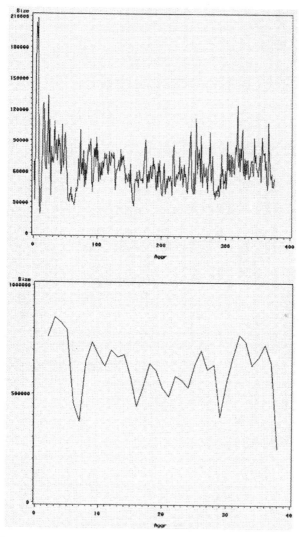

Fig. 26 Low quality MPEG-4 aggregate size vs. number, 50 minute trace, 240:1 and 2400:1 aggregation.

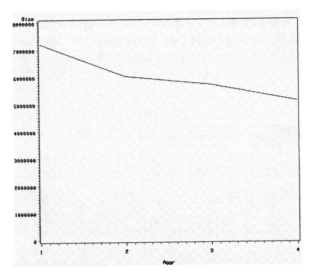

Fig. 27 Low quality MPEG-4 aggregate size vs. number, 50 minute trace, 24000:1 aggregation.

Aggregation can occur in several ways. One might aggregate frame sizes into a constant time interval (e.g., 1, 10, 100 seconds), or aggregate frame sizes into scenes of a few seconds to several minutes in length. There is theory that describes the temporal aggregation properties of a particular ARIMA sub-family [2, 3, 12], although it is not clear whether these results can be extended to ARIMA models, SARIMAs, fractional ARIMAs, etc. that are more typical of encoded multimedia traffic. (Results for ARMA temporal aggregation are summarized in Appendix III.)

3.9 An Empirical Approach

It is likely that one can use a 'characteristic signatures' approach to build a representative set of parametric traffic models for standard multimedia types. The general approach is to: **(1)** select a number of representative multimedia traces with different multimedia types, encodings, time scales, etc.; **(2)** characterize them statistically; **(3)** tabulate the characterizations; and **(4)** build realistic, parametric traffic generators for these models. SARIMA processes and models are one approach, but others may be also be used; Results might be tabulated as follows:

Table 2. Tabulation of Models for Various Multimedia Traces.

Multimedia type	Encoding	Aggregation	Timescale	Model
Video	MPEG-4	None	0.033 sec	SARIMA(s,p,d,q)
Video	MPEG-4	12:1	0.400 sec	...
Video	H.263	...	...	...
Audio	...	...	...	...
...	...	...	...	...

Characterization (step 2) could be very labor-intensive. However, a semi-automated heuristic for characterizing models might be implemented as follows:

- Define a practical upper bound d_{max} for the degree of differencing required to stabilize a nonstationary series mean; $d \in \{0,1,2\}$ in practice.
- Define practical ranges for the empirical SARIMA orders (s, p, q). E.g., s $\in \{1, 2, 3, 12\}$, $p < 5$, and $q < 5$ in practice.
- For each empirical series:
  - Test for a white noise process using a Box-Pierce or Ljung-Box statistic; if white noise, then stop.
  - Test the series for empirical stationarity. If the dth-differenced time series is not mean-stationary ($d=0,1,...,d_{max}-1$), repeat the test using the precomputed $d+1^{st}$-differenced series up to d_{max}. Perform logarithmic or Box-Cox transformations if the series is not variance-stationary.
  - Fit all possible candidate SARIMA models using a statistical package (e.g., SAS, R, S-Plus).
  - Select among the candidate models using some information criterion[10].
  - Perform a final test for stationarity (based on Schur's theorem) after orders (s, p, d, q) and parameters $\{\alpha_i\}$ and $\{\beta_j\}$ are determined.

---

[10] The Akaike information criterion (AIC) [1] is defined as AIC = -2(ln L) + 2P where L is the exact maximum likelihood and P is the number of parameters. The AIC tends to overfit, so some prefer the Schwartz Bayesian information criterion defined as SBC = -2(ln L) + P × (ln R) where R is the number of computed residuals. Others include the Hannan-Quinn information criterion, Sawa's criterion, Akaike's reformulated AIC, Chow's criterion, Amemiya's PC, etc. Regardless, one selects the adequate model with the smallest information criterion.

Poorly fit models can arise in some situations. However, models derived from over-parameterized or over-differenced time series are usually valid (though not optimal).

3.10 Preliminary Conclusions

Using SARIMA processes to model empirical 25-30 frame per second MPEG-4 multimedia traces confirms that analytical multimedia traffic models can provide reasonable approximations. We have developed several multiplicative SARIMA(s,p,d,q) models and have examined the behavior of these models with different sample sizes and at different time scales (with aggregation). Preliminary results suggest that the seasonal effect induced by MPEG GOP encoding is the dominant factor at time scales up to a few tens of seconds, while scene length is the dominant factor at longer time scales. It also appears feasible to automate the modeling step, which could be used to analyze relatively large trace libraries. Note that while the MPEG-4 traces we have used have GOPs of 12, this is not a requirement.

4. SUMMARY

We believe the preliminary results outlined herein address COST Action 285's Task 2.2 – multilayer traffic modeling in which models from several protocol layers are combined; and Task 1.3 – multimedia traffic modeling at different time scales. We believe the research directions described herein will lead to additional results of value to the COST Action.

APPENDIX I – TCP Reno Model [19]

Table 3. Parameters Used in the TCP Reno Model.

Input	p	Probability that a packet is lost, typically 0.01 to 0.05
	RTT	Round trip time, typically 0.1 to 4.0 seconds
	T_0	Timeout interval (i.e., wait upon loss), typically 1 to 7 seconds
	W_m	Maximum congestion window size (i.e., largest window size)
Computed	$G(p)$	Algebraic function of p thru six timeout intervals
	$W(p)$	Window size when a loss occurs
	$Q(p, W(p))$	Probability that a loss in a window of size $W(p)$ is due to a timeout
	$Q(p, W_m)$	Probability that a loss in a window of size W_m is due to a timeout
	$T(p)$	TCP throughput (packets received per unit time)

$$G(p) = 1 + p + 2p^2 + 4p^3 + 8p^4 + 16p^5 + 32p^6 \qquad (1)$$

$$W(p) = \frac{2}{3} + \sqrt{\frac{4(1-p)}{3p} + \frac{4}{9}} \qquad (2)$$

$$Q(p, W(p)) = \min\left(1, \ \frac{(1-(1-p)^3)(1+(1-p)^3(1-(1-p)^{W(p)-3}))}{1-(1-p)^{W(p)}}\right) \qquad (3)$$

$$Q(p, W_m) = \min\left(1, \ \frac{(1-(1-p)^3)(1+(1-p)^3(1-(1-p)^{W_m-3}))}{1-(1-p)^{W_m}}\right) \qquad (4)$$

$$\text{if } W(p) < W_m \text{ then } T(p) = \left(\frac{\dfrac{1-p}{p} + \dfrac{W(p)}{2} + Q(p,W(p))}{RTT(W(p)+1) + \dfrac{Q(p,W(p))G(p)T_0}{1-p}} \right) \quad (5)$$

$$\text{else} \qquad T(p) = \left(\frac{\dfrac{1-p}{p} + \dfrac{W_m}{2} + Q(p,W_m)}{RTT(\dfrac{W_m}{4} + \dfrac{1-p}{pW_m} + 2) + \dfrac{Q(p,W_m)G(p)T_0}{1-p}} \right) \quad (6)$$

APPENDIX II – ARMA, ARIMA, and SARIMA Models

An autoregressive moving average process of order (p, q), denoted ARMA(p, q), is defined as

$$(Y_t - \mu) - \alpha_1(Y_{t-1} - \mu) - \ldots - \alpha_p(Y_{t-p} - \mu) = \varepsilon_t - \beta_1\varepsilon_{t-1} - \ldots - \beta_q\varepsilon_{t-q} \tag{7}$$

or more commonly, following a centering transformation, as

$$X_t - \alpha_1 X_{t-1} - \ldots - \alpha_p X_{t-p} = \varepsilon_t - \beta_1\varepsilon_{t-1} - \ldots - \beta_q\varepsilon_{t-q} \equiv X_t = \sum_{i=1}^{p}\alpha_i X_{t-i} + \varepsilon_t - \sum_{j=1}^{q}\beta_j\varepsilon_{t-j} \tag{8}$$

where Y_t is the measured frame size at time t, μ is the mean of the time series, $\{\varepsilon_i\} \sim \text{iid}(0, \sigma^2_\varepsilon)$ is a series of uncorrelated random disturbances, X_t is the mean-centered frame size at time t, and $\{\alpha_i\}$ and $\{\beta_j\}$ are parameter sets. In general, the frames sizes $\{Y_i\}$ exhibit serial autocorrelation and thus are not independent. The series of uncorrelated random disturbances $\{\varepsilon_i\} \sim \text{i.i.d.}(0, \sigma^2_\varepsilon)$ can also be viewed as a time series of residuals, where

$$\varepsilon_i = (X_i)_{actual} - (X_i)_{predicted} \tag{9}$$

If d denotes the number of differencing transformations needed to induce a stationary ARMA(p, q) process, then an autoregressive integrated moving average (ARIMA) process of order (p, d>0, q) is simply an ARMA(p,q) representation of the series of dth order differences $\{D_i\}$ as

$$D_t - \alpha_1 D_{t-1} - \ldots - \alpha_p D_{t-p} = \varepsilon_t - \beta_1\varepsilon_{t-1} - \ldots - \beta_q\varepsilon_{t-q} \equiv D_t = \sum_{i=1}^{p}\alpha_i D_{t-i} + \varepsilon_t - \sum_{j=1}^{q}\beta_j\varepsilon_{t-j} \tag{10}$$

ARMA(p, q) processes incorporate short-term trends and cycles; ARIMA(p, d>0, q) processes can also model longer term (but local) trends in the realization of the time series.

A seasonal process is a weighted sum of periodic values whose cycles reflect strong harmonic relationships. E.g., a seasonal process may include a term that reflect MPEG-4's GOP structure with strong harmonics at periods of 3 and 12 for P and I frames, respectively (e.g., Figure 19, left):

$$X_t - X_{t-12} = \varepsilon_t - 0.70\varepsilon_{t-3} \tag{11}$$

Seasonal effects are common in multimedia traffic where frame sizes are cyclically related to frame sizes in earlier time periods. Seasonal ARIMA processes, denoted SARIMA(p, d, q, s) and seasonal multiplicative ARIMA processes, denoted ARIMA(p, d, q) × SARIMA(P, D, Q, S) are not uncommon for some types of traffic at longer time scales. (S and s represent the periods of the seasonal components.)

APPENDIX III – Aggregating ARMA Processes

If frame sizes $\{Y_i\}$ follow an ARMA(p, q) process, then an aggregate of the $\{Y_i\}$ over an interval that is several hundred times larger can be predicted using results from econometrics [2, 3, 12]].

Case 1. An ARMA(p, q) process consisting of consecutive and non-overlapping sums of m terms is estimated by an ARMA(p, floor(((p+1) · (m-1)+q)/m)) temporally aggregated process. For sufficiently large m,

if q ≥ p+1 then ARMA(p, q) → ARMA(p, p+1)   (12)
if q < p+1 then ARMA(p, q) → ARMA(p, p)

The implications for ARMA traffic streams are that aggregates of ARMA streams produce streams in which the orders (p, q) are predictable.

One might also sample every k^{th} observation without aggregation. If frame sizes $\{Y_i\}$ follow an ARMA(p, q) process, then a systematic (sparse, non-aggregated) sampling of the $\{Y_i\}$ over an interval that is several hundred times larger can be predicted using additional results from econometrics [2, 3].

Case 2. An ARMA(p, q) process which is generated at some arbitrary time unit t and is consecutively sampled every $k \cdot t$ time units ($k > 1$) is estimated by an ARMA(p, floor((p · (k-1)+q)/k)) process. The limit for large k depends on p and q. When sampled at every k^{th} interval,

if q ≥ p then ARMA(p, q) → ARMA(p, p)   (13)
if q < p then ARMA(p, q) → ARMA(p, p-1).

The implications for ARMA streams are that sparse-sampled ARMA streams produce streams in which the orders (p, q) are predictable.

REFERENCES

[1] Akaike H. "A new look at the statistical model identification." *IEEE Trans. Automatic Control.* Vol AC-19. No 6. December 1974.

[2] Amemiya T. and R. Wu. "The effect of aggregation on prediction in the autoregressive model." *Jour. American Statistical Assoc. - Theory and Methods Section.* Vol 67. No 339. September 1972.

[3] Brewer K. "Some consequences of temporal aggregation and systematic sampling for ARMA and ARMAX models." *Jour. Econometrics.* Vol 1. No 2. June 1973.

[4] Bragg A. and W. Chou. "Traffic analysis for high-speed networks." *Proc. IEEE Globecom '91.* December 1991.

[5] Bragg A. and W. Chou. "Analytic models and characteristics of video traffic in high-speed networks." *Proc. IEEE Second Intl. MASCOTS '94.* January 1994.

[6] Bragg A. and W. Chou. "The locality and transitional behavior of ARIMA network traffic models." *Proc. IEEE Third ICCCN '94.* September 1994.

[7] Bragg A. and W. Chou. "Real-time forecasting of bandwidth demand in high-speed communications networks." *Proc. 29th Conf. Info. Sciences and Systems (CISS '95).* March 1995.

[8] Bragg A. and W. Chou. "Real-time computation of empirical autocorrelation, and detection of non-stationary traffic conditions in high-speed networks." *Proc. IEEE Fourth ICCCN '95.* September 1995.

[9] Chimento P. "A review of video sources in ATM networks." in Viniotis Y. and R. Onvural (Eds.). *Asynchronous Transfer Mode Networks.* Plenum Press. New York. 1993.

[10] TCP Performance over ESNet, http://www-unix.mcs.anl.gov/~bester/historical/dsl/esnet.html

[11] MPEG-4 and H.261/263 Video Compression, http://www.apl.jhu.edu/Notes/Geckle/525759/ lecture11.pdf

[12] Granger C. and M. Morris. "Time series modelling and interpretation." *Jour. Royal Statistical Society* (A). Vol 139. Part 2. 1976.

[13] Grünefelder R. et al. "Characterization of video CODECs as autoregressive moving average processes and related queueing system performance." *RACE Traffic Performance in IBCN.* Munich. July 1990.

[14] Grünefelder R. et al. "Characterization of video CODECs as autoregressive moving average processes and related queueing system performance." *IEEE J. Selected Areas in Communic. JSAC* Vol 9. No 3.

[15] Heyman D. et al. "Statistical analysis and simulation study of video teleconference traffic in ATM networks." *IEEE Trans. Circuits and Systems for Video Technology.* Vol 2. No 1. March 1992.

[16] Koga H. et al., "Performance Comparison of TCP Implementations in QoS Provisioning Networks,", *Proc. INET 2000,* July 2000, Japan.

[17] Maglaris B. et al. "Performance models of statistical multiplexing in packet video communications." *IEEE Trans. on Communications.* Vol 36. No 7. July 1988.

[18] Nomura M. et al. "Basic characteristics of variable rate video coding in ATM environment." *IEEE Jour. Selected Areas in Communications.* Vol 7. No 5. June 1989.

[19] Padhye, J. et al., "Modeling TCP Reno Performance: A Simple Model and Its Empirical Validation," *Proc. IEEE/ACM Trans. Networking,* Vol. 8 No. 2, April 2000.

[20] Padhye J. and S. Floyd, "On inferring TCP behavior", *Proc. ACM SIGCOMM '01,* August 2001.

[21] Padhye J. et al., "Modeling TCP throughput: a simple model and its empirical validation", *Proc. ACM SIGCOMM '98,* August 1998.

[22] Rodriguez-Dagnino R. et al. "Prediction of bit rate sequences of encoded video signals." *IEEE J. Selected Areas in Communications. JSAC* Vol 9. No 3. April 1991.

[23] Sen P. et al. "Models for packet switching of variable-bit-rate video sources." *IEEE Jour. Selected Areas in Communications.* Vol 7. No 5. June 1989.

[24] Telecommunication Networks (TKN) Group, Department of Electrical Engineering, Technical University Berlin (http://www.tkn.tu-berlin.de/research/trace/trace.html) Verbiest W. et al. "The impact of the ATM concept on video coding." *IEEE Jour. Selected Areas in Communications.* Vol 6. No 9. December 1988.

CHAPTER 4

Multimedia Traffic Behavior: Analysis and Implications

Rachid El Abdouni Khayari, Axel Lehmann

Universitaet der Bundeswehr München, Deutschland
Institut fur Technische Informatik
WernerHeisenberg Weg 39, 85577 Neubiberg, Deutschland

Abstract. *System Performance strongly depends on the incoming workload. To improve the user perceived perfomance, it is mandatory to analyze this observed workload with the aim to choose the most adequate system configuration and processing methods of the incoming requests. In this work, we will examine, whether some characterisctis, like heavy taildeness, temporal locality, and frequency of references are restricted to certain document types, or are present for all of them. We will see, that this analysis is helpful for further studies, especially for developing new approaches to improve the system performance (e.g. for caches)*

Keywords: Workload analysis, heavy-tailed distributions, WWW, proxy server, multimedia, caching.

1. INTRODUCTION

The phenomenal growth of the World-Wide Web is caused by a permanent increase of web services and applications resulting in an increasing demand for web servers and proxies limiting web service performance. These limitations are mostly caused by a limited bandwidth or by web server access conflicts. It has been shown that considering special characteristics of the workload can help in developing new methods for improving the perceived system performance [3, 4, 6]: from measurements, ideas about typical request patterns are gained. These obtained insights are then used to develop methods for improving the system performance. For validating these new approaches, simulations can be used.

It is also known from different studies, that the presence of properties, such as self similarity, fractality and long-range dependency, in network traffic has significant implications on the system performance; ignoring such properties can lead to underevaluation of important system performance measures. For instance, we simulated a $.|G|n$-FCFS queue with two different arrival processes: an "empirical" arrival process obtained by logging requests to the technical university of Aachen from Germany (RWTH) web server and a Poisson arrival process with the same mean interarrival time. Some performance measures are shown in Table 1 for $n = 30$ and $n = 35$. As can be seen, the Poisson arrivals return a much too optimistic estimation of the performance measures. Furthermore, we can see that the bursty nature of the empirical arrival process leads to a strong increase of the response time when the number of servers is decreased from 35 to 30 (thus increasing the load on the servers)

Table 1. Performance Measures for a $-|G|N$-FCFS   Queue for Different Arrival Processes (RWTH Trace)

	arrival process	empirical (trace)	Poisson process
35	mean queue length	21.25	0.02
35	mean response time (ms)	5026	1871
35	cv^2 response time	12.02	22.66
30	mean queue length	384.14	0.29
30	mean response time (ms)	58958	1911
30	cv^2 response time	9.4	21.73

In many studies, some characteristics of traces of web requests have been studied. These traces comprise the access logs to some (proxy-) servers for requesting some documents. The analysis of the trace can help to achieve a detained workload characterization. This analysis is done for all the stored documents types (text, image, video, etc.). Therefore the results obtained can be seen as 'averaged' for all these stored and requested document types. In this work, we will examine, whether some characterisctis, like heavy-taildeness, temporal locality, and frequency of references are restricted to certain document types, or are applicable for all of them. We will see, that this analysis is helpful for further studies, especially for developing new approaches to improve the system performance (e.g. for scheduling or caching)

Caching has been recognized as one of the most important techniques to reduce the Internet bandwidth. Different caching algorithms have been developed to achieve this task. These caching algorithms have been developed for specific contexts and show different performance results depending on their application domains. This is caused by the fact that each caching algorithm rates document characteristics more important than others. For more details, we rely to [3, 4, 6]. In this paper we compare the performance of some well-known caching strategies with respect to different document types. For this purpose, we use trace-driven simulation and consider the resulting hit rates as well as the byte hit rates. For that purpose, we dealt with the following well-known caching strategies: LRU (Least Recently Used), SLRU (Segmented LRU), LRU-k, LFU (Least Frequently Used), LFU-Aging, LFF (Largest File First), GDS (Greedy Dual-Size), GDS-Hit, GDS-Byte, GDSF (GDS with Frequency). Further information about these caching strategies can be found in [5].

2. TRACE ANALYSIS

For our purpose, we used three differents traces. Table 2 shows informations about these. For our experiments, we focus on caching requests for static documents. Dynamic generated objects have been removed from the three traces. First, we have analysed the statistics of the whole three traces. Thereafter, we have extracted files with respect to some applications (document types) and collected the statistics of each file. In the second step, we analysed the performance of different caching algorithms for each extracted file using trace-driven simulation. In our analysis, we have considered the following document types audio, application, images, text, and video.

Table 2. Requesting Periode and Number of Requests for the Used Traces

	RWTH 2000	RWTH 2002	RWTH 2004
Begin	17.02.2000	4.09.2002	22.06.2004
End	02.03.2000	02.10.2002	13.07.2004
Number of requests	9 Mio	9 Mio	39,2 Mio

2.1   Analysis of the Whole Traces

Table 3 presents some important statistics for all the three traces. For all these traces, the heavy-tailedness of the object-size distribution is clearly visible: high squared coefficients of variation, and very small medians compared to the means.

Table 3. Statistics of the Three Traces

	RWTH 2000	RWTH 2002	RWTH 2004
Cacheable requests	8,278,346	8,567,655	37,620,430
Cachable byte	88.31 GB	110.31 GB	347.22 GB
Average object size	11,454.3 Byte	13,827.8 Byte	9,910.1 Byte
Median	2297 Byte	2365 Byte	335 Byte
Smallest object	118 Byte	17 Byte	2 Byte
Largest object	68.7 MB	714.7 MB	2047.9 MB
Squared coeff. of variation	275.33	2776.59	2495.39
number of different objects	4,668,689	4,819,490	15,748,853
Different Byte	72.28 GB	92.91 GB	281.41 GB

Due to the voluminous results, we only present the results for the RWTH2004 trace. We want here to mentione that similar results for both other traces have been observed.

2.2   Distribution of the Object Sizes

Figures 1 shows the complementary log-log plot of the object-size distribution. As can be seen, this distribution decays more slowly than an exponential distribution, thus showing heavy-tailedness. This becomes even more clear from the histogram of object sizes in Figure 2. The heavy-tailedness is also present when looking at the request frequency as a function of the object size (see Figure 3) It shows that small objects are not only more numerous but also that they are requested more often than large objects (this inverse correlation between file size and file popularity has also been stated in [7]). Thus, caching strategies which favor small objects are supposed to perform better. However, the figure also shows that large objects cannot be neglected.

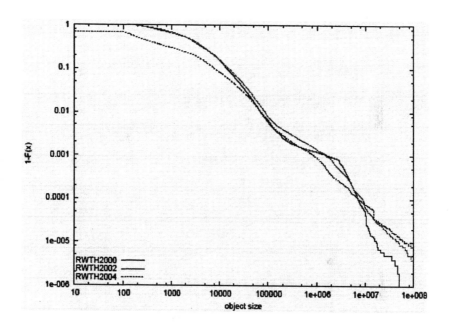

Fig. 1. Complementary Log-Log plot of document size distribution for the three traces: RWTH2000, RWTH2002 and RWTH2004

2.3   Recency of References

Another way to determine the popularity of objects is the temporal locality of their references [1]. However, recent tests have pointed out that this property decreases [2], possibly due to client caching. We performed the common LRU stack-depth methode [1] to analyze the temporal locality of references. The results are given in Figure 4. The positions of the requested objects within the LRU stack are combined in 5000 blocks. The figure shows that about 45% of all requests have a strong temporal locality, thus suggesting the use of a recency-based caching strategy.

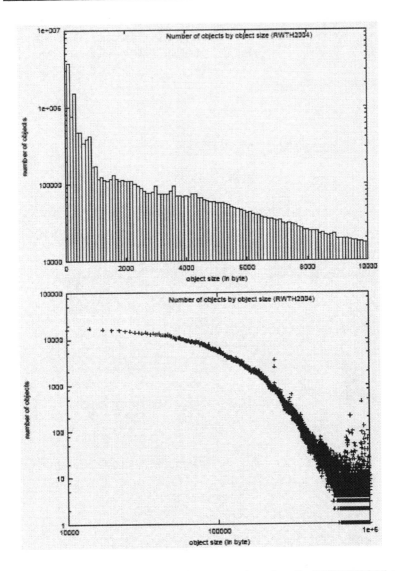

Fig. 2 Distribution of the document sizes for the RWTH2004 trace; top: document sizes until 10,000 byte, bottom: document sizes up 10.000 Byte (log-log scale)

2.4   Frequency of References

Objects which have been often requested in the past, are probably popular for the future too. This is explained by Zipf's law: if one ranks the popularity of words in a given text (denoted ρ) by their frequency of use (denoted P), then it holds $P \sim 1/\rho$.

Studies have shown that Zipf's law also holds for WWW objects. Figure 5 shows a log-log plot of all 8.3 million requested objects of the RWTH-trace. As can be seen the slope of the log-log plot is nearly -1, as predicted by Zipf' law, suggesting the use of frequency-based strategies. It should be mentioned that there are many objects which have been requested only once. Frequency-based strategies have the advantage that "one timers" are poorly valued, so that frequently requested objects stay longer in the cache, and thereby cache pollution can be avoided.

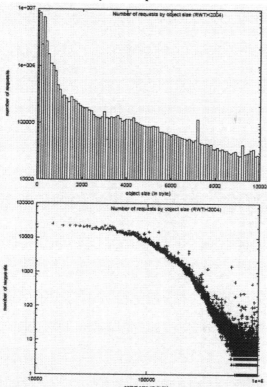

Fig. 3. Number of objects as function of object size for the RWTH2004 trace; top objects until 10,000 Bytes, bottom: log-log scale for objects larger than 10kb

2.5   Analysis of The Extracted Files

As for the whole trace, we study the extracted files with respect to the specific document types. Altogether, we found the same characteristics for the extracted files. In the foil wing are listed:

- In Table 4, some statistics of the different document types are listed.
- In Figure 6, the distribution of the object sizes are shown; for all the document types, the heavy-taildeness can be shown.

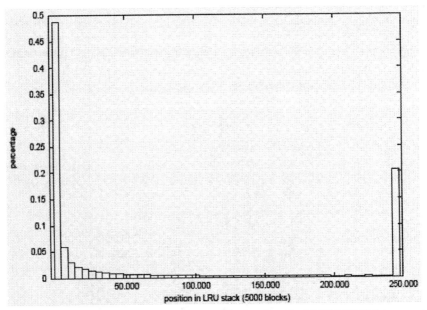

Fig. 4   Temporale locality chacteristics for the RWTH2004 trace

- In Figure 7, the temporale locality of the extracted files for the RWTH004 trace is shown (for images, text, audio, video and applications).
- In Figure 8, the Zipf law for the extracted files (images, text, audio, video and application) is illustrated.

Table 4. Statistics of the Extracted Traces for the RWTH2004 Trace

	audio	App.	image	Text	Video
Requests	25,328	462,652	16,143,840	7,053,820	28,779
Byte total	25.2 GB	54.87 GB	79.37 GB	91.88 GB	80.77 GB
average document	1,043.25 KB	124.37 KB	5.15 KB	1,365.86 KB	2,942.95 KB
size Median	325,595 Byte	3,418 Byte	867 Byte	828 Bte	848,192 Byte
smallest object	152 Byte	176 Byte	126 Byte	115 Byte	209 Byte
largest object	318.33 MB	466 MB	23.93 MB	700 MB	2,047.94 MB
squared coeff. of var.	15.29	399.94	29.35	4,116.96	3.33
different objects	20,960	134,578	7,579,474	2,349.021	26,841
different Byte	22.7 GB	45.84 GB	62.46 GB	61.39 GB	6.65 GB

3. IMPLICATIONS FOR CACHING

The cache size is a decisive factor for the performance of the cache, hence, we want to choose the caching strategy that provides the best result for a given cache size. To compare the caching strategies, we have performed the evaluation with different cache sizes. We performed the trace-driven simulations using our own simulator, written in C++. To obtain reasonable results, the simulator has to run for a certain amount of time without hits or misses being counted. The so-called *warm-up* phase was set to 8% of all requests.

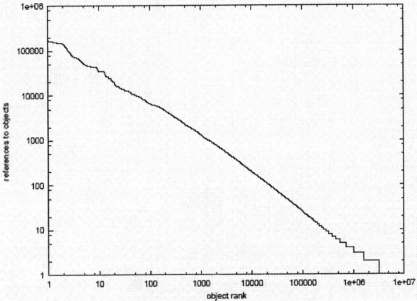

Fig. 5 Zipf Law: Frequency of reference as a function of object rank for the RWTH2004 trace

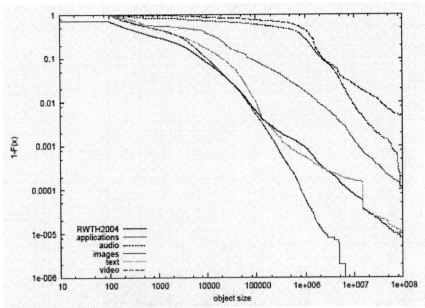

Fig. 6 Complementary log-log plot of document size distributions for the different document types

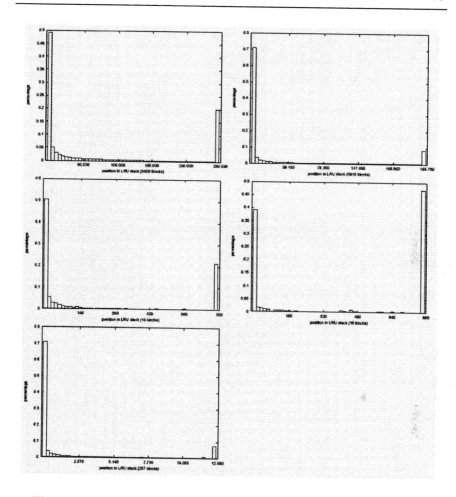

Fig. 7 Temporale locality of the extacted files for the RWTH2004 trace: images, text, audio, video and applications

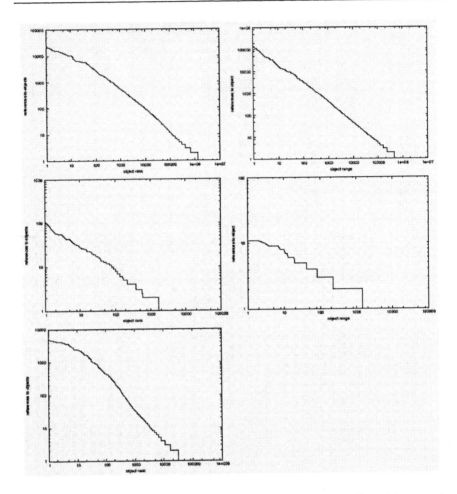

Fig. 8   Zipf law for the extracted files: image, text, audio, video   and applications

In Figure 9, we show the simulation results of the whole RWTH2004 trace for the different caching strategies with respect to the hit rate and byte hit rate. Respect to the hit rate, the simulations show that GDS-Hit (and LFF) provides the best performance. For the byte hit rate, one observes that the performance of all strategies is nearly equal, except for LFF which yields the worst results.

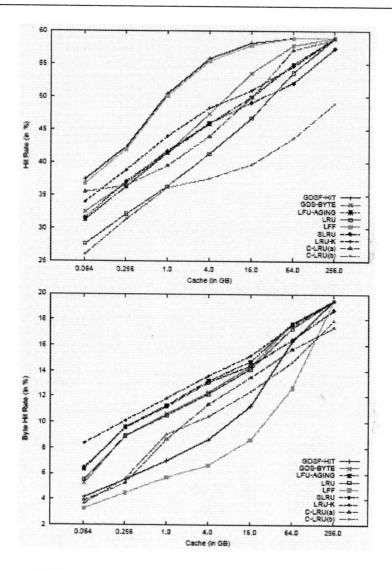

Fig. 9 Hit rate and byte hit rate for the whole RWTH2004 trace

We also performed trace-driven simulations using the same simulator for these different extraced files by considering the hit rate as well as the byte hit rate. Figures 10 to 14 show the obtained results for the extracted files respect to the document types. All algorithms have nearly the same behaviour according to the graphs with the byte hit rate. Only LFF performs significantly worse. Regarding the hit rate, GDSF-Hit caching algorithm offers the best results.

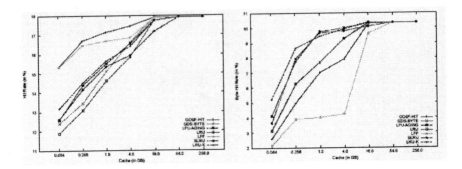

Fig. 10   RWTH2004 trace: hit rate and byte hit rate for audio documents

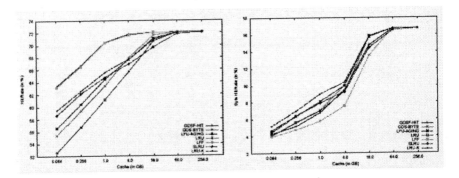

Fig. 11 RWTH2004 trace: hit rate and byte hit rate for application documents

The results of these trace driven simulations complied with those obtained in [4,5], namely that the GDS-Hit delivers the best results for the hit rate, and for the byte hit rate no notable difference could be detected, except for LFF which performs worse.

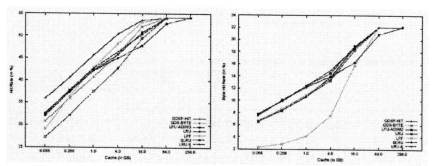

Fig. 12 RWTH2004 trace: hit rate and byte hit rate for image documents

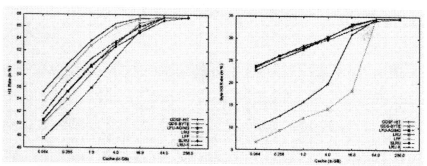

Fig. 13 RWTH2004 trace: hit rate and byte hit rate for text documents

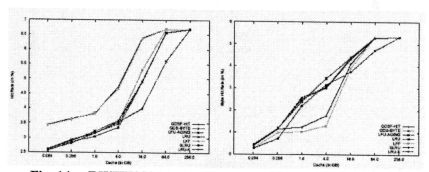

Fig. 14   RWTH2004 trace: hit rate and byte hit rate for video document

4. CONCLUSIONS AND FURTHER WORK

The analyses of the different document types have shown that some specific workload characteristics, heavy-tailedness of document sizes, temporale locality and Zipf's law, are maintained for the whole trace as well as in extracted files related to different document types. This fact is responsable for the obtained similar results for the performance resutls of caching algorithms. One of the conclusions that can be made here ist that it is better to concentrate the researches on developing performant methods (e.g. for caching) than to differentiate between document types.

REFERENCES

[1] M. F. Arlitt and C. L. Williamson. Internet Web servers: Workload Characterization and Performance Implications. *IEEE/ACM Transactions on Networking* 5(5)631-645, October 1997.

[2] P. Barford, A. Bestavros, A. Brdley and M. Crovell Changes in Web Client Access Patterns. *WWW Journal* 2(1)-16, 1999.

[3] B. R. Haverkort, R. El Abdouni Khayari, and R. Sadre. A Class- Based Least Recently Used Caching Algorithm for WWW Proxies. In *Computer Performance Evaluation-Modling Techniques and Tools,* pages 273-290. Lecture Notes in Computer Science 2324, Springer Verlag, Berlin, 2003.

[4] R. El Abdouni Khayari R. Sadre, and B. R. Haverkort. Class-Based Least-Recently Used Caching Algorithm for WWW Proxies. In *Proceedings of the 2nd Polish-German Teletraffic Symposium,* pages 295-306, Gdansk, Poland, September 2002.

[5] Rachid El Abdouni Khayari. *Workload-Driven Design and Evaluation of Web-Based Systems.* IBN: 3-89959-073-2. Der andere Verlag, Osnabrueck, Germany 2003.

[6] Rachid El Abdouni Khayari. *Workload-Driven Design and Evaluation of Web-Based Systems.* PhD thesis, Technical University of Aachen, Germany, February 2003.

[7] J. Robinson and M. Devrakond. Data Cache Management Using Frequency-Based Replacement. In *Proceedings of ACM SIGMTRICS,* pages 13442, Boulder, M 1990.

CHAPTER 5

Traffic Modeling and Prediction using ARIMA/GARCH Model

Bo Zhou, Dan He, and Zhili Sun

Centre for Communication System Research
University of Surrey
Guildford, Surrey
United Kingdom

Abstract. *The predictability of network traffic is a significant interest in many domains such as congestion control, admission control, and network management. An accurate traffic prediction model should have the ability to capture prominent traffic characteristics, such as long-range dependence (LRD) and self-similarity in the large time scale, multifractal in small time scale. In this paper we propose a new network traffic prediction model based on non-linear time series ARIMA/GARCH. This model combines linear time series ARIMA model with non-linear GARCH model. We provide a parameters estimation procedure for our proposed ARIMA/GARCH model. We then evaluate a scheme for our models' prediction. We show that our model can capture prominent traffic characteristics, not only in large time scale but also in small time scale. Compare with existing FARIMA model, our model have better prediction accuracy.*

Keywords: Traffic prediction, traffic modeling, ARIMA, GARCH

1. INTRODUCTION

The predictability of network traffic is of significant interest in many domains, including adaptive applications [1], congestion control [2], admission control [3], wireless [4] and network management [5]. An accurate traffic prediction model should have the ability to capture the prominent traffic characteristics, e.g. short- and long- range dependence, self-similarity in large-time scale and multifractal in small-time scale. For these reasons, time-series models which are widely used in financial research field are introduced in network traffic modeling and prediction. In the earliest work, network traffic prediction models use linear time series models, e.g. Auto Regressive (AR) and Auto Regressive Integrated Moving Average (ARIMA) [6][7]. The exponential decay of the autocorrelation function of these models gives them the ability to capture the short-range dependence (SRD) characteristics only. However, it has been shown that the traffic data exhibited a high degree of long-range dependence (LRD) characteristics in addition to SRD [8]. Thus, such models cannot characterize the network traffic well, and unable for traffic prediction [6][8][9]. Another related model, known as the Fractional Auto Regressive Integrated Moving Average (FARIMA) model, can capture both SRD and LRD and has been used to model and predict traffic data [9][10]. However, this model cannot capture the multifractal which has been found in the network traffic in small time scale. For this reason, another Multifractal Wavelet model (MWM) has been introduced to solve this problem. MWM model can capture multifractal but cannot predict traffic [11]. On the other hand, it has been found that traffic exhibited non-stationary and non-linear properties and threshold autoregressive (TAR) model [12] has been proposed to model such properties. In [12], the authors have developed the first network measurement system which integrated prediction and they have also proposed running multiple predictors simultaneously and forecasting one which exhibiting the smallest prediction error produced on its measurements. Another significant prediction research work has been introduced in [6], which analyzed the prospects for multi-step prediction of network traffic using ARMA and MMPP models. Their analysis is based on continuous time ARMA and MMPP models driven by Gaussian noise sources. Making the assumption that such models are appropriate, they have developed analytic expressions on how far into the future prediction is possible before errors would exceed a bound, and they also showed how traffic aggregation and smoothing monotonically can help to increase prediction accuracy. Apart from the above mentioned model-based prediction schemes, [19] has reported that non-model-based prediction provides better prediction than model-based prediction as long as the traffic is LRD or self-similar. However, the authors only compared their non-model-based prediction model with the FARIMA and FBM models. Both these two models cannot capture traffic bursty very well and this bursty characteristic affects traffic prediction accuracy.

In this paper, we introduce ARIMA/GARCH model to our network traffic prediction research field. The Auto Regressive Integrated Moving Average (ARIMA) with Generalized Auto Regressive Conditional Heteroscedasticity (GARCH) model is a non-linear time series model which combined the linear ARIMA with conditional variance GARCH. It captures both SRD and LRD characteristics and observes self-similarity and multifractal. The ARIMA/GARCH has been applied to model many financial time series [13][14][15]. ARIMA model has been proposed in literatures for both modeling and prediction purposes. However in such models, the variance is constant, comparing with the bursty which has been found in data traffic, therefore these models cannot capture such characteristic well. For this reason, we introduced new GARCH model to explain bursty characteristics. The most important contribution of GARCH model is its dynamic variance, where the variance varies over time. Using this conditional variance property, we can explain our network's bursty characteristic.

First, we present a parameter estimation procedure for the ARIMA/GARCH model. After that, an adaptive prediction scheme is used to allow capturing the non-stationary characteristic and react to any possible growth in the future traffic. We apply our models to model and predict real traffic data. In the paper, the experimental data are processed to present the bytes per unit time, time unit is from 100ms to 10s.

The remaining of this paper is organized as follows. Section 2 presents the ARIMA/GARCH model, its parameter estimation procedure and the adaptive prediction scheme. The results of applying the ARIMA/GARCH to model and predict actual traffic data are given in section 3 and section 4 concludes this paper.

2.   ARIMA/GARCH MODEL

2.1   Introduction to ARIMA/GARCH model

ARIMA/GARCH is a combination which combined linear time series ARIMA with GARCH conditional variance. We call this the conditional mean and conditional variance model. This model can be repressed in the following mathematical expressions.

The general ARIMA (r,d,m) model for the conditional mean applies to all variance models. Let X_t to be the time series we want to model.

$$\left(\nabla^d X_t\right) = \sum_{i=1}^{r} \phi_i \left(\nabla^d X_{t-i}\right) + Z_t + \sum_{j=1}^{m} \theta_j Z_{t-j}, Z_t \sim WN(0, \sigma_t^2) \tag{1}$$

We can use the other concise form of this expression,
$$\phi(B)(1-B)^d X_t = \theta(B) Z_t$$
where $\phi(.)$ and $\theta(.)$ are the rth and m^{th} degree polynomials, and B is the backward shift operator ($B^j X_t = X_{t-j}$, $B^j Z_t = Z_{t-j}$, $j = 0, \pm 1, ...$), d is differenced d times. We define that X_t is the time series and Z_t is the innovation of the original time series. We also define the lag-1 difference operator ∇ by
$$\nabla X_t = X_t - X_{t-1} = (1-B) X_t$$
The conditional variance of the innovations Z_t, σ^2, is by definition,
$$V_{t-1}(X_t) = E_{t-1}(Z_t^2) = \sigma_t^2$$

The key insight of GARCH lies in the distinction between conditional and unconditional variances of the innovation process Z_t. The term conditional implies explicit dependence on a past sequence of observations. The term unconditional is more concerned with long-term behavior of a time series and assumes no explicit knowledge of the past. The various GARCH models characterize the conditional distribution of Z_t by imposing alternative parameterizations to capture serial dependence on the conditional variance of the innovations.

The general GARCH (p,q) model for the conditional variance of innovations Z_t is,

$$\sigma_t^2 = \alpha_0 + \sum_{i=1}^{p} \gamma_i \sigma_{t-i}^2 + \sum_{j=1}^{q} \alpha_j Z_{t-j}^2, \quad \alpha_0 > 0, \quad \alpha_j \geq 0, \quad \gamma_i \geq 0 \tag{2}$$

with constraints,

$$\sum_{i=1}^{p} \gamma_i + \sum_{j=1}^{q} \alpha_j < 1 \quad \alpha_0 > 0 \quad \gamma_i \geq 0 \quad i = 1, 2, \ldots, p \quad \alpha_j \geq 0 \quad i = 1, 2, \ldots, q$$

Combined with two models together, then get the whole ARIMA(r, d, m)/GARCH (p, q) model in mathematical expression,

$$\left(\nabla X_t\right)^d = \sum_{i=1}^{r} \phi_i (\nabla X_{t-i})^d + Z_t + \sum_{J=1}^{m} \theta_j Z_{t-j}, Z_t \sim WN(0, \sigma_t^2) \tag{3}$$

$$\sigma_t^2 = \alpha_0 + \sum_{i=1}^{p} \gamma_i \sigma_{t-i}^2 + \sum_{j=1}^{q} \alpha_j Z_{t-j}^2$$

2.2 ARIMA/GARCH Parameter Estimation

Parameter estimation is the first step in fitting an ARIMA/GARCH model to observed data. In a standard ARIMA(r, d, m)/GARCH (p, q) model, there are five parameters to be estimated. The whole procedure to estimate the ARIMA/GARCH parameters are as following:

The first step is to compute differenced parameter d. This parameterdetermines the time series' stationary. For the request of GARCHparameters' estimation [14], it can only work if the time series is stationary. For the network traffic data that behave non-stationary, we can use differenced operation to change it from non-stationary to stationary. Using the estimation method in [16], the differenced parameter d can be estimated and tested by the autocorrelation function (ACF).

Assume we define the differenced parameter d already, in the other hand, we can find the time series become stationary after differenced operation. The ARIMA model is a linear time series model that the mean is conditional changed but the variance is constant. The ARIMA back shift parameters' order of r and m can be determined by the autocorrelation function (ACF) and its partial autocorrelation function (PACF) [16]. From the characteristics of the ACF, it describes the correlation between the current states of the time series with the past. Using ACF, we can determine the moving average (MA) parameter's order m straight. From the characteristics of the PACF, it describes the correlation between the current states' innovation of the time series with the past. Using PACF, we can determine the auto regressive (AR) parameter's order r directly. The following Table 1 shows how the procedure works:

Table 1 Using ACF and PACF Detect ARMA (R,M) Order

	AR (r)	MA (m)	ARMA (r,m)
ACF	Tails off to 0	Cuts off to 0 after lag m	Tails off to 0 after m lags
PACF	Cuts off to 0 after lag r	Tails off to 0	Tails off to 0 after r lags

The initial GARCH parameters p and q should be estimated independently comparing with the ARIMA parameters r and m. we first assume that the three parameters of the ARIMA/GARCH model have been estimated, and the order of r and m are set to be zero. That means we only need to use the simplified GARCH model to model the time series. In other words, we only want to estimate the value of the parameter p and q in the GARCH model. In this model, the mean is constant, but the variance is conditional changed. The order of the parameters of GARCH can be determined by the ACF and PACF again, but normally these orders are not very accurate. From research experience [13][15], setting the order of p and q to one is already good enough to capture the variance's performance.

The parameter estimation procedure of ARIMA and GARCH models are both based on Box-Cox methodology. In such model, we use maximum likelihood method (MLE) to estimate initial parameters. Initial parameter estimates of general ARMA (r, m) conditional mean models are estimated by the three step methods outlined in [20]. At first, we estimate the autoregressive coefficients, φ, by computing the sample autocovariance matrix and solving the Yule-Walker equations. Then, using these estimated coefficients, we filter the observed series to obtain a pure moving average process. Finally, we compute the autocovariance sequence of the moving average process, and use it to iteratively estimate the moving average coefficients, θ. This last step provides an estimation of the unconditional variance of the innovations.

For GARCH models, we assume that the sum of the γ_p and the α_q is close to 1. Specifically, for a general GARCH (p,q) model, we assumes that $\gamma_1 + K + \gamma_p + \alpha_1 + K + \alpha_q = 0.9$. If $p>0$, then we allocate 0.85 out of the available 0.9 to the p GARCH coefficients, and allocate the remaining 0.05 among the q ARCH coefficients. $p=0$ specifies an ARCH (q) model in which we allocate 0.90 to the q ARCH terms. In our experimental case, we assume the GARCH model we use is (1,1) model, and initial estimates are expressed as follows: $\sigma_t^2 = k + 0.85\sigma_{t-1}^2 + 0.05\varepsilon_{t-1}^2$. Finally, we estimate the constant k of the conditional variance model by first estimating the unconditional, or time-independent, variance of ε. $\sigma^2 = \frac{1}{T}\sum_{t=1}^{T}\varepsilon_t^2$. This can also be expressed as $\sigma^2 = \dfrac{k}{1-\sum_{i=1}^{p}\gamma_i - \sum_{j=1}^{q}\alpha_j} = \dfrac{k}{1-(0.85+0.05)}$. And so,

$k = \sigma^2(1-(0.85+0.05)) = 0.1\sigma^2$.

The whole process of ARIMA/GARCH parameter estimation is shown in the Figure 1.

2.3 ARIMA/GARCH Prediction Scheme

The ARIMA/GARCH model can be used to predict one-step-ahead value of time series, which can be extended to k-step-ahead. We can use the mathematical expression to explain what is one-step-ahead and k-step-ahead prediction. Let X_t to be the time series that we want to predict its performance. The k-step-ahead prediction can be defined with $\hat{X}_{t+k}$. This means $\hat{X}_{t+k}$ denotes the k-step forecast made at origin t of X_{t+k} at some future time $t+k$, we call the k is the lead time. When $k=1$, then we call it is one-step-ahead prediction.

We use the one-step-ahead prediction to explain our prediction scheme, and then extend it to be the k-step-ahead prediction. The basic prediction scheme is given as Figure 2.

In one-step-ahead prediction scenario, we first set the forecast step to be one, which means each time we only forecast value at one time unit. According the ARIMA/GARCH modeling, we have known all the parameters from the historical actual time series, and we also know the last one of the historical time series data. From the basic idea of the forecasting [17], we can use the minimum mean square error (MMSE) forecast method to forecast its performance at one time unit. Because each time we will get a forecast range of the prediction result, we propose to provide the

upper probability limit as done in normal prediction method techniques
[17] to decide our forecast value. This value is the M_{t+1} value that we
expect. Once the actual traffic value becomes available, we update the
historical traffic data, and estimate the new value of parameters.

To extend the horizon of time series, the k-step-ahead prediction value
can be computed recursively. For example, to obtain the two-step-ahead
prediction value, the one-step-ahead predicted value is computed first.

Then it is used with the other lagged traffic values to compute the two-
step-ahead predicted value. This procedure is repeated to generate
subsequent k predicted values. This is similar as the one-step-ahead
prediction.

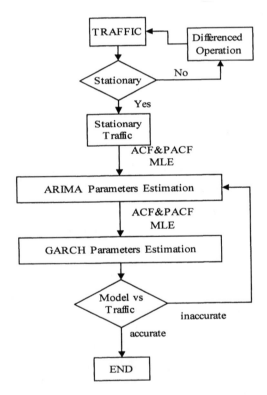

Fig. 1 Process of ARIMA/GARCH Parameter Estimation

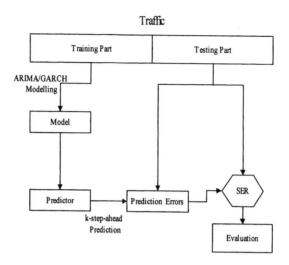

Fig. 2 ARIMA/GARCH Prediction Scheme

3.   EXPERIMENTAL RESULTS

In this paper, we apply the ARIMA/GARCH to model and predict the real network traffic data: *LBL-TCP-3*. It is available in the public domain at the Internet Traffic Archive site [18]. The traffic data are processed to present the amount of traffic in bytes per unit of time. This trace contains two hours' worth of all wide-area TCP traffic between the Lawrence Berkeley Laboratory and the rest of the world. Then they are aggregated at three different time scales: 100 ms for intermediate time scale, 1 s and 10 s for large time scale. The data we try to use contains two hours traffic, we use the first hour's traffic to estimate model's related parameters, after then use the second hour's traffic to test our model's prediction accuracy.

In the prediction procedure, we use one-step-ahead prediction and extend it to *k*-step-ahead prediction, where "step" means the time unit. For instance, if the time unit equals to 100ms, then one-step-ahead prediction will predict traffic in 100ms in future. This approach could operate on intermediate time scale and large time scale from our experiments. When the time scale becomes large, the approach's prediction approach becomes more accurate.

3.1   Traffic Modeling

The first step is to use ACF and PACF to test our time series' stationary, and to determine the model's differenced parameter d. After we estimate the parameter of d, we use the ACF and PACF again to determine the order of parameters of r and m. Then we can use MLE method to estimate the models parameters' values. Using the parameters estimation procedure we proposed above, we can compute the corresponding order and values of parameters p and q for GRACH. From the estimation procedure, we can determine all the details of parameters, and then we perform the traffic modeling.

In the original trace, we notice that there appears to be no long-run average level about which the series evolves. This is evidence of a non-stationary time series. After difference operation, the differenced trace appears to be quite stable over time, and the differenced operation has produced a stationary time series. We can also find this from its ACF, stationary ACF is balanced more or less evenly around zero correlation while non-stationary ones usually decline continuously with increasing lag. We also observe that the orders of r and m are all equal to one from ACF and PACF. Considered the property of GARCH model we mentioned before, the orders of GARCH can be defined as (1,1). Thus, the "ARIMA (1,1,1) / GARCH (1,1)" model is determined to model the *LBL-TCP-3* traffic.

Figure 3 shows the results of ACF and PACF operation.
Figure 3(a) is ACF without any difference operation, it can show non-
    stationary clearly;
Figure 3(b) is ACF after one difference operation, and
Figure 3(c) is PACF after one difference operation.

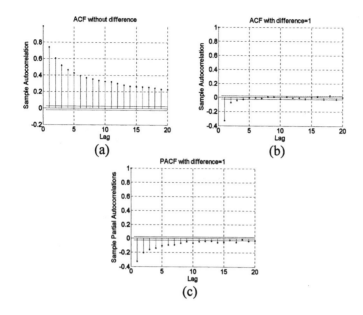

Figure 3 ACF and PACF

The mathematical expression of this model is given as following:

$$Y_t = X_t - X_{t-1}, \quad Y_t = C + \phi_1 Y_{t-1} + Z_t + \theta_1 Z_{t-1}, \quad \sigma_t^2 = \alpha_0 + \gamma_1 \sigma_{t-1}^2 + \alpha_1 Z_{t-1}^2$$

$$\text{Where } \alpha_0 > 0, \quad \gamma_1 > 0, \quad \alpha_1 > 0 \quad \gamma_1 + \alpha_1 < 1 \tag{4}$$

In equation 4, X_t represents the process of the actual traffic trace, and Y_t represents the process that differenced the original non-stationary trace to be stationary. Z_t is the error of the modeled Y_t' trace comparing with the actual Y_t trace (innovation), and σ^2 represents the conditional variance of Z_t. Notice that the C is a constant value.

After we get all coefficients' orders of the model, we can use the Maximum Likelihood Estimation (MLE) method to estimate their values, all parameters initial values are in Table 2.

Table 2 Values of Coefficients

Time scale	C	ϕ_1	θ_1
100ms	109.3	0.962	-0.684
1s	1485	0.935	-0.566
10s	1.243×10^5	0.593	0.039

Time scale	α_0	γ_1	α_1
100ms	8.95×10^5	0.642	0.272
1s	5.74×10^7	0.516	0.483
10s	5.5172×10^9	0.694	0.211

From Table 2, it is easily to find two constant value C and α_0 are always get the huge values for our aggregate traffic, and the related parameter $\phi_1, \theta_1, \gamma_1$ and α_1 are very small. That means our model is very like the linear time series model, but its variance is not stable, variance will vary on time. Although the values of variance and mean change very small in our model, it can behave bursty of our traffic. This is a major traffic character which has been found on the small time scale.

3.2   Model Analysis

To prove our model is useful and significant, we compare its features with original traffic and Multifractal Wavelet Model (MWM) [11] in two scenarios. The first one is comparing their large-time scale characteristics, and the second is comparing their small-time characteristics Because we get three traffic models on three time scale, and the performances of different traffic model are very similar, we just use the traffic model on intermediate time scale (time unit = 100ms) to study our models.

The LRD and multifractal are two important characteristics which can be found in large-time scale and small-time scale. We use the variance-time plot on measure the LRD characters on large time scale [8][11] and use the multifractal spectrum to detect its multifractal character on small time scale [11]. Figure 4(a) shows the variance-time plot results and the multifractal characters are shown in the Figure4(b). Comparing the results shown in the variance-time plot, it can be seen that all models can capture the LRD in large time scale very well. In the multifractal characteristics comparison graph, we observe that both MWM and ARIMA/GARCH model cannot capture the actual traffic's multifractal very well. However, the ARIMA/GARCH model's multifractal spectrum is more closely match the actual data's than the MWM models. The spectrum of ARIMA/GARCH model matches the better of the real data, whereas the spectrum of MWM model's showed a divergence on the small alpha part. It indicates that the probability of observing small alpha in the MWM data is fairly too small. For this reason, the burst estimated by MWM is less frequent than that in the real condition. Moreover, the spectrum of ARIMA/GARCH model behaves a little worse when alpha become large.

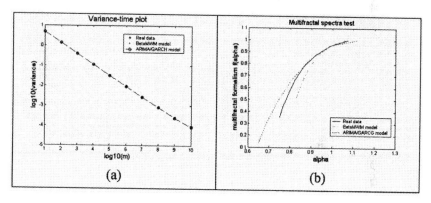

Fig. 4 Traffic Characters Comparison

3.3 Traffic prediction

The estimated parameters are used to build the ARIMA/GARCH prediction model. The problem of one-step-ahead traffic value prediction at different timescales and extension of one-step-ahead to k-step-ahead prediction are considered. The prediction quality is quantified by the prediction signal to error ratio (SER). The SER is defined as following:

$$SER = 10\log_{10}\left(\frac{E\left[X_t^2\right]}{E\left[\left(X_t - \tilde{X}_t\right)\right]}\right) dB \tag{5}$$

Where E (.) is the mean value of the expected value, X (t) is the actual traffic value, and $\tilde{X}(t)$ is the predicted traffic value. As the prediction accuracy increases, the prediction SER becomes higher. This is a good measure for comparing the performances over different databases and single database at different timescales.

In our prediction scenario, we use two ways to test our models prediction accuracy. The first one is to compare the traffic prediction performance in different time scales by one-step-ahead prediction, and the second one is to compare the traffic prediction performance in same time scale by different k-step-ahead prediction.

ARIMA/GARCH model uses the lagged actual traffic values to predict one-step-ahead traffic value at the three timescales 100ms, 1s and 10s. In Fig. 5, we show the prediction performance at different time scales by using ARIMA/GARCH model. The Fig. 5(a) shows results at intermediate time scale (100ms), the Fig. 5(b) shows results at middle time scale (1s), and the Fig. 5(c) shows results at large time scale (10s).

The real traffic data at different time scale are shown as solid lines. The corresponding predicted values are superimposed as dotted lines. The prediction results can capture the actual traffic very well. However there bound to be some prediction errors that cannot be avoided by the prediction model. This is because the real trace dynamic variance is out of expectation of GARCH model variance prediction.

We compare prediction accuracy of ARIMA/GARCH with FARIMA model. We use the prediction SER to quantify their prediction results. In Table 4, we show the prediction result of ARIMA/GARCH and FARIMA model using prediction SER measure.

Table 4 shows that the ARIMA/GARCH prediction model's prediction performances are better than the FARIMA models. In each time scale, it can be found the prediction SER of ARIMA/GARCH model is 5dB higher than the FARIMA models'. From equation (5), if the prediction accuracy increases, the SER prediction will be higher. From this table, it can be seen that if the time scale becomes larger, the prediction SER will become higher too. In other words, as the timescale increases, the prediction performance increases. This result can be explained in term of the possibility of bursty decreases when the timescale becomes larger. It can be seen that the bursty will affect the time series prediction accuracy seriously.

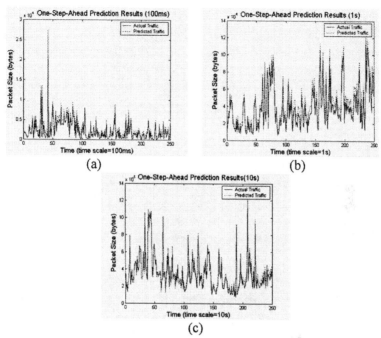

(a)   (b)

(c)

Fig. 5 One-Step-Ahead Prediction Results at Different Time Scales

After discussed the one-step-ahead prediction scheme of ARIMA/GARCH model, we show how we can extend the one-step-ahead prediction to k-step-ahead prediction. However, we would like to note that as the number of k increases, the prediction errors increase as well. Therefore, one critical question that needs to be discussed is "How far should we extend the prediction step length". Based on the acceptable accuracy, we can predict the required k-step-ahead prediction. In this paper, we study our prediction model which time scale is defined as 100ms, and then increase the value of k, find the values of prediction SER.

Fig. 6 shows the prediction results of k-step-ahead ARIMA/GARCH prediction model. Fig. 6(a) shows the results for one-step-ahead prediction ($k=1$), Fig. 6(b) shows the results for ten-step-ahead prediction ($k=10$), and Fig. 6(c) shows the results for hundred-step-ahead prediction ($k=100$). The actual traffic values at different time scale are shown as solid lines; the corresponding predicted values are superimposed as dotted lines. It can be seen that when k=1, the prediction performance can capture the actual traffic performance well. As k increases from 1 to 10, we observe that the prediction errors start to increase. When k=100, the prediction performance is totally different from the actual traffic. In other words, it means the prediction errors are bigger than expected. Table 3 shows the prediction results comparing ARIMA/GARCH with FARIMA model by prediction SER measure. In this table, when the SER is equal to zero, that means the value of SER is too small to take. It should be 0.01 or less. In other words, the prediction result is too far away from the actual network traffic that we want to predict.

From Table 3 results, it can be seen that when k increases, the value of SER decreases. That means longer prediction step reduces the prediction accuracy. Comparing our ARIMA/GARCH model with FARIMA model, our model can obtain better prediction performance not only in one-step-ahead prediction scheme but also in k-step-ahead prediction scheme.

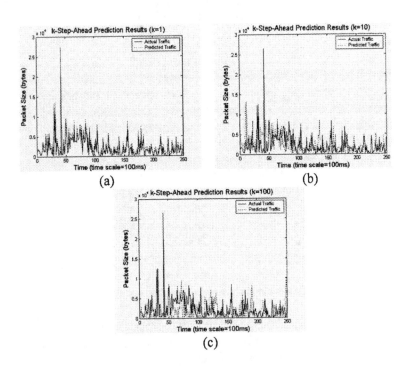

(a)

(b)

(c)

Fig. 6 *K*-Step-Ahead Prediction Results at the Small Time Scale (Time Scale=100ms)

In Fig. 7, the Fig. 7(a) shows the SER of prediction results for traffic in different timescales and the Fig. 7(b) shows the SER of predictionresults for different k-step-ahead. From the graph results, it is clearly shown that the time scale and the prediction scale both affect the prediction accuracy. When the time scale increase, (i.e., the traffic get more aggregated), some nature of traffic has been ignored (e.g. some bursty points have been lost). In this case, we can treat the variance of the time series as stable and the trend as linear. When the time series become linear, its prediction results can get more accuracy. In Fig. 6(b), it shows the   relationship between the prediction accuracy with the prediction step length. We understand that the network traffic behaves bursty (e.g. its   variance varies over time). When the prediction step length increases, the variation of the variance is far away from our model's expectation, which affects prediction.

Table 3 SER of Prediction Results for Traffic Various from k

Prediction Model	K=1	K=10	K=100
FARIMA (dB)	13.54	0.42	0
ARIMA/GARCH (dB)	18.91	3.89	0

Table 4 SER of Prediction Results for Traffic Various Timescales

Prediction Model	100ms	1s	10s
FARIMA (dB)	13.54	14.19	15.21
ARIMA/GARCH (dB)	18.91	19.48	20.69

The two scenarios prediction test results indicated the performance of the ARIMA/GARCH model outperforms the traditional FARIMA model. From Table 4 and Table 3, we notice that the timescale and the value of prediction scale k play significant roles in the traffic prediction performance. From our studying, it can be found that the prediction nature of the network traffic. For the network traffic, as the timescale increases, the prediction performance increases. This result can be explained in term of the user's habit of using the network. For example, during the day (longer time scale) it is less challenging to predict when and how long the user will access the network, and how much traffic is expected over this period from user's traffic history. Another critical question has already discussed in this section: the relationship between the prediction accuracy with how many step-ahead prediction we need. We wish our model can use less actual traffic to capture more prediction traffic performance. That means we want to use k-step-ahead prediction scheme to predict the traffic. The prediction results show that number of k increases, the prediction error increases.

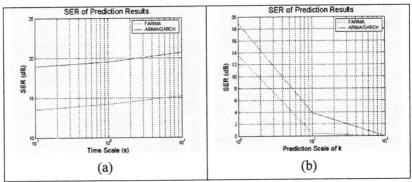

Fig. 7 SER of Prediction Results

4.   CONCLUSION

Network traffic exhibits high degree of self-similarity at large time scale and high degree of multifractal at small time scale. The traditional traffic models are unable to capture all these characteristics, while ARIMA/GARCH model provides such flexibility when it is applied to model the network traffic.

In this paper, we present a new procedure to estimate the parameters of the ARIMA/GARCH model, and an adaptive prediction scheme. The prediction performance is tested on real network traffic data using three different time scales and from one-step-ahead prediction to k-step-ahead prediction. The results show that the performance of the ARIMA/GARCH model outperforms traditional FARIMA model. The traffic predictability depends on the traffic nature, the considered time scales and prediction step length.

The parameters of the ARIMA/GARCH model capture statistical traffic characteristics and provide detailed descriptions about the data in time domain. The study of their effects on the network performance could be useful for the traffic measurements and network management and control. The ARIMA/GARCH traffic prediction model can be used to build effective congestion control schemes, e.g. dynamic bandwidth allocation and admission control.

From this study, it shows that non-linear time series model can model and forecast better then the classical linear time series models, even the linear time series model can also behave self-similarity. The basic idea to introduce the non-linear time series model instead of the linear time series model is based on the network's property: bursty. In the statistical theory, if the time series behaves bursty, that means its variance vary time. The ARIMA/GARCH model is a non-linear time series model which can capture the conditional variance (variance vary time) very well.

The non-linear time series model ARIMA/GARCH can model and forecast the network traffic better than the traditional linear time series model. However, its prediction methodology is more complex and unstable. How to determine its prediction accuracy and prediction scale is a challenge that should be solved in future. Considered our network traffic behave different characteristics in different time scale. How to model and predict the network traffic in multi-time-scale is another challenge that should be solved in future. We will study the wavelet multi-resolution method by the next step to solve this question. For rapid changed of applications and   topologies of computer network, current network traffic should change its behavior comparing with the data that we study for ARIMA/GARCH model. Try to use more current traffic to test our ARIMA/GARCH model's modeling and prediction skill is also considered in future work.

REFERENCES

[1] M.Stemm, S.Seshan and R.H.Katz, "A network measurement architecture for adaptive applications", *Proceedings of INFOCOM 2000*, vol.1, pp.285-294.

[2] V.Jacobson, "Congestion avoidance and control", *Proceedings of the ACM SIGCOMM 1988 Symposium*, pp.314-329.

[3] S.Jamin, P.Danzig, S.Schenker and L.Zhang, "A measurement-based admission control algorithm for integrated services packet networks", *Proceedings of ACM SIGCOMM'95*, pp.56-70.

[4] M.Kim and B.Noble, "Mobile network estimation", *Proceedings of the Seventh Annual International Conference on Mobile Computing and Networking*, pp.298-309.

[5] S.F.Bush, "Active virtual network management prediction", *Proceedings of the 13th Workshop on Parallel and Distributed Simulation (PADS'99)*, pp. 182-192.

[6] A.Sang and S.Li, "A predictability analysis of network traffic", *Computer networks*, vol. 39, pp. 329-345, 2002.

[7] A.Adas, "Traffic models in broadband networks", *IEEE Comm. Mag.*, vol. 35, pp. 82-89. July 1997.

[8] W.E.Leland, M.S.Taqqu,W.Willinger and D.V.Wilson, "On the self-similar nature of Ethernet traffic (Extended Version)", *IEEE/ACM Trans. Net.*, vol.2, no.1, pp. 1-15, Feb 1994.

[9] Y.Shu, Z.Jin, L.Zhang, L.Wang and O.W.W.Yang, "Traffic prediction using FARIMA models", *ICC'99*, vol.2, pp. 891-895, 1999.

[10] M.Corradi, R.G.Garroppo, S.Giordano and M.Pagano, "Analysis of f-ARIMA processes in the modeling of broadband traffic", *ICC'01*, vol.3, pp.964-968, 2001.

[11] Rudolf H. Riedi, Vinay J. Ribeiro, Matthew S. Crouse, Richard G. Baraniuk, "Network Traffic Modeling Using a Multifractal Wavelet Model," *DSPCS'99*, Perth, 1999.

[12] C.You and K.Chandra, "Time series models for Interent data traffic", *Proceedings of the 24th Conference on Local Computer Networks (LCN 99)*, pp. 164-171.

[13] Bollerslev,T.,R.Y.Chou, and K.F.Kroner, "ARCH Modeling in Finance: A Review of the Theory and Empirical Evidence," Journal of Econometrics, Vol.52, 1992.

[14] C.Gourieroux, "ARCH Models and Financial Applications", Springer Series in Statistics, Springer-Verlag New York, 1997.

[15] T.Nakatsuma and H.Tsurumi, "ARMA-GARCH models: Bayes Estimation Versus MLE, and Bayes Non-stationary Test", department working papers with number 199619, department of Economics,   Retgurs University, 1996.

[16] P.Brockwell and R.Davis, "Introduction to Time Series and Forecasting", Springer, 1996.

[17] S.Makridakis, S.C.Wheelwright and V.McGee, "Forecasting Methods and Applications", second edition, Wiley, 1983.

[18] http://ita.ee.lbl.gov/html/contrib/LBL-TCP-3.html

[19] Y.Gao, G.He and J.C.Hou, "On Leveraging traffic predictability in active queue management", Proc. IEEE INFOCOM 2002, June 2002.

[20] Box, G.E.P., G.M.Jenkins, and G.C. Reinsel, "Time Series Analysis: Forecasting and Control", 3rd edition, Prentice Hall, 1994.

CHAPTER 6

On the Scalability of Fluid Models of IP Networks Loaded by Long-lived TCP Flows

M.Ajmone Marsan, G.Carofiglio, M.Garetto, P.Giaccone,
E.Leonardi, E.Schiattarella, A.Tarello

Dipartimento di Elettronica
Politecnico di Torino, Italy

Abstract. *Fluid models of IP networks have been recently proposed, to break the scalability barrier of traditional performance evaluation approaches, both simulative (e.g., ns-2) and analytical (e.g., queues and Markov chains). Fluid models adopt a deterministic description of the average source and network dynamics through a set of (coupled) ordinary differential equations that are solved numerically, obtaining estimates of the time-dependent behavior of the IP network.*

The most attractive property of fluid models resides in the fact that they are scalable, i.e., their complexity is independent of the number of TCP flows and of link capacities. In this paper we provide a theoretical investigation of the origins of the scalability of fluid models. We show that the set of differential equations defining the network dynamics under both drop-tail and AQM buffering exhibits a nice invariance property, that allows an equivalence relation to be established among different systems. The validity of the invariance property is verified in realistic network scenarios with ns-2 simulations.

1. INTRODUCTION

Traditional approaches to performance evaluation of telecommunication networks in general, and of packet networks in particular (we specifically refer to IP networks in this paper), have normally relied on attempts to describe as closely as possible the dynamics of network elements over a discrete state-space. We shall refer to these approaches as 'discrete models'.

Discrete stochastic models are quite a natural choice in light of the fact that the operations of traffic sources, of switches, and of protocols, are normally governed by finite-state machines, whose dynamics determine the IP network performance. However, discrete models, requiring the description of the dynamics of the different network elements over their discrete state-spaces (accounting for the dependencies induced by network control algorithms, by end-to-end protocols, and by the traffic flowing through several network elements), suffer from limited scalability, thus allowing only the performance analysis of rather small networking setups. This is one of the reasons why only toy topologies are normally considered in IP network performance studies, and models almost invariably concentrate on a very limited subset of the network protocol stack.

Today, IP networks have become extraordinarily large and complex, and, in order to predict the effects on performance of new topologies or protocols, scalable modeling approaches are a must. To obtain scalability, researchers often model one network element (a protocol layer or a router) with a high degree of accuracy, whereas the remaining network elements are either neglected, or modeled in a very coarse manner. This approach may sometimes produce acceptable results, but many cases exist where the interaction among network elements is a focal point for performance issues (for example, in a satellite IP network, the interaction between TCP and the ARQ protocol at layer 2 is crucial).

The above comments apply to analytical models as well as simulation. Indeed, traditional discrete models for simulation are based on a detailed packet-level description of the network, and consequently suffer the scaling problems resulting from the growth of CPU times and memory requirements beyond the capability of available machines. On the analytical side, the application of discrete-state probabilistic models to the analysis of sizable portions of the Internet appears prohibitive, although such models (continuous-time or discrete-time Markov chains and queuing models in particular) have traditionally been the mathematical tool of choice in the networking field.

A new class of semi-analytical models has recently been introduced in the networking arena, and today appears to be the most promising approach for scalable and accurate performance analysis of large IP networks. These new models, that are often called 'fluid models', adopt an abstract deterministic description of the average network dynamics through a set of differential equations [1–5], thus neglecting the short-term, packet-by-packet description of the stochastic network dynamics. The resulting set of differential equations is then solved numerically, obtaining estimates of the time-dependent network behavior.

The most attractive property of fluid models resides in the fact that their complexity (i.e., the number of differential equations to be solved) is independent of the number of TCP flows and of link capacities, when considering traffic scenarios comprising only long-lived TCP flows (commonly called 'elephants'). In addition, fluid models have been recently proved to capture the limiting behavior of TCP elephants in single bottleneck topologies when the number of TCP flows, the bottleneck capacity and the buffer-size jointly grow linearly to infinity [1, 7–10].

Recently [11, 12] we have developed a fluid model approach in which the sources dynamics are described by *partial* differential equations. This new approach allows the description of the evolution of TCP window size distributions, instead of averages, with no sacrifice in the scalability of the model. By doing so, the model provides better accuracy in the performance predictions, and does not require the approximations and "correction factors" that have to be introduced in the other models based on the average window dynamics.

In this paper we provide a theoretical investigation on the structure of the equations defining fluid models. We show that the set of differential equations defining the network dynamics under both drop-tail and AQM queuing disciplines exhibits a nice invariance property, that allows an equivalence relation to be established among different systems, thus providing a theoretical foundation to the scalability of fluid models of IP networks. The validity of the invariance property is verified in realistic network scenarios with *ns-2* simulations.

This paper is organised as follows. Section 2 discusses the traditional fluid models based on the average window dynamics. Sections 3 and 4 describe our fluid model, based on partial differential equations. The main contribution of this paper is presented in Section 5, which shows the invariance property of fluid models. The insight provided by the invariance is then validated by the scenarios studied in Section 6.

2. THE MGT FLUID MODEL OF IP NETWORKS

In [3–5], Misra, Gong and Towsley presented simple differential equations to describe the behavior of TCP elephants over networks of IP routers adopting a RED (Random Early Detection [13]) active queue management (AQM) scheme. Their approach (that we name MGT) spurred several research efforts aiming at the application of various kinds of fluid models to the performance analysis of packet networks. It is important to note that the equations of the MGT model heavily rely on the assumptions mentioned above (all TCP connections are elephants, and all IP routers adopt RED), and that the extension to mice and drop-tail routers may be not simple. Note that the model we proposed in [11, 12] has been inspired by the MGT model, and it is able to overcome those limitations, supporting gracefully both TCP mice and drop-tail routers.

We now introduce the notation to describe the MGT fluid model. Consider a network comprising K router output interfaces, equipped with FIFO buffers, and interfacing data channels at rate C (the extension to non-homogeneous data rates is straightforward). The network is fed by I classes of long-lived TCP flows; all the elephants within the same class follow the same route through the network, thus experiencing the same round-trip time (RTT), and the same average loss probability (ALP). At time t = 0 all buffers are assumed to be empty. Buffers drop packets according to their average occupancy, as dictated by a RED AQM scheme.

2.1 TCP Source Evolution Equations

Consider the ith class of elephants; the temporal evolution of the average window of TCP sources in the class, $W_i(t)$, is described by the following differential equation:

$$\frac{dW_i(t)}{dt} = \frac{1}{R_i(t)} - \frac{W_i(t)}{2}\lambda_i(t) \qquad (1)$$

where $R_i(t)$ is the average RTT for class i, and $\lambda_i(t)$ is the loss indicator rate experienced by TCP flows of class i.

The differential equation is obtained by considering the fact that elephants can be assumed to always be in congestion avoidance (CA) mode, so that the window dynamics are close to AIMD (Additive Increase, Multiplicative Decrease). The window increase rate in CA mode is approximatively linear, and corresponds to one packet per RTT. The window decrease rate is proportional to the rate with which congestion indications are received by the source, and each congestion indication implies a reduction of the window by a factor two.

2.2   Network Evolution Equations

$Q_k(t)$ denotes the (fluid) level of the packet queue in the kth buffer at time t; the temporal evolution of the queue level is described by:

$$\frac{dQ_k(t)}{dt} = A_k(t)\left[1 - p_k(t)\right] - D_k(t) \tag{2}$$

where $A_k(t)$ represents the fluid arrival rate at the buffer, $D_k(t)$ the departure rate from the buffer (which equals C, provided that $Q_k(t) > 0$) and the function $p_k(t)$ represents the instantaneous loss probability at the buffer, which depends on the queuing discipline (as explained later in Section 4).

If $T_k(t)$ denotes the instantaneous delay of buffer k at time t, we can write

$$T_k(t) = Q_k(t)/C \tag{3}$$

If $\mathbf{F}_k$ indicates the set of flows traversing buffer k, $A^i_k(t)$ and $D^i_k(t)$ are respectively the arrival and departure rates at buffer k referred to elephants in class i, we get:

$$A_k(t) = \sum_{i \in F_k} A^i_k(t)$$

$$\int_0^{t+T_k(t)} D_k(a)da = \int_0^t A_k(a)[1 - p_k(a)]da$$

$$\int_0^{t+T_k(t)} D'_k(a)da = \int_0^t A'_k(a)[1 - p_k(a)]da$$

which means that the total amount of fluid arrived up to time t at the buffer leaves the buffer by time $t + T_k(t)$, since the buffer is FIFO. By differentiating the last equation:

$$D_k^i(t + T_k(t))\left(1 + \frac{dT_k(t)}{dt}\right) = A_k^i(t)[1 - p_k(t)]$$

2.3 Source-Network Interactions

Consider elephants in class i. Let $k(h, i)$ be the hth buffer traversed by them along their path P_i of length L^i. The RTT $R_i(t)$ perceived by elephants of class i satisfies the following expression:

$$R_i\left(t + g_i + \sum_{h=1}^{L_i} T_{k(h,i)}(t_{k(h,i)})\right) = g_i + \sum_{h=1}^{L_i} T_{k(h,i)}(t_{k(h,i)}) \tag{4}$$

where g_i is the total propagation delay[1] experienced by elephants in class i, and $t_{k(h,i)}$ is the time when the fluid injected at time t by the TCP source reaches the hth buffer along its path P_i. We have:

$$t_{k(h,i)} = t_{k(h-1,i)} + T_{k(h-1,i)}(t_{k(h-1,i)})$$

The loss indicator rate is instead given by:

$$\lambda_i(t + R_i(t)) = \alpha \frac{W_i(t)}{R_i(t)} p_i^F(t) \tag{5}$$

---

[1] Equation (4) comprises the propagation delay gi in a single term, as if it were concentrated only at the last hop. This is just for the sake of easier reading, since the inclusion of the propagation delay of each hop would introduce just a formal modification in the recursive equation of tk(h,i).

where $W_i(t)/R_i(t)$ is the instantaneous emission rate of TCP sources, α is a calibration parameter, and $p^F_i(t)$ is the instantaneous loss probability experienced by elephants in class i:

$$p_i^F(t) = 1 - \prod_{h=1}^{L^i}\left[1 - p_{k(h,i)}(t_{k(h,i)})\right]$$

(6)

Finally;

$$A_k(t) = \sum_i \sum_q r_{qk}^i D_q^i(t) + \sum_i e_k^i \frac{W_i(t)}{R_i(t)} N_i$$

(7)

where $e_k^i = 1$ if buffer k is the first buffer traversed by elephants of class i, and 0 otherwise; r_{qk}^i is derived by the routing matrix, being $r_{qk}^i = 1$ if buffer k immediately follows buffer q along P_i; N_i is the number of class i active flows.

3.   PARTIAL DIFFERENTIAL EQUATION BASED DESCRIPTION OF TCP SOURCE DYNAMICS

The class of fluid models that we consider in this paper was first proposed in [11] and differs from previous proposals (with the exception of [10]) because, instead of describing just the evolution of the average window size of TCP sources, the model describes the evolution of the window size *distribution* for the whole TCP flow population. In other words, rather than just describing the average TCP connection behavior, the model describes statistically the dynamics of the entire population of TCP flows sharing the same path. This approach leads to systems of partial derivatives differential equations, and produces more flexible models, while preserving scalability with respect to the population size.

In this section we first revise the basic model for the TCP flow population. This basic model can be extended by adding several features, which permit a progressively more accurate description of the behavior of TCP sources.

3.1 Basic TCP Sources

Consider a fixed number of TCP elephants. We use $P_i(w, t)$ to indicate the number of elephants of class i whose window is $\leq w$ at time t. For the sake of simplicity, we consider just one class of flows, and omit the index i from all variables. The source dynamics are approximately described by the following equation, for $w \geq 1$:

$$\frac{\partial P(w,t)}{\partial t} = \int_{w}^{2w} \lambda(\alpha,t)\frac{\partial P(\alpha,t)}{\partial \alpha}d\alpha - \frac{1}{R(t)}\frac{\partial P(w,t)}{\partial w} \tag{8}$$

where $\lambda(w, t)$ is the loss indication rate. A formal derivation of (Eq.8) is reported in Appendix A, to provide a deeper insight in the model. Note that the above equation is equivalent to the deterministic transport equation reported in Corollary 1 of [10], which was obtained by applying Mean Field Analysis.

The intuitive explanation of the formula is the following. The time evolution of the population described by $P(w, t)$ is governed by two terms: (i) the integral accounts for the growth rate of $P(w, t)$ due to sources with window between w and $2w$ that experience losses; (ii) the second term accounts for the decrease rate of $P(w, t)$ due to sources increasing their window with rate $1/R(t)$.

The quantity $\lambda(w, t)$ can be computed by recalling (Eq.5):

$$\lambda(w,t) = \frac{wp^F(t)}{R(t)}$$

in which the current window of the sources that emitted the lost fluid approximates the window value at which those sources emitted this fluid. Intuitively, this loss model distributes the lost fluid over the entire population, proportionally to the window size.

3.2 Accounting for the Maximum Window Size

We now extend the basic model of (Eq.8) to account for the maximum window size of TCP sources, that we denote by W^{max}. We have:

$$\frac{\partial P(w,t)}{\partial t} = \int_{w}^{min(2w,W^{max})} \lambda(\alpha,t) \frac{\partial P(\alpha,t)}{\partial \alpha} d\alpha +$$

$$+ \lambda(W^{max},t)P_{max}(t)u(w-W^{max}/2) - \frac{1}{R(t)} \frac{\partial P(w,t)}{\partial w}$$

(9)

for $1 \leq w < W^{max}$, where $u(\cdot)$ is the unit step function, and $P_{max}(t)$ is the number of TCP flows whose window is exactly equal to W^{max}.

For $P_{max}(t)$ we can write :

$$\frac{dP_{max}(t)}{dt} = \frac{1}{R(t)} \lim_{w \uparrow W^{max}} \frac{\partial P(w,t)}{\partial w} - \lambda(W^{max},t)P_{max}(t)$$

(10)

with the boundary conditions: $P(1^{-}, \quad t) \quad = \quad 0$ and $\lim_{w \uparrow W^{max}} P(w.t) + P_{max}(t) = N$. The derivation of (Eq.9) is very similar to that of (Eq.8). The first term in (Eq.9) is the contribution of all TCP sources which experience losses at window size between w and 2w (W^{max} if 2w exceeds it). The second term of (Eq.9) is the contribution of all TCP sources at maximum window size that experience losses; note that this contribution exists only for windows greater than $W^{max}/2$.

The growth rate of $P_{max}(t)$ is obtained as the limit of the usual growth rate $(\partial P(w,t)/\partial w)/R(t)$ of $P(w, t)$. The decrease rate of $P_{max}(t)$ is simply $\lambda(W^{max}, t)$.

3.3 Considering Fast Recovery

Newer versions of TCP (such as NewReno - see RFCs 2001, 2581, 2582) avoid halving the window more than once for RTT, even in the case of multiple losses. To model this fact, we divide the population $P(w, t)$, representing the number of TCP flows whose congestion window is $\leq w$ at time t, in two classes: class L comprises all those sources that experienced losses during the last RTT, while class O is composed by remaining sources[2]

---

[2] For the sake of simplicity, the equations in this section and in the rest of the paper do not consider the effect of the maximum window size. However, in all numerical results that are presented in this paper the effect of the maximum window size is always accounted for.

$(P(w,t) = P_L(w,t) + P_O(w,t)).$

We can write:

$$\frac{\partial P_O(w,t)}{\partial t} = -\int_1^w \lambda(\alpha,t) \frac{\partial P_O(\alpha,t)}{\partial \alpha} d\alpha - \frac{1}{R(t)} \frac{\partial P_O(w,t)}{\partial w} + \frac{1}{R(t)} P_L(w.t) \qquad (11)$$

$$\frac{\partial P_L(w,t)}{\partial t} = +\int_1^{2w} \lambda(\alpha,t) \frac{\partial P_O(\alpha,t)}{\partial \alpha} d\alpha - \frac{1}{R(t)} P_L(w,t) - \frac{1}{R(t)} \frac{\partial P_L(w,t)}{\partial w} \qquad (12)$$

A formal derivation of (Eq.11) and (Eq.12) is reported in [11]. An intuitive explanation of the two equations can be provided as follows. In the right hand side of (Eq.11), the first two terms account for the decrease rate of the number of elephants of class O whose window is $\leq w$ at time t, due to: (i) sources in class O experiencing losses and moving to class L, (ii) sources in class O increasing their window. The third term refers to the sources moving to class O from class L after experiencing a RTT without losses. In the right hand side of (Eq.12), the first term accounts for the growth rate of the number of elephants of class L whose window is $\leq w$ at time t, due to sources in class O experiencing losses. The second and third terms account for the decrease rate due to: (i) sources moving to class O from class L after a RTT without losses, (ii) sources in class L increasing their window.

4.   COMPUTATION OF THE PACKET LOSS PROBABILITY

In this section we describe how different queuing disciplines implemented in network routers are accounted for in the model. In particular, we need to characterize the temporal evolution of the packet loss probability given as input to the TCP source model. Equation (Eq.2) describes the evolution of a fluid queue of infinite capacity. In practice, router buffers are finite, and react to congestion by dropping and/or marking incoming packets. The two most popular queuing disciplines implemented today are drop-tail and RED. Our fluid approach allows us to characterize in a simple way the instantaneous loss process introduced by both queuing disciplines, as well as to describe networks comprising a mixture of drop-tail and RED buffers. The packet loss probabilities introduced by these queuing disciplines are described separately in the next Sections, but they can be combined at will using (Eq.6) to obtain the overall packet loss probability along a network path.

4.1 Modeling RED Buffers

In RED queues and other variants of AQM, the drop probability is computed as a function of the buffer level, according to a given loss "profile". In particular, RED maintains a moving average $Q_a(t)$ of the instantaneous queue size $Q(t)$, updated whenever a packet arrives, according to the rule

$$Q_a(t) \leftarrow (1 - w_q)Q_a(t) + w_q Q(t)$$

where w_q is the weight of the exponentially moving average. The instantaneous drop probability is then computed as piecewise, increasing function of $Q_a(t)$ [13]. In order to describe a RED queue in a fluid model, we need to characterize the temporal evolution $Q_a(t)$ of the moving average as a continuous function of time, relating it to the actual packet-by-packet behavior. This was originally done in [3], where the authors have shown that the evolution of $Q_a(t)$ is related to $Q(t)$ through the differential equation

$$\frac{dQ_a(t)}{dt} = \frac{log(1-w)}{\delta} Q_a(t) - \frac{log(1-w)}{\delta} Q(t) \tag{13}$$

where δ is the sampling period. In RED, the estimation of $Q_a(t)$ is updated at each packet arrival. Following [3], the actual sampling period can be approximated in two possible ways: (i) δ is the instantaneous interarrival time of packets at the queue: $\delta(t) = 1/A(t)$; (ii) δ is constant and given by the queue capacity: $\delta = 1/C$. Differently from [3], in our model we have chosen the first approximation, which is more accurate. From $Q_a(t)$ one can immediately derive the instantaneous packet loss probability $p(t)$ from the RED profile. Networks comprising only RED buffers are well described by fluid models [3], since the loss probability depends only on the average queue level, which is a smooth function of time.

4.2 Modeling Drop-Tail Buffers

The case of drop-tail buffers is more difficult to describe with fluid models, since in this case the loss probability is a discontinuous function of the instantaneous queue size. Many studies have shown that the behavior of networks carrying TCP traffic is pulsing: congestion epochs in which some buffers are overloaded (and overflow) are interleaved to periods of time in which traffic is lighter, buffers are not saturated, and no loss is experienced. Light traffic periods are the result of losses at the previous congestion epochs, that force TCP sources to reduce their emission rate. As a consequence, the loss processes experienced by TCP flows traversing drop-tail buffers are quite bursty. This burstiness induces a high degree of correlation (synchronization) among the dynamics of TCP sources sharing the same buffer. In addition, during congestion epochs, losses are not evenly distributed among TCP flows, but are more likely to affect TCP sources with larger window size. In this context, it is necessary to distinguish among sources with different instantaneous window size, while at the same time accounting for the effects of the TCP fast recovery mechanism, which prevents TCP sources from halving their window several times within one round trip time.

The level of detail in the description of the TCP sources dynamics adopted in this paper allows an easy description of the time-dependent behavior of the packet loss probability introduced by a drop-tail queue:

$$p(t) = \frac{max(0, A(t) - C)}{A(t)} 1\!I_{\{Q(t) = B\}} \tag{14}$$

that is, the loss probability $p(t)$ equals $(A(t) - C)/A(t)$ (the relative difference between the instantaneous arrival rate and the service rate) only when the buffer is full, being B the capacity of the buffer, and $1\!I_{\{\cdot\}}$ the indicator function.

We note that a different approach is used in [1] and [2] to describe the dynamics of the average window size for TCP flows traversing a network with drop-tail buffers. In those papers, the loss indicator rate is obtained by applying queuing theory results which are not "internal" to the fluid model. That approach is probably difficult to generalize to networks including both drop-tail and AQM buffers.

5.   THE INVARIANCE PROPERTY OF FLUID MODELS

In this section we discuss a structural property of the system of differential equations which defines the proposed fluid model for IP networks loaded by TCP elephants. Structural properties of fluid models are interesting on the one hand to gain deeper insight into the underlying system, on the other hand to better understand the validity of fluid models. Invariance properties are specially important because they allow us to extrapolate the behavior of large systems from the behavior of a scaled-down version of the same system. In this sense, they provides a way to dramatically reduce the computational requirements of simulations and the cost of experiments. Indeed, the time/memory complexity of packet-level simulators (like *ns-2*) scales badly with both the link capacity and the number of active connections (each packet triggers one or more events, and each source requires the allocation of a class object).

Table 1. Correspondence of Input Parameters and Dynamic Variables in the Original Fluid Model and in the Transformed Model

Input Parameters	Original Network	Transformed Network
Propagation delay	g	g
Path length	L	L
RED loss probability	p_{max}	p_{max}
RED thresholds	min_{th}, max_{th}	$\eta min_{th}, \eta max_{th}$
RED filter constant	w_q	w_q/η
Number of flows	N	ηN
Buffer size	B	ηB
Link capacity	C	ηC

Dynamic Variables	Original Network	Transformed Network
RTT	$R(t)$	$R(t)$
Loss indication rate	$\lambda(w,t)$	$\lambda(w,t)$
Drop probability	$p(t)$	$p(t)$
Window size	$W(t)$	$W(t)$
Flow loss probability	$p^F(t)$	$p^F(t)$
Queuing delay	$T(t)$	$T(t)$
Queue length	$Q(t)$	$\eta Q(t)$
Departure rate	$D(t)$	$\eta D(t)$
Arrival rate	$A(t)$	$\eta A(t)$
Flow CDF	$P(w,t)$	$\eta P(w,t)$
Flows at W_{max}	$P_{max}(t)$	$\eta P_{max}(t)$
Flow CDF-FR mechanism	$P_O(w,t)$	$\eta P_O(w,t)$
Flow CDF-FR mechanism	$P_L(w,t)$	$\eta P_L(w,t)$
Slow start flow CDF	$P_s(w,t)$	$\eta P_s(w,t)$

In particular, we are interested in verifying the invariance properties of the proposed fluid model with respect to linear transformations, following an approach similar to the one in [14]. Consider an IP network loaded by

TCP elephants, referred as the *original network*, whose dynamics are described by the fluid model. Consider then a different network, whose model can be derived from the one of the original network through simple linear transformations. We refer to this new network with the term *transformed network*.

Table 1 summarizes the transformations which map the original network parameters into those of the transformed network, being $\eta \in R^+$ the multiplicative factor applied to the model parameters.

Basically, the transformed network is obtained from the original network by multiplying by a factor η the number of elephants, as well as all transmissive capacities, and buffer dimensions. The basic idea is that loss rates and round trip times are not modified by the transformation. Table 1 also reports the transformations which relate the modeled behavior of the original network to the one of the transformed network.

The verification of the invariance properties with respect to the considered linear transformation of the fluid model of TCP elephants can be easily carried out by direct inspection. Indeed, look first at the window dynamics of TCP sources. For simplicity, we consider only the simplest case, described by (Eq.8); however, the same arguments can be immediately extended to the other cases as well.

The behavior of sources depends only on the feedback received by the network, expressed in terms of loss rate $\lambda(w, t)$ and round trip time $R(t)$. As already mentioned, both $\lambda(w, t)$ and $R(t)$ are left unchanged in the transformed system (this is proved below). Suppose the original system satisfy (Eq.8), under the initial condition $P(\infty, 0) = N$. Since (Eq.8) is linear in $P(w, t)$, $\eta P(w, t)$ satisfies equation (Eq.8) under the initial condition $P(\infty, 0) = \eta N$. We can conclude that (Eq.8) is invariant with respect to the linear transformation of parameters and dynamics variables reported in Table 1. It can be easily shown that also the equations modeling the maximum window (Eq.9)-(Eq.10) and the fast recovery mechanism (Eq.11)-(Eq.12) are satisfied as well. Notice that, in the transformed system, sources have exactly the same *distribution* of window size as in the original system, thus the same average window size $\overline{w}(t)$.

Now consider the network dynamics. In order to maintain the same round trip time, both queuing delays and propagation delays need to remain the same. Propagation delays are fixed parameters, which are left unchanged by our transformation. The instantaneous queuing delay at any queue is also maintained equal by the transformation, because in (Eq.3) both $Q(t)$ and C are multiplied by η. The link capacity C is an input parameter, which is scaled by the factor η according to Table 1. Therefore, we only need to show that the instantaneous buffer level $Q(t)$ in the transformed network evolves as a scaled replica of $Q(t)$ as observed in the original network. This follows immediately from (Eq.2), because both the arrival rate $A_k(t)$ and the departure rate $D_k(t)$ are multiplied by η, thus variations in the amount of fluid in the queue are also multiplied by η: if we start from the initial condition in which all buffers are empty, it follows that the entire trajectory of $Q(t)$ is multiplied by η. The fact that arrival rates $A_k(t)$ in the transformed network are scaled by η follows from (Eq.7): service rates of queues are indeed scaled accordingly, while the fluid emitted by sources depends on the window size distribution (unchanged) and is proportional to the number of flows (multiplied by η), thus it scales as well.

Finally, we need to show that packet loss probabilities remain equal in the transformed network. In the case of drop-tail queues, this derives immediately from equation (14), because in the ratio $(A(t) - C)/A(t)$ the common factor η is simplified, while the evolution of $Q(t)$ guarantees that the buffer overflows during the same intervals. In the case of RED, according to Table 1, the queue thresholds defining the loss profile (min_{th}, max_{th}) are scaled by η, whereas p_{max} is the same. Following the scaling rules introduced so far, whether $\delta(t) = 1/A(t)$ or $\delta = 1/C$, the sampling period in the transformed network is obtained by dividing δ by η. In order to have the same instantaneous packet loss probability, besides $Q(t)$ also the moving average $Q_a(t)$ has to scale up by η. Looking at (Eq.13), observe that this can be achieved by *dividing* the weight parameter w_q by η (this is an exception, not reported in Table 1), i.e., by slowing down the dynamics of the filter. Indeed, when $w_q \ll 1$ (as it is usually the case), from (13) we have $\log(1 - w_q) \approx -w_q$, thus the factor $1/\eta$ applied to w_q cancels with an equal reduction of the sampling period δ, therefore the evolution of $Q_a(t)$ is coupled to $Q(t)$ as in the original network.

In conclusion, the complete system of equations defining the fluid model is invariant with respect to the linear transformation of parameters and dynamics variables reported in Table 1. Hence, the behavior of the transformed network can be easily derived from the one of the original network.

6.   VALIDATION OF THE INVARIANCE PROPERTY

In the previous Section we described an invariance property of fluid models of IP networks that was proved by simply inspecting the structure of the differential equations defining the fluid model. Besides verifying this property analytically, the question naturally arises about the actual invariance of the behavior of IP networks. In order to answer this question, we performed simulation experiments with *ns-2* in different network scenarios comprising drop-tail or RED buffers to validate our fluid models, and to assess the validity of the invariance property in real IP networks.

6.1   Experiments with Drop-Tail

In [14] the authors claim that the scaling property fails in networks comprising drop-tail buffers. In [15] they argue that this is due to the fact that drop-tail buffers introduce bursty, correlated losses, thus the loss process cannot be modeled as Poisson, hence the invariance property in not satisfied. This conclusion is not confirmed by our study. The loss process is indeed bursty, and cannot be described by a homogeneous Poisson process. However, it is well described by an inhomogeneous point process with time-varying intensity given by (5), which is left unchanged under the transformation of system parameter specified in Section 5. Therefore, the scaling property indeed holds in networks comprising drop-tail buffers.

To prove our claim, we consider the same network scenario adopted in [14], which is illustrated in Figure 1. A tandem network of two congested drop-tail queues is loaded by three groups of elephants denoted by *grp1*, *grp2*, *grp3*. Group *grp1* crosses only the first link, group *grp2* crosses only the second link, whereas *grp3* traverses both links. The link speeds are equal to 100 Mbps, and the buffer capacities are equal to 8000 packets. Propagation delays for the three groups are 150ms, 200ms, and 250ms, respectively. The number of active elephants in each group varies in time, as flows switch on and off as shown in the timing diagrams reported in Figure 1.

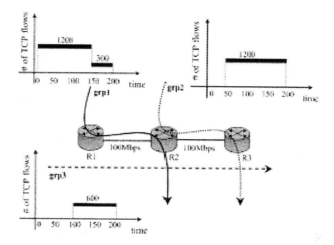

Fig. 1 Network topology with two congested links and three classes of elephants switching on and off (reproduced from [14]).

The plot in Fig. 2 reports the time average of the queuing delay (evaluated from the beginning of the simulation at t=0) for both queues (note the different scales on the left and right vertical axes). All the considered invariant versions of the original system ($\eta = 1/3$, 1/10, 1/30) show extremely similar behaviors, showing that fluid models of IP networks are invariant also in presence of drop-tail buffers. Of course, the plots cannot exactly overlap, due to the intrinsic randomness of the simulator.

Fig. 3 shows the instantaneous queuing delay at the first queue during the first 10 seconds of simulation. Again, the behavior of the queue is very similar for all scaled versions of the system. The plot proves that the detailed trajectories of the number of packets in the queue, which is directly related to instantaneous queuing delays by (3), have nearly the same shape.

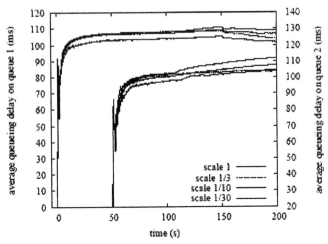

Fig. 2 Average queuing delay at the first queue (left Y axis), and at the second queue (right Y axis), for different invariance factors, evaluated by NS simulations.

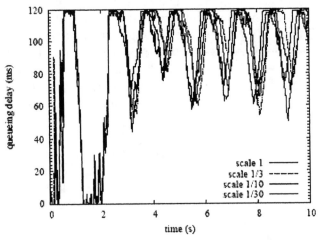

Fig. 3 Queuing delay at the first queue, for different invariance factors, evaluated by NS Simulations

We observe that the simulation results for $\eta = 1/30$ are obtained with a CPU time which is about 4%of the time required to simulate the original network, and with a memory occupation which is about 9%. This proves the advantage that can be gained by simulating a transformed version of the network.

However, when the scaling factor becomes very small (for example, η = 1/100 for this scenario), the approximation is no longer reliable. The reason is that the TCP flow population size becomes very small, and the approximation of large population underlying our fluid model does not hold any more.

6.2 Experiments with RED

We now consider a simple scenario comprising just one RED queue. The RED queuing discipline is more amenable to analysis by fluid models, thanks to the smooth variation of the moving average of the buffer occupancy. However, RED queues can exhibit very complex behavior due to instability in the control loop of the source-network interaction, as well documented in the literature. Here we are interested in studying how networks comprising RED buffers scale, also with regards to stability. Our basic scenario comprises a single link of capacity 100 Mbps, with round trip propagation delay of 50ms. Initially, 30 elephants are active. Starting from $t = 10$s, 3 elephants are added to the original group every 10s, up to $t = 90$s. At this point, 57 elephants are active. Symmetrically, 3 elephants leave the group every 10s starting at $t = 110$s. At time $t = 190$s, the system is loaded again only by the initial 30 elephants. A gentle version of RED is implemented with $min_{th} = 100.0$, $max_{th} = 200.0$, $p_{max} = 0.1$. Before scaling the system, we consider the stability of the queue for two different values of w_q, 0.0001 and 0.0005.

Figures 4 and 6 report, respectively, the evolutions of queue length and average window size, in the case $w_q = 10^{-5}$, obtained with *ns-2*. Figures 5 and 7 plots the corresponding results obtained with the fluid model. Similarly, Figures 8 and 10 show the results obtained with *ns-2* when $w_q = 5 \cdot 10^{-5}$, and the corresponding fluid model predictions are reported in Figures 9 and 11.

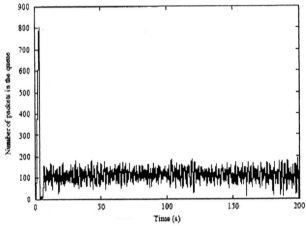

Fig. 4 Buffer evolution obtained with NS-2, $w_Q = 10^{-5}$.

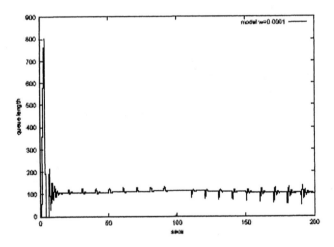

Fig. 5 Buffer evolution obtained with Fluid Model, $w_Q = 10^{-5}$.

In all cases, the fluid model predictions match simulation results very well. When $w_q = 0.0001$, the queue exhibits small fluctuations at any time (the RED queue is stable). When $w_q = 0.0005$, the queue exhibits wild fluctuations during the intervals [0, 50]s and [150, 200]s, in which the number of active flows is smaller than 45. The queue stabilizes when more than 45 flows are active (Figure 9). The fact that RED is more stabilizes when the number of flows increases is well known in the literature [6]. The model predicts this behavior, and this demonstrates its ability to capture complex source-network dynamics.

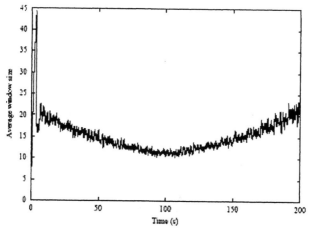

Fig. 6 Average window size obtained with NS-2, $w_Q = 10^{-5}$.

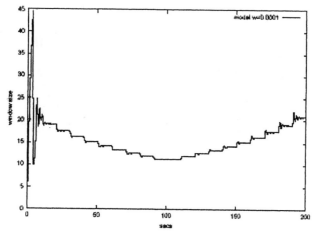

Fig. 7 Average window size obtained with Fluid Model, $w_Q = 10^{-5}$.

We now scale up the system by a factor $\eta = 10$. Correspondingly, the link capacity is set to 1 Gbps, the number of elephants and its variations are multiplied by 10, whereas the weight parameter of the RED filter has to be divided by 10. Thus, we consider the two cases of $w_q = 10^{-6}$ and $w_q = 5 \cdot 10^{-6}$.

Figures 12 and 13 report, respectively, the evolution of queue length and average window size, in the case $w_q = 10^{-6}$, obtained with *ns-2*.

Similarly, Figures 14 and 15 show the simulation results when $w_q = 5 \cdot 10^{-6}$. The model predictions are not reported because the curves perfectly match those obtained in the unscaled system (as for the queue length, the only difference is that it is scaled up by $\eta = 10$). Simulations confirm that the transformed system behave exactly as the original system. Again, in the case $w_q = 5 \cdot 10^{-6}$, the queue stabilizes at about $t = 50s$ when the number of active flows grows sufficiently large. Interestingly, in the stable regime the residual oscillations of the queue (and, correspondingly, of the average window size of the sources) reduce with respect to the original network. This is due to the increased number of concurrent flows. Indeed, fluid models capture the limiting behavior of the system when the number of flows goes to infinity.

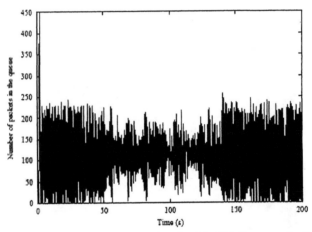

Fig. 8 Buffer evolution obtained with NS-2, $w_Q = 5 \cdot 10^{-5}$.

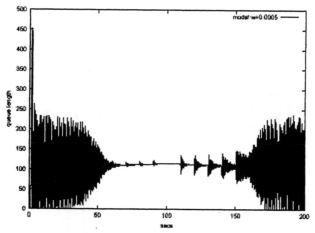

Fig. 9 Buffer evolution obtained with Fluid Model, $w_Q = 5 \cdot 10^{-5}$.

7.   CONCLUSIONS

In this paper we have provided a theoretical investigation of the structural properties of the differential equations defining fluid models of IP networks. We have shown that the set of differential equations defining the network dynamics under both drop-tail and AQM queuing disciplines exhibits a nice invariance property, which allows an equivalence relation to be established among different systems, thus providing a theoretical foundation to the scalability of fluid models of IP networks. The validity of the invariance property has been verified in realistic network scenarios with *ns-2* simulations.

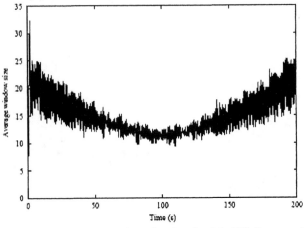

Fig. 10 Average window size obtained with NS-2, $w_Q = 5 \cdot 10^{-5}$.

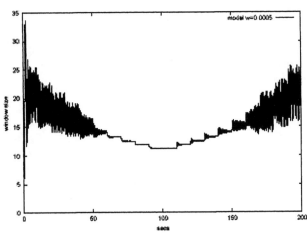

Fig. 11 Average window size obtained with Fluid Model, $w_Q = 5 \cdot 10^{-5}$.

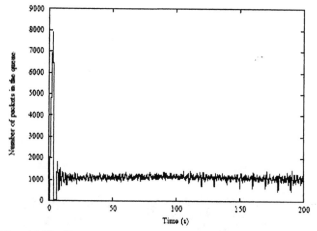

Fig. 12 Buffer evolution obtained with NS-2, $w_Q = 10^{-6}$.

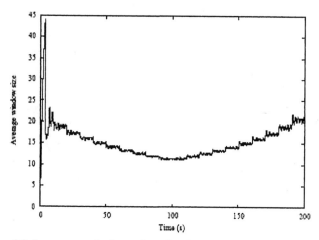

Fig. 13 Average window size obtained with NS-2, $w_Q = 10^{-6}$.

A Proof of Eq. (8) - Basic Sources

We wish to estimate the evolution of $P(w, t)$; we define $v(w, t) = \partial P(w,t)\partial w$ as the probability density of the window distribution at time t. Consider a small enough Δt such that $R(t) \approx \mathbf{R}(t + \Delta t)$. Let ΔP^- be the number of sources with window $\leq w$ at time t, but with window $> w$ at time $t + \Delta t$. All the sources which do not experience any loss indication during the interval $[t, t + \Delta t)$ increase their window with rate $1/R(t)$. Among these sources, ΔP^- includes only the ones with initial window $>$

$w - \Delta t / R(t)$, since they will exceed w by time $t + \Delta t$. If we assume to model (locally) the loss indication process with a Poisson process with rate $\lambda(w, t)$, the probability that no losses are experienced

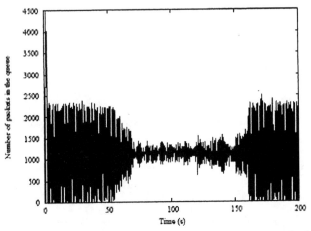

Fig. 14 Buffer evolution obtained with NS-2, $w_Q = 5 \cdot 10^{-6}$.

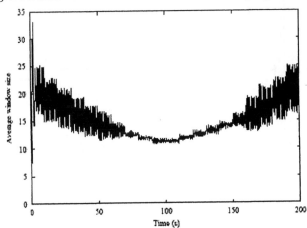

Fig. 15 Average window size obtained with NS-2, $w_Q = 5 \cdot 10^{-6}$.

during Δt is $(1 - \lambda(\alpha,t)\Delta t + o(\Delta t))$, then;

$$\Delta P^{-} = \int_{w-\Delta t/R(t)}^{w} (1 - \lambda(\alpha,t)\Delta t + o(\Delta t))v(\alpha,t)d\alpha$$

$$\frac{\Delta P^{-}}{\Delta t} \rightarrow \frac{1}{R(t)} v(w,t)$$

Let now ΔP^+ be the number of sources with window $> w$ at time t, but with window $\leq w$ at time $t + \Delta t$. ΔP^+ include only the sources (i) with window in the range $(w, 2w - \Delta t/R(t)]$ at time t, and (ii) receiving a loss indication in the interval $[t, t + \Delta t)$. Note that the probability of receiving multiple loss indications is $o(\Delta t)$, hence negligible. Hence,

$$\Delta P^+ = \int_w^{2w-\Delta t-R(t)} \lambda(\alpha,t)\Delta tv(\alpha,t)d\alpha + o(\Delta t)$$

$$\frac{\Delta P^+}{\Delta t} \to \int_w^{2w} \lambda(\alpha,t)v(\alpha,t)d\alpha$$

Since $P(w, t + \Delta t)=P(w,t)+\Delta P^+-\Delta P^-$, we can find (8):

$$\frac{\partial P}{\partial t}(w,t)= \lim_{\Delta t \to 0} \frac{\Delta P^+ - \Delta P^-}{\Delta t} = \int_w^{2w} \lambda(\alpha,t)v(\alpha,t)d\alpha - \frac{1}{R(t)}v(w,t)$$

REFERENCES

[1]   F.Baccelli, D.Hong, "Interaction of TCP Flows as Billiards", *IEEE Infocom 2003,* San Francisco, CA, March 2003.

[2]   F.Baccelli, D.Hong, "Flow Level Simulation of Large IP Networks", *IEEE Infocom 2003,* SanFrancisco, CA, March 2003.

[3]   S.Misra, W.B.Gong, D. Towsley, "Fluid-Based Analysis of a Network of AQM Routers Supporting TCP Flows with an Application to RED", *ACM SIGCOMM 2000,* Stockholm, Sweden, August 2000.

[4]   C.V.Hollot, V.Misra, D.Towsley, W.B.Gong, "On the Design of Improved Controllers for AQM Routers Supporting TCP Flows", *IEEE Infocom 2001,* Anchorage, Alaska, USA, April 2001.

[5]   Y.Liu, F.Lo Presti, V.Misra, D.Towsley, "Fluid Models and Solutions for Large-Scale IP Networks", *ACM Sigmetrics 2003,* San Diego, CA, June 2003.

[6]   C.V.Hollot, V.Misra, D.Towsley, W.B.Gong, "A Control Theoretic Analysis of RED", *IEEE Infocom 2001,* Anchorage, Alaska, USA, April 2001.

[7]   S.Deb, S.Shakkottai, R.Srikant, "Stability and Convergence of TCP-like Congestion Controllers in a Many-Flows Regime", *IEEE Infocom 2003,* San Francisco, CA, March 2003.

[8]   S.Shakkottai, R.Srikant, "How Good are Deterministic FluidModels of Internet Congestion Control?" *IEEE INFOCOM 2002,* New York, June 2002.

[9]   P.Tinnakornsrisuphap, A.Makowski, "Limit Behavior of ECN/RED Gateways Under a Large Number of TCP Flows", *IEEE Infocom 2003,* San Francisco, CA, March 2003.

[10] F.Baccelli, D.R.McDonald, J.Reynier, "A Mean-Field Model for Multiple TCP Connections through a Buffer Implementing RED", *Performance Evaluation,* vol. 49 n. 1/4, pp. 77-97, 2002.

[11] M.Ajmone   Marsan,   M.Garetto,   P.Giaccone,   E.Leonardi, E.Schiattarella, A.Tarello, "Using Partial Differential Equations to Model TCP Mice and Elephants in Large IP Networks", *Infocom 2004,* Hong Kong, March 2004.

[12] M.Ajmone   Marsan,   G.Carofiglio,M.Garetto,   P.Giaccone, E.Leonardi, E.Schiattarella, A.Tarello, "Of Mice and Models", *QoS IP 2005,* Catania, February 2005.

[13] S.Floyd, V.Jacobson, "Random Early Detection Gateways for Congestion Avoidance", *IEEE/ACM Transactions on Networking,* vol. 1, n. 4, pp. 397-413, August 1993.

[14] R. Pan, B. Prabhakar, K. Psounis, D.Wischik, "SHRiNK: A method for scalable performance prediction and efficient network simulation", *IEEE Infocom 2003,* San Francisco, CA, March 2003

[15] R. Pan, B. Prabhakar, K. Psounis, M. Sharma, "A study of the applicability of a scaling hypothesis", *in Proc. ASCC 2002,* Singapore, September 2002

CHAPTER 7

The Optimal Dimensioning Of Multi-Service Links

Iversen, V.B.[1] and Stepanov, S.N.[2]

[1] COM.DTU, Technical University of Denmark,
DK-2800 Kgs. Lyngby, Denmark

[2] Sistema Telecom
125047, Moscow, 1st Tverskay-Yamskaya 5, Russia

Abstract. *In this paper we describe an approach for optimal dimensioning of multi-service lines. The solution presented is based on the method to convert recursions for global state probabilities of multi-service models into stable form. The suggested approach is numerically stable because it deals with normalised values of global state probabilities used for estimation of main stationary performance measures.*

Keywords. dimensioning, multi-service links, stable recurrence, soft blocking, trunk reservation, complexity.

1. PROBLEM FORMULATION

One of the main problems that often has to be solved by operators is to determine the volume of telecommunication resources that is enough for serving given input flows of demands with prescribed characteristics of QoS (Quality-of-Service). Usually, the resource considered is the number of circuits, the value of bandwidth represented by integer number of units, the number of radio channels, etc. The adequacy of volume is determined by comparison of suitably chosen performance measure with prescribed level of system functioning.

A multi-service link is a basic element of transport network in realization of NGN (Next Generation Network) concept. In teletraffic theory the multi-service link is considered as a single link traffic model, where the link transmission capacity is represented by k basic bandwidth units. In the model we have some number of flows of demands for bandwidth which we for simplicity will name by calls. Let $p(i)$ be the mean proportion of time when exactly i service units are occupied. The main stationary performance measures of practical interest can be expressed as a function of $p(i)$, $i = 0, 1, \ldots, k$. Let B denote a measure used for dimensioning. A traditional dimensioning problem related with system planning is formulated as follows.

For given input flow, find a minimum value of k satisfying the inequality:

$$B \leq \pi, \tag{1}$$

where π is a prescribed level of system functioning. Typically it will be the proportion of calls lost or the proportion of time the system is blocked for call servicing. Usually only a limited number of $p(i)$'s are used in estimation of B and the number of $p(i)$'s used does not depend on the number of service units. For example, if each call of the input flow needs b units for its servicing then the proportion of time the system is blocked for call servicing can be estimated by the expression:

$$B = p(k - b + 1) + p(k - b + 2) + \ldots + p(k).$$

Let us suppose that performance measure B used for dimensioning of the multi-service line is calculated according to expression:

$$B = B(p(k - b + 1), p(k - b + 2), \ldots, p(k)), \tag{2}$$

where b is some integer number independent of k. The function $B(.)$ depends on the parameters of the model and the form of the criteria used for dimensioning. Specific types of the function $B(.)$ will be presented in the following sections where we will consider examples of teletraffic models.

Solving the problem formulated by inequality (Eq.1) is usually going on according to the following recursive scheme. Let $P(i)$ denote unnormalised values of $p(i)$, $i = 0, 1, \ldots, k$, and suppose that the following relation is known for estimation of $P(i)$.

$$P(i) = \begin{cases} 0 & \text{if} \quad i<0 \\ a & \text{if} \quad i=0 \\ f\big(P(i-1), P(i-2),\ldots, P(i-m)\big) & \text{if} \quad i=1,2,\ldots k \end{cases} \qquad (3)$$

In (Eq.3) m is an integer. We suppose that its value is independent of the number of service units k. In some cases m is just equal to the maximum number of service units needed for call servicing. The parameter a is any positive real number chosen as starting value of recurrence. The function $f(.)$, depends on the parameters of the model and specific examples of the function $f(.)$ will be presented later.

Using (Eq.3) sequentially, we can express all $p(i)$, $i = 1, 2, \ldots\ldots, k,$, by $P(0)$. Then we find the normalising constant $Q = \sum_{i=0}^{k} P(i)$ and the normalised values of state probabilities.

$$p(i) = \frac{P(i)}{Q}, i = 0,1,2,\ldots,k \qquad (4)$$

Let us suppose that for any choice of a we obtain the same set of probabilities $p(i)$, $i = 0,1,\ldots\ldots, k$, after normalising. Next we calculate the performance measure B used in criteria (Eq.1). If number of service units k does not satisfy (Eq.1), then the value of k is increased and we repeat the described actions once more. The numerical complexity of this approach is estimated by the numerical complexity of finding the performance measure for given input parameters multiplied by the number of searches. We call the described scheme of dimensioning a traditional procedure.

The implementation of traditional procedure has at least two negative aspects. They are clearly observed for a large number of service units k. For practically cases of interest almost all probability mass is located in states with large number of occupied service units (see for example results of estimation of $P(i)$ presented at Fig.4). It means that states with a small number of occupied service units have very small probabilities of existence. The state with number of occupied service units equals to zero has the smallest probability different from zero for bigger systems. So when we start to express all $P(i)$ through $P(0)$ we very quickly experience problems of overflow. Another negative aspect is the increase of computational efforts. To solve the formulated dimensioning problem we need to perform a run of (Eq.3) for each value of k serving as a candidate for desired solution. Each time we need to start calculation from $i = 0$.

The significance of solving the formulated problem for optimal distribution of telecommunication resources has forced researchers to look for ways to decrease the amount of computational work. One direction is to realize the approach implemented for optimization of dimensioning procedure based on the Erlang-B loss formula.

Let us consider the details of the optimization scheme. The Erlang model is a particular case of the model considered. It has one Poissonian flow of demands for connections with intensity λ. For simplicity we suppose that each call uses one bandwidth unit for the time of connection having mean value equal to one. Erlang formula for estimation of the mean proportion of time when all k bandwidth units are occupied looks as follow.

$$E_k(\lambda) = \frac{\dfrac{\lambda^k}{k!}}{1 + \lambda + \dfrac{\lambda^2}{2} + ... + \dfrac{\lambda^k}{k!}} \tag{5}$$

For optimization of dimensioning based on (Eq.5) the expression for $E_k(\lambda)$ is written in the form of recurrence on number of service units.

$$E_k(\lambda) = \frac{\dfrac{\lambda}{k}.E_{k-1}(\lambda)}{1 + \dfrac{\lambda}{k}.E_{k-1}(\lambda)}, k = 1,2,...,E_0(\lambda) = 1 \tag{6}$$

The relation (Eq.6) is obtained by dividing upper and lower parts of (Eq.5) by $1 + \lambda + \dfrac{\lambda^2}{2!} + ... + \dfrac{\lambda^{k-1}}{(k-1)!}$ and using the definition of the Erlang-B formula. The recursion (Eq.6) works with normalised values $E_k(\lambda)$ so its implementation does not suffer from numerical problems such as overflow or loss of precision that very often occur for large values of k. Using (Eq.6) consecutively for $k = 1, 2, \ldots$,it is easy to find the value of link transmission capacity k that is minimum for solving the given traffic flow with prescribed characteristics of QoS. It is easy to see that computational work for solving the formulated dimensioning problem equals computational work for estimating $E_k(\lambda)$.

The aim of this paper is to suggest a unified approach for multi-service links to convert recursions for global state probabilities of the type (Eq.3) into stable form where recursions are made on the number of service units. This type of recursions is well suited for solving the dimensioning problem based on form of criteria (Eq.1) and performance measure of type (Eq.2). The proposed method is a direct generalisation of optimal dimensioning scheme used for Erlang model. The general form of solution will be presented in Section 2. The usage of the unified approach will be shown for a number of cases in Section 3, where alternative and very simple solutions will be given for some already known results, and in Section 4 where new solutions will be presented. The paper is an extended and updated version of [1].

2.   BASIC ALGORITHM

Let us formulate a condition on implementation of function $f(.)$ where we can solve the dimensioning problem in one run as was done by recurrence (Eq.6) for service systems whose performance measures numerically are estimated by means of Erlang's B-formula. We indicate when necessary the number of available service units by lower index for the corresponding set of probabilities. Let us assume that for any $k > 1$ the implementation of recurrence (Eq.3) for number of service units $k-1$ and k and the same starting value gives sets of unnormalised probabilities $P_{k-1}(0) = a$, $P_{k-1}(1)$, $\ldots$, $P_{k-1}(k-1)$, and $P_k(0) = a$, $P_k(1)$, $\ldots$, $P_k(k-1)$, $P_k(k)$ satisfying the following relations:

$$P_k(i) = P_{k-1}(i), i = 0, 1, \ldots, k - 1. \tag{7}$$

Property (Eq.7) allows us to rewrite (Eq.3) in a form dealing with normalised probabilities.

Let us assume that we know the normalised set of probabilities $p_{k-1}(0)$, $p_{k-1}(1),\ldots,p_{k-1}(k-1)$ for number of service units $k-1$, and that the performance measure B calculated for this number of service units doesn't satisfy the criteria (Eq.1). Let us assume that the number of service units is increased by one to k. Then taking as starting value $P_k(0) = p_{k-1}(0)$ and applying (Eq.3) we obtain according to property (Eq.7) the following set of unnormalised probabilities.

$$
\begin{aligned}
&P_k(0) = p_{k-1}(0), \\
&P_k(1) = p_{k-1}(1),
\end{aligned}
\tag{8}
$$

$$
\ldots \qquad \ldots
$$

$$
\begin{aligned}
&P_k(k-1) = p_{k-1}(k-1), \\
&P_k(k) = f(\, p_{k-1}(k-1), p_{k-1}(k-2),\ldots, p_{k-1}(k-m)).
\end{aligned}
$$

For normailsed probability $P_k(k)$ we have:

$$
\begin{aligned}
p(k) &= \frac{P_k(k)}{P_k(0) + P_k(1) + \ldots + P_k(k)} \\
&= \frac{P_k(k)}{p_{k-1}(0) + p_{k-1}(1) + \ldots + p_{k-1}(k-1) + p_{k-1}(k)} \\
&= \frac{P_k(k)}{1 + P_k(k)}
\end{aligned}
\tag{9}
$$

Using (Eq.8) we rewrite (Eq.9) in the form.

$$
p_k(k) = \frac{f(p_{k-1}(k-1), p_{k-1}(k-2),\ldots, p_{k-1}(k-m))}{1 + f(p_{k-1}(k-1), p_{k-1}(k-2),\ldots, p_{k-1}(k-m))}
\tag{10}
$$

At this step we need to calculate the performance measure B used for solving the dimensioning problem. Because the number of service units is increased by one we need to normalise probabilities $P_k(k-1) = p_{k-1}(k-1)$, $P_k(k-2) = p_{k-1}(k-2),\ldots, P_k(k-b+1) = p_{k-1}(k-b+1)$, used for estimation of B (the normalised value $P_k(k)$ has already been obtained by relation (10)). We can do this by means of equalities.

$$
p_k(i) = \frac{p_{k-1}(i)}{1 + f(p_{k-1}(k-1), p_{k-1}(k-2),\ldots, p_{k-1}(k-m))}, i = k-1, k-2,\ldots, k-b+1.
\tag{11}
$$

If the calculated value of B satisfies criteria (Eq.1), then the desired number of service units will be k. Otherwise it is necessary to increase the number of service units k by one and make a new estimation of B for $k+1$. According to (Eq.8)-(Eq.10) to perform this step it is necessary to have normalised values of $P_k(k-1)$, $P_k(k-2)$, . . . , $P_k(k-m)$. It is obvious that these normalisations and (Eq.11) can be done in one run in the form.

$$p_k(i) = \frac{p_{k-1}(i)}{1 + f(p_{k-1}(k-1), p_{k-1}(k-2), \dots, p_{k-1}(k-j))}, i = k-1, k-2, \dots, k-j \qquad (12)$$

where $j = \max(b - 1, m)$. A one-run algorithm giving at each step normalised values of state probabilities which are necessary for solving the problem of dimensioning based on (Eq.1)-(Eq.3) looks as follows.

- *Step 1.* Let $p_0(0) = 1$.
- *Step 2.* Let $j = \max(b - 1, m)$. For fixed $k = 1, 2, \dots,$ find normalised value of $p_k(i)$ by using relations (10),(12).

$$p_k(k) = \frac{f(p_{k-1}(k-1), p_{k-1}(k-2), \dots, p_{k-1}(k-j))}{1 + f(p_{k-1}(k-1), p_{k-1}(k-2), \dots, p_{k-1}(k-j))}, \qquad (13)$$

$$p_k(i) = \frac{p_{k-1}(i)}{1 + f(p_{k-1}(k-1), p_{k-1}(k-2), \dots, p_{k-1}(k-j))},$$
$$i = k-1, k-2, \dots, \max(k-j, 0)$$

- *Step 3.* Here we calculate the performance measures, check the dimensioning criteria and either stop or continue the process of estimating the number of service units needed.

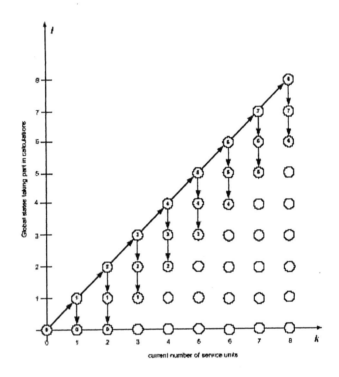

Fig. 1 The Order of calculation of the normalised stationary probabilities in accordance with suggested approach ($j = 2$)

Fig. 1 shows the order of calculation of the normalised values of stationary probabilities in accordance with suggested approach. For the simplicity $j = 2$. We see that for fixed number of service units k it is necessary to find only $\min(k,j)+1$ normalised probabilities to be able to calculate the performance measure B and proceed if necessary to the next step of recursion defined by (Eq.13). Computational efforts for current number of service units k is estimated by the total number of operations N_f that are needed for estimation of function $f(.)$ (once) with the number of operations N_B that are needed for estimation of performance measure B (once), and $j+1$ operations of normalisation. So computational efforts to reach number of service units k starting from 1 is estimated by $O\{(N_f +N_B+j+1)k\}$ Storage requirements are estimated by $O\{j+1\}$.

Because stable recursions are obtained directly from recursions for global state probabilities we can easily rewrite them into other forms which in some cases are more suitable for calculations than basic form. We consider two modifications of the basic version of the formulated algorithm. Both allow to start stable recursions not from $k = 1$ but from some intermediate value chosen according to availability of some additional information concerning model functioning. The first modification gives exact results and the second gives approximate values, but with very good convergence to exact results.

2.1 Modification Of Basic Algorithm Taking Account Of The Known Values Of State Probabilities

We now consider the first modification of the basic algorithm. It is clear that we can start stable recursions from some intermediate value of $k = k_0$ if in some way (for example by means of (Eq.3)) we know normalised values of $p_{k0} (k_0)$, $p_{k0} (k_0 - 1)$, $\ldots\ldots$, $p_{k0} (k_0 - j)$. This follows from previous stated algorithm (Eq.7)-(Eq.12) and Fig. 1. For simplicity we assume that $k_0 \geq j$. The algorithm looks as follows.

- $Step\ 1$. Using (3) find normalised values $p_{k0} (k_0)$, $p_{k0} (k_0 -1)$, $.\ .$, $p_{k0}(k_0-j)$.
- $Step\ 2$. Let $j = \max(b-1,m)$. For fixed $k = k_0+1$, k_0+2, $.\ .\ .\ .$, find normalised value of $p_k(i)$ by using relations (10),(12).

$$p_k(k) = \frac{f(p_{k-1}(k-1), p_{k-1}(k-2),..., p_{k-1}(k-j))}{1 + f(p_{k-1}(k-1), p_{k-1}(k-2),..., p_{k-1}(k-j))}. \qquad (14)$$

$$p_k(i) = \frac{p_{k-1}(i))}{1 + f(p_{k-1}(k-1), p_{k-1}(k-2),..., p_{k-1}(k-j))},$$

$$i = k-1, k-2,..., k-j.$$

- $Step\ 3$. The same as formulated in the basic version of algorithm.

Fig. 2 shows the order of calculation of the normalised stationary probabilities in accordance with modified version of basic algorithm. For simplicity $j = 2$, $k_0 = 5$. The computational efforts are in accordance with modified versions of basic algorithm is estimated by $O\{(N_f+(N_B+j +1)(1-\frac{k_0}{k}\ )) k\}$. Storage requirements are estimated by $O\{j + 1\}$.

2.2 Modification Of Basic Algorithm With Truncations Of The State Space

We now suppose that the algorithm is applied for solving the dimensioning problems for large value of number of service units k. In this case the distribution of probability mass for global state probabilities has a specific form. Almost all probabilistic mass is located in the area close to the state (k), when all available service units are occupied.

Heuristics based on analysis done for particular tele-traffic models allow us to make the following suggestions. Let us choose some truncation level` and suppose that model functioning is located in global states (ℓ), $(\ell+1)$,..., (k). If the value of ℓ is chosen so that $p(\ell)$ is quite small, then recurrence (Eq.3) for $P(i)$ can rewritten in the form

$$P^*(i) = \begin{cases} 0, & \text{if} \quad i < \ell \\ a, & \text{if} \quad i = \ell \\ f\big(P^*(i-1), P^*(i-2),..., P^*(i-m)\big) & \text{if} \quad i = \ell+1, \ell+2,...,k \end{cases} \tag{15}$$

In (15) we use asterisk to show that $P^*(i)$ gives approximate value of $P(i)$. The accuracy of estimation should be studied separately for each specific model. We will show details on the example of one link multi-rate Poissonian model in Section 4.1. The modified version of stable recurrence based on (Eq.15) becomes.

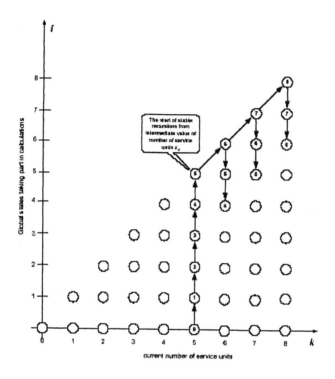

Fig. 2 The order of calculation of the normalised stationary probabilities in accordance with the modified version of basic algorithm when the start of recursions was done fromsome intermediate value of service units $k_0 = 5$ ($j = 2$, $k_0 = 5$)

- *Step 1*. Let $p_\ell^* (\ell) = 1$.
- *Step 2*. Let $j = \max(b-1, m)$. For fixed $k = \ell+1, \ell+2, \ldots$, find normalised value of $p_k^*(i)$ by using relations (Eq.10), (Eq.12) and (Eq15).

$$p_k^*(k) = \frac{f(p_{k-1}^*(k-1), p_{k-1}^*(k-2), \ldots, p_{k-1}^*(k-j))}{1 + f(p_{k-1}^*(k-1), p_{k-1}^*(k-2), \ldots, p_{k-1}^*(k-j))}; \quad (16)$$

$$p_k^*(i) = \frac{p_{k-1}^*(i)}{1 + f(p_{k-1}^*(k-1), p_{k-1}^*(k-2), \ldots, p_{k-1}^*(k-j))},$$
$$i = k-1, k-2, \ldots, \max(k-j, \ell)$$

- *Step 3*. The same as formulated for basic version of algorithm.

Fig. 3 shows the order of calculation of the normalised stationary probabilities in accordance with modified version of the basic algorithm. For simplicity $j = 2$, $\ell = 4$. The computational efforts to reach number of service units k starting from ` is estimated by $O\{(N_f + N_B + j + 1)\ (k-\ell)\}$. Storage requirements are estimated by $O\{j + 1\}$.

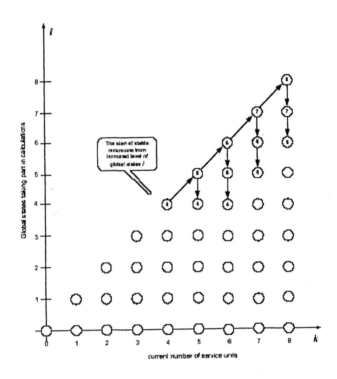

Fig. 3 The order of calculation of normalised stationary probabilities in accordance with the modified version of basic algorithm when global states are truncated by level ℓ ($j = 2$, $\ell = 4$)

2.3   Discussions and Previous Results

The results obtained are based on ideas from [2],[3]. If we implement this approach to specific teletraffic models we obtain numerically stable versions of recursive algorithms designed for estimation of performance measures. For one link multi-rate models with product form solutions a number of numerically stable algorithms for calculation of main performance measures are known. [4],[5],[6]. If we compare the best results in this field [5],[6] with algorithms obtained after usage of this approach we can conclude that they have the same storage and computational requirements, but we derive stable algorithms in a much simpler and way by one general scheme. All we need is just to take the recurrence for global state probabilities in the form (Eq.3) and only put it into relations (Eq.13) or its modifications (Eq.14), (Eq.16). This is illustrated in Sections 3 and 4.

The approach suggested has a number of other positive features. Firstly, the possibility of increasing the efficiency by truncation of the global state space. We have already discussed the general aspects for realization of this idea in this section. Examples are presented in Section 4.1. Another advantage is the possibility in a unified way to treat a wide class of teletraffic models with/without product form solution. We illustrate this for a couple of models in Sections 4.2-4.3, but first we show the application of the approach for a link models with product form solutions which is a multi-service analogue of the Erlang model.

3   ONE-LINK MULTI-SERVICE MODELS WITH PRODUCT FORM SOLUTIONS

3.1   Multi-Rate Poissonian Models

Let us consider a single link traffic model, where the link transmission capacity is represented by k basic bandwidth units, and let us suppose that we have n incoming Poisson flows of calls with intensities λ_s, $s = 1, \ldots, n$. A call of flow s uses b_s bandwidth units during all time of connection. Without loss of generality we shall assume that the holding times all are exponentially distributed with the same mean value chosen as time unit, but it is known that the model considered is insensitive to the distribution of the holding time, and each flow may furthermore have individual mean holding times.

Let $i_s(t)$ denote the number of calls of the flow s served at time t. The model is described by an n-dimensional Markovian process of type $r(t) = \{i_1(t), i_2(t), \ldots, i_n(t)\}$ with state space S consisting of vectors $(i_1, \ldots, i_n)$, where i_s is the number of calls of the s'th flow being served by the link under stationary conditions. The state space S is defined as follows. $(i_1, \ldots, i_n) \in S$, $i_s \geq 0$, $s = 1, \ldots, n$, $\sum_{s=1}^{n} i_s b_s \leq k$. Let $P(i_1, \ldots, i_n)$ denote the unnormalised values of stationary probabilities of $r(t)$. After normalisation, $p(i_1, \ldots, i_n)$ denotes the mean proportion of time when exactly $\{i_1, \ldots,$

i_n}connections are established. For state $(i_1, \ldots, i_n)$ the value i denotes the total number of occupied bandwidth units $i = i_1 b_1 + \ldots + i_n b_n$.

The process of transmission of s'th flow is described by blocking probability π_s, $s = 1, \ldots, n$. Their formal definition through values of state probabilities are as follows (here and further, summations are for all states $(i_1, \ldots, i_n) \in S$ satisfying the formulated condition, and by small characters we denote the normalised values of state probabilities).

$$\pi_S = \sum_{i+b_s > k} p(i_1, \ldots, i_n). \tag{17}$$

There are many algorithms for estimation of π_s. All of them are based on product form relations valid for $P(i_1, \ldots, i_n)$.

$$P(i_1, \ldots, i_n) = P(0, \ldots, 0) \cdot \frac{\lambda_1^{i_1}}{i_1!} \cdot \frac{\lambda_2^{i_2}}{i_2!} \cdot \ldots \cdot \frac{\lambda_n^{i_n}}{i_n!}, (i_1, \ldots, i_n) \in S.$$

The most efficient calculation scheme is the recurrence algorithm first obtained in [7] and later also derived in [8], [9]. This algorithm exploits the fact that the performance measures (Eq.17) can be found if we for process $r(t)$ know probabilities $p(i)$ of being in the state where exactly i bandwidth units are occupied.

$$p(i) = \sum_{i_1 b_1 + \ldots + i_n b_n = i} p(i_1, \ldots, i_n).$$

The corresponding formulas are as follows.

$$\pi_S = \sum_{i=k+b_s+1}^{k} p(i), s = 1, 2, \ldots, n. \tag{18}$$

The unnormalised values of $P(i)$ are found by the recurrence.

$$P(x) = \begin{cases} 0, & \text{if} \quad i < 0 \\ a, & \text{if} \quad i = 0 \\ \dfrac{1}{i} \sum_{s=1}^{n} \lambda_s b_s P(i - b_s), & \text{if} \quad i = 1, 2, \ldots, k. \end{cases} \tag{19}$$

This relation gives a specific function $f(.)$ defined by (Eq.3) for the model considered. Using the general framework formulated in Section 2 we obtain a one-run algorithm that at each step gives the normalised values of state probabilities which are necessary for solving the problem of dimensioning based on the form of function $B(.)$ specified by (Eq.18).

- *Step 1.* Let $p_0(0) = 1$.
- *Step 2.* Let $b = \max\limits_{0 \le s \le n} (b_s)$. For fixed $k = 1, 2, \ldots$, find normalised value of $p_k(i)$ using (Eq.13).

$$p_k(k) = \frac{\dfrac{1}{k} \sum\limits_{s=1}^{n} \lambda_s b_s p_{k-1}(k - b_s)}{1 + \dfrac{1}{k} \sum\limits_{s=1}^{n} \lambda_s b_s p_{k-1}(k - b_s)} ; \tag{20}$$

$$p_k(i) = \frac{p_{k-1}(i)}{1 + \dfrac{1}{k} \sum\limits_{s=1}^{n} \lambda_s b_s p_{k-1}(k - b_s)}, i = k - 1, k - 2, \ldots, \max(k - b, 0)$$

- *Step 3.* Here we calculate the performance measures defined by (Eq.18), check the dimensioning criteria (for example, $B = \max\limits_{1 \le s \le n} \pi_s$) and either stop or continue the process of estimating the number of service units needed.

When implementing this version of the recurrence algorithm we need to keep a vector of size $O\{b\}$ in computer memory. Computational efforts are estimated by $O\{(n + b)k\}$.

3.2 Multi-Rate State-Dependent Models

We consider a single link traffic model, where the link capacity is represented by k basic bandwidth units, and let us consider n incoming flows of calls. For s'th flow $\{s = 1, \ldots ,n\}$, the time interval between successive call arrivals has a state-dependent exponential distribution with parameter $\lambda_s(i_s) = \alpha_s + i_s \beta_s$, where i_s is the number of calls of flow s served by the link. A call of flow s uses b_s bandwidth units for the time of connection. Without loss of generality we shall assume that the holding times all are exponentially distributed with the same mean value chosen to as time unit, but it is known that the model considered is insensitive to the distribution of the service time, and that each flow may have individual mean holding times.

Let $i_s(t)$ denote the number of calls of the s'th flow being served at time t. The model is described by an n-dimensional Markovian process of the following type $r(t) = \{i_1(t), i_2(t), \ldots, i_n(t)\}$ with state space S consisting of vectors $(i_1, \ldots, i_n)$, where i_s is the number of calls of the s'th flow served by the link under stationary conditions which impose some restrictions on values of α_s and β_s. The state space S is defined as follows. $(i_1, \ldots, i_n) \in S$, $i_s \geq 0$, $s = 1, \ldots, n, \sum_{s=1}^{n} i_s b_s \leq k$. Let $P(i_1, \ldots, i_n)$ denote the unnormalised values of stationary probabilities of $r(t)$. After normalisation the value $p(i_1, \ldots, i_n)$ denotes the proportion of time when exactly $(i_1, \ldots, i_n)$ connections are established. Assume that for state $(i_1, \ldots, i_n)$ the value i denotes the total number of occupied bandwidth units $i = i_1 b_1 + \ldots + i_n b_n$.

The process of transmission of the s'th flow is described by time blocking probability π_s and by call blocking probability ω_s, $s = 1, \ldots, n$. Their formal definition by state probabilities are as follows.

$$\pi_s = \sum_{i+b_s > k} p(i_1, \ldots, i_n), \omega_s = \frac{\sum_{i+b_s > k} \lambda_s(i_s) P(i_1, \ldots, i_n)}{\sum_{(i_1, \ldots, i_n) \in S} \lambda_s(i_s) P(i_1, \ldots, i_n)}. \tag{21}$$

This model like the previous one has product form representation for $P(i_1 \ldots, i_n)$.

$$P(i_1, \ldots, i_n) = P(0, \ldots, 0) \cdot \frac{\prod_{m=0}^{i_1 - 1} (\alpha_1 + m\beta_1)}{i_1!} \cdot \ldots \cdot \frac{\prod_{m=0}^{i_n - 1} (\alpha_n + m\beta_n)}{i_n!}. \tag{22}$$

Based on (Eq.22) a number of algorithms are constructed for estimating the performance measures (Eq.21). Let us formulate an algorithm based on results from [10]. This algorithm exploits the fact that the performance measures (Eq.21) can be found if we for the process $r(t)$ know auxiliary characteristics defined for the state where exactly i bandwidth units are occupied.

$$p(i) = \sum_{i_1 b_1 + \ldots + i_n b_n = i} p(i_1, \ldots, i_n),$$

$$m_s(i) = \sum_{i_1 b_1 + \ldots + i_n b_n = i} i_s \cdot p(i_1, \ldots, i_n), s = 1, 2, \ldots, n.$$

The corresponding formulas are as follows.

$$\pi_s = \sum_{i=k-b_s+1}^{k} p(i), \omega_s = \frac{\sum_{i=k-b_s+1}^{k} \{\alpha_s p(i) + \beta_s m_s(i)\}}{\sum_{i=0}^{k} \{\alpha_s p(i) + \beta_s m_s(i)\}}, s = 1, 2, \ldots, n \qquad (23)$$

We introduce auxiliary characteristic.

$$m_{s,k} = \sum_{i=0}^{k} \beta_s m_s(i).$$

The unnormalised values of $P(i)$ and $M_s(i)$, $s = 1, 2, \ldots, n$ are found by recurrence.

$$P(i) = \frac{1}{i} \sum_{s=1}^{n} b_s M_s(i), i = 1, 2, \ldots, k, \qquad (24)$$

where

$$M_s(i) = \alpha_s P(i - b_s) + \beta_s M_s(i - b_s),$$

with starting value $P(0) = 1$, $M_s(0) = 0$, $s = 1, 2, \ldots, n$. This relation gives the type of function $f(.)$ defined by (Eq.3). When necessary, for $M_s(i)$ or $m_s(i)$ the number of available service units k is denoted by lower index k. Using general framework formulated in Section 2 we obtain a one-run algorithm that at each step gives the normalised values of the model's characteristics which are necessary for solving the dimensioning problem based on performance measures (Eq.23).

- *Step 1.* Let $p_0(0) = 1$.

- *Step 2.* Let $b = \max\limits_{0 \le s \le n}(b_s)$. For fixed $k = 1, 2, \ldots$, find normalised value of

$p_k(i)$, $i = k, k\text{-}1, \ldots$, $\max(k - b, 0)$ by using relations (Eq.13).

$$p_k(k) = \frac{\dfrac{1}{k}\sum_{s=1}^{n}(\alpha_s b_s p_{k-1}(k - b_s) + \beta_s b_s m_{s,k-1}(k - b_s))}{1 + \dfrac{1}{k}\sum_{s=1}^{n}(\alpha_s b_s p_{k-1}(k - b_s) + \beta_s b_s m_{s,k-1}(k - b_s))} ; \tag{25}$$

$$p_k(i) = \frac{p_{k-1}(i)}{1 + \dfrac{1}{k}\sum_{s=1}^{n}(\alpha_s b_s p_{k-1}(k - b_s) + \beta_s b_s m_{s,k-1}(k - b_s))} ,$$

$$i = k - 1, k - 2, \ldots, \max(k - b, 0).$$

Also for all $s = 1,2,\ldots,n$ we need to normalize:

$$m_{s,k}(i) = \frac{m_{s,k-1}(i)}{1 + \dfrac{1}{k}\sum_{s=1}^{n}(\alpha_s b_s p_{k-1}(k - b_s) + \beta_s b_s m_{s,k-1}(k - b_s))} \tag{26}$$

$$i = k, k - 1, \ldots, \max(k - b, 0)$$

$$m_{s,k} = \frac{m_{s,k-1}}{1 + \dfrac{1}{k}\sum_{s=1}^{n}(\alpha_s b_s p_{k-1}(k - b_s) + \beta_s b_s m_{s,k-1}(k - b_s))} + \beta_s m_{s,k}(k).$$

- *Step 3.* Here we calculate the performance measure defined by (Eq.23), check the dimensioning criteria (for example, $B = \max\limits_{1 \le s \le n} \pi_s$ or $B = \max\limits_{1 \le s \le n} \omega_s$ and either stop or continue the process of estimating the desired number of service units.

When implementing this version of recurrence algorithm we need to keep a vector of size $O\{b\}$ in computer memory with computational efforts estimated by $O\{(n + b) k\}$ if we only use π_s for solving dimensioning problem. We need to keep a vector of size $O\{n\, b\}$ in computer memory with computational efforts estimated by $O\{n\, b\, k\}$ if we use both π_s and ω_s.

4. STABLE ALGORITHMS FOR SOME TELETRAFFIC MODELS

4.1 Multi-rate Poissonian model with truncations

To illustrate the necessity of truncation for large values of service units we consider a numerical example. Fig. 4 shows the distribution of relative values of $P(i)$.

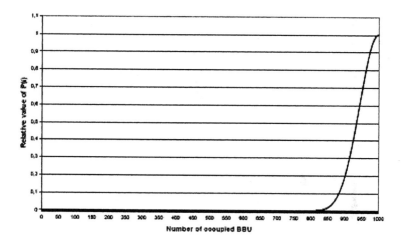

Fig. 4 The distribution of relative value of $P(i)$ (normalised by $\max\limits_{0 \leq i \leq k} P(i)$) with respect to number of basic service units (BBU)$_i$

The model input parameters are $k = 1000$, $n = 5$, $b_s = s$, $\lambda_s = k / (nb_s)$, $s = 1, 2, 3, 4, 5$. The results show that approximately only 10% of global state probabilities are important for estimation of the performance measures.

Let us consider the details of implementing the stable recurrence with truncation according to the scheme suggested in Section 2.3. The results will be presented for the one link multi-rate Poissonian model studied in Section 3.1.

First, we derive the modified version of recurrence for global state probabilities suggested by relation (Eq.15). Let us choose some truncation level ` and construct the auxiliary model to approximate the behaviour of the one-link multi-rate model considered in Section 3.1. The auxiliary model behaves as the initial model in states $k = \ell, \ell + 1, \ldots, k$, except for the cases when a transition moves the initial model to the states $(i_1, \ldots, i_n)$ with $i_1 b_1 + \ldots + i_n b_n < \ell$. In auxiliary model we prevent this transition by immediately putting a call of corresponding type into service (type of the call depends on the number of bandwidth units occupied by the call just finishing service). This fictitious call is served the same way as any call in the initial model.

The auxiliary model dynamic is described by a Markov process $r^*(t)$ with the same components as $r(t)$ used for description of the initial model in Section 3.1. Process $r^*(t)$ is defined on a state space S^* that is smaller than S and consists of vectors $(i_1, \ldots, i_n)$ with nonnegative components satisfying conditions. $(i_1, \ldots, i_n) \in S^*$, $\ell \le \sum_{s=1}^{n} i_s b_s \le k$. Let $P^*(i)$ denote unnormalised probabilities of being in the global state (i) when exactly i bandwidth units are occupied. Probabilities $P^*(i)$ will serve as estimates for the corresponding probabilities $P(i)$ of the initial model.

In the same way as for the initial model we can prove that a product form solution is valid for stationary probabilities and corresponding recursion scheme for estimation of unnormalised value of $P^*(i)$

$$P^*(i) = \begin{cases} 0, & \text{if} \quad i < \ell \\ a, & \text{if} \quad i = \ell \\ \dfrac{1}{i}\displaystyle\sum_{s=1}^{n} \lambda_s b_s P^*(i - b_s), & \text{if} \quad i = \ell+1, \ell+2, \ldots k \end{cases} \tag{27}$$

As result we prove the possibility of using the modification of recurrence (Eq.3) in the form suggested by relation (Eq.15) when deriving stable recursion with truncation of global state space.

Using the general framework formulated in Section 2 we obtain a one-run algorithm that at each step gives the normalised values of the model's characteristics that are necessary for solving the dimensioning problem based on performance measures (Eq.18).

- *Step 1.* Let $p_\ell^*(\ell) = 1$.

- *Step 2.* Let $b = \max\limits_{0 \le s \le n} (b_s)$. For fixed $k = \ell + 1, \ell + 2, \ldots$, find normalised value of $p_k^*(i)$ by using relations (Eq.27).

$$p_k^*(k) = \frac{\frac{1}{k}\sum\limits_{s=1}^{n} \lambda_s b_s p_{k-1}^*(k - b_s)}{1 + \frac{1}{k}\sum\limits_{s=1}^{n} \lambda_s b_s p_{k-1}^*(k - b_s)}; \qquad (28)$$

$$p_k^*(i) = \frac{p_{k-1}^*(i)}{1 + \frac{1}{k}\sum\limits_{s=1}^{n} \lambda_s b_s p_{k-1}^*(k - b_s)},$$

$$i = k - 1, k - 2, \ldots, max(k - b, \ell).$$

- *Step 3.* The same as formulated for basic version.

The accuracy of calculations with truncation of global state space greatly depends on possibilities of finding the proper truncation level ℓ. Let us consider this problem in more detail. The problem of choosing a proper value of ℓ depends on many factors (type of model and performance measure used for dimensioning, proposed accuracy of estimation, the number of service units, etc). It is obvious that relative error of estimation the performance measures depends on the value of $p^*(\ell)$, the proportion of time the model spends in truncation state ℓ. We can easily find strict bounds for relative error as a function of $p^*(\ell)$ if some propositions are true.

Let M_s for the initial model denote the mean number of occupied service units by demands of s flow. It can be proved that the following relation is true.

$$M_s = \lambda_s b_s (1 - \pi_s), s = 1, \ldots, n. \qquad (29)$$

Let us for the auxiliary model denote the performance measures in the same way as for initial model, only with an additional asterisk. Then we obtain for π_s^* and M_s^* (for simplicity we suppose that $k - b > \ell$).

$$\pi_s^* = \sum\limits_{i=k-b_s+1}^{k} p^*(i) \qquad (30)$$

$$M_s^* = \lambda_s b_s (1 - \pi_s^*) + \alpha_{\ell,s} b_s,$$

Where

$$a_{\ell,s} = \sum_{i_1b_1+...+i_nb_n=\ell} i_s p^*(i_1,...,i_n) \le \frac{\ell}{b_s} p^*(\ell). \tag{31}$$

Let us suppose that the following inequalities are true.

$$M_s^* \ge M_s, \pi_s^* \ge \pi_s, s = 1,...,n. \tag{32}$$

This is not true in general, but for most cases of practical interest it is so because the auxiliary model serves more traffic than the initial model. Then using (Eq.29)-(Eq.32) we have the relative error estimations

$$0 \le \frac{\pi_s^* - \pi_s}{\pi_s^*} \le \frac{\ell p_k^*(\ell)}{\lambda_s b_s \pi_s^*}, \tag{33}$$

$$0 \le \frac{M^* - M}{M^*} \le \frac{\ell p_k^*(\ell)}{M^*}. \tag{34}$$

The value of $p_k^*(\ell)$ can be found when performing the recursion (Eq.28). It is easy to show that.

$$p_k^*(\ell) = \frac{p_{k-1}^*(\ell)}{1 + \frac{1}{k}\sum_{s=1}^{n} \lambda_s b_s p_{k-1}^*(k-b_s)}, k = \ell+1, \ell+2,... \tag{35}$$

To illustrate the computational savings obtained when using stable recurrence with or without truncations of global state space we consider a numerical example. In Table 1 relative efforts for three schemes of calculation are compared to reach values of $k = 100, 500, 1000, 5000, 10000$. Other model parameters are $n = 4$, $b_1 = 1$, $b_2 = 2$, $b_3 = 3$, $b_4 = 5$,

$$\lambda_s = \frac{k}{nb_s}, s = 1, 2, 3, 4.$$

Table 1. Relative Values of Computational Efforts to Reach the Value k for Three Schemes of Calculation Normalised by Efforts Needed for Usage of Stable Recurrence in the Form Defined by (Eq.20)

No. of service units k	Full search	Stable recurrence	Stable recurrence & truncation	Truncation level ℓ
100	50	1	0.37	63
500	250	1	0.18	410
1000	500	1	0.13	870
5000	2500	1	0.06	4700
10000	5000	1	0.042	9580

Table 2. Comparison Between Exact and Approximate (Eq.33) Estimation of Relative Error in Realization of Truncated Version of the Recursive Algorithm

Trun-cation ℓ	Relative error							
	1-st flow		2-nd flow		3-rd flow		4-th flow	
	Exact	Appr.	Exact	Appr.	Exact	Appr.	Exact	Appr.
676	1.0E-06	1.3E-05	1.0E-06	2.5E-05	1.0E-06	3.1E-05	1.0E-06	6.3E-05
705	9.8E-06	1.1E-04	9.8E-06	2.3E-04	9.8E-06	2.8E-04	9.8E-06	5.7E-04
727	4.7E-05	5.1E-04	4.7E-05	1.0E-03	4.7E-05	1.3E-03	4.7E-05	2.5E-03
738	9.8E-05	1.0E-03	9.8E-05	2.0E-03	9.8E-05	2.5E-03	9.8E-05	5.1E-03
764	4.9E-04	4.6E-03	4.9E-04	9.2E-03	4.9E-04	1.1E-02	4.9E-04	2.3E-02
776	9.6E-04	8.7E-03	9.6E-04	1.7E-02	9.6E-04	2.2E-02	9.6E-04	4.3E-02
808	4.9E-03	3.9E-02	4.9E-03	7.8E-02	4.9E-03	9.7E-02	4.9E-03	1.9E-01
823	9.7E-03	7.2E-02	9.7E-03	1.4E-01	9.7E-03	1.8E-01	9.7E-03	3.6E-01
865	5.0E-02	3.0E-01	5.0E-02	6.0E-01	5.0E-02	7.5E-01	5.0E-02	1.5E+00
886	9.8E-02	5.3E-01	9.8E-02	1.1E+00	9.8E-02	1.3E+00	9.8E-02	2.6E+00

In Table 1 the results presented in the column "Full search" are based on step-by-step usage of recurrence (Eq.19), the results presented in the column "Stable recurrence" are based on usage of stable recurrence as defined by (Eq.20), the results presented in the column "Stable recurrence & truncation" are based on usage of stable recurrence with truncation as defined by (Eq.28), the results presented in the column "Truncation level" give the value of ` that is chosen to provide relative error of one percent in estimating weighted blocking probabilities P defined as

$$P = \sum_{s=1}^{n} \lambda_s b_s \pi_s / \sum_{s=1}^{n} \lambda_s b_s$$

For fixed value of k, computational efforts are normalised by efforts needed for using stable recurrence in the form defined by (Eq.20). The results show considerable savings of computational efforts. To illustrate the accuracy of estimation the error caused by truncation we consider the numerical example. Table 2 shows the comparison between exact and approximate (Eq.33) estimation of relative error in realization of truncated version of the recursive algorithm for multi-rate model with $k = 1000$, $n = 4$, $\lambda_1 = 25$, $b_1 = 10$, $\lambda_2 = 50$, $b_2 = 5$, $\lambda_3 = \dfrac{125}{2}$, $b_3 = 4$, $\lambda_4 = 125$, $b_4 = 2$. The results show that the accuracy of estimation is acceptable for practical purposes.

The results show considerable savings of computational efforts when applying the derived algorithms for solving dimensioning problems for teletraffic models.

Among other positive features is the simplicity in derivation of stable recursions for new models. All we need is to take recurrence for global state probabilities in the form (Eq.3) and put it into the three steps algorithm formulated in Section 2. We illustrate this for two models in the next two subsections

4.2   One-Link Model With Batch Arrivals

We consider a single link traffic model, where the link transmission capacity is represented by k basic bandwidth units. In the model we have one Poissonian flow of batches for connections with intensity λ. Let f_i denote the probability that batch size is i. Each call from the batch uses for its transmission only one bandwidth unit for exponentially distributed connection time with parameter equal to one. Let us suppose that index i for f_i varies from 1 to b. If size of the batch exceeds the number of free bandwidth units, then the available bandwidth units are occupied by calls from the batch and excess calls are lost.

Let $i(t)$ be number of calls being served at time t. The dynamics of the model is described by a Markov process $r(t)$ having one component $r(t) = i(t)$. The process $r(t)$ takes values in the finite set of states $S = \{0, 1, \ldots, k\}$ defined in accordance with link capacity. Let $P(i)$ denote the unnormalised values of stationary probabilities of $r(t)$. After normalisation, the value $P(i)$ denotes the mean proportion of time when exactly i

connections are established or i bandwidth units are occupied. From the system of state equations it is easy to derive recurrence for $P(i)$.

$$
P(i) = \begin{cases} 0, & \text{if} \quad i < 0 \\ a, & \text{if} \quad i = 0 \\ \dfrac{\lambda}{i}\left\{ P(i-1)\sum_{s=1}^{b} f_s + \dots + P(i-b)\sum_{s=b}^{b} f_s \right\} & \text{if} \quad i = 1,2,\dots,k. \end{cases} \tag{36}
$$

The performance measure for solving dimensioning problem for this model can be taken as traffic congestion π_c, defined as a the fraction of offered traffic that is not carried.

$$
\pi_c = \frac{\lambda\sum_{s=1}^{b} f_s s - \sum_{i=1}^{k} ip(i)}{\lambda\sum_{s=1}^{b} f_s s} \tag{37}
$$

An alternative relation for carried traffic is obtained after summation of (Eq.6) over i.

$$
\sum_{i=1}^{k} ip(i) = \lambda(1-p(k))\sum_{s=1}^{b} f_s + \lambda(1-p(k)-p(k-1))\sum_{s=2}^{b} f_s + \dots + \tag{38}
$$
$$
+ \lambda(1-p(k)-p(k-1)-\dots-p(k-b+1))\sum_{s=b}^{b} f_s
$$

The recurrence (Eq.36) and the definition of performance measure used for link dimensioning (Eq.37), (Eq.38) satisfy conditions formulated in Section 2. So we can implement a one-run three-step algorithm that at each step gives the normalised values of the model's characteristics which are necessary for solving the dimensioning problem based on performance measure (Eq.37) and (Eq.38).

4.3 Multi-Rate Poissonian Model With Soft Blocking

We consider multi-rate Poissonian model studied in Section 3.1 and assume now that demand of s flow $s = 1, 2, \dots, n$ arriving in when exactly i bandwidth units are occupied is blocked with probability $\varphi_{s,i}$. Let us assume that a $b > 0$ exists so that for all $s = 1, 2, \dots, n$, $\varphi_{s,i} = 0$ for $i = 0$,

$1, \ldots, k - b$. In practice the proper choice of $\varphi_{s,i}$ allows us to introduce reservation schemes, to take into account losses caused by interference in mobile network and so on. The process of serving demands of s flow is characterised by the proportion of calls lost.

$$\pi_s = \sum_{i=k-b+1}^{k} p(i)\varphi_{s,i} \qquad s = 1,2,\ldots,n. \tag{39}$$

Heuristics based on the analysis done for particular teletraffic models allows us to use the modified version of recurrence for approximate evaluation of $P(i)$ unnormalised probabilities of the state when i bandwidth units are occupied (Eq.19)

$$P(i) = \begin{cases} 0, & \text{if} \quad i < 0 \\ a, & \text{if} \quad i = 0 \\ \dfrac{1}{i}\sum_{s=1}^{n} \lambda_s b_s (1 - \varphi_{s,i-b_s})P(i - b_s) & \text{if} \quad i = 1,2,\ldots,k \end{cases} \tag{40}$$

The recurrence (Eq.40) and the definition of performance measure used for link dimensioning (Eq.39) satisfy conditions formulated in Section 2. So we can implement a one-run three-step algorithm that gives at each step the normalised values of the model's characteristics that are necessary for solving the dimensioning problem based on performance measure (Eq.39).

5. CONCLUSIONS

In this paper we describe an approach for optimal dimensioning of multi-service lines. The presented solution is based on a method to convert recursions for global state probabilities of the type (Eq.3) into stable form where recursions are made on number of service units. This type of recursions is well suited for solving the dimensioning problem based on the form of criteria (Eq.1) and performance measure of the type (Eq.2). The results obtained are based on ideas from [2],[3] and generalise for multi-rate models the well known recurrence formula for the Erlang-B formula. In addition to being numerically stable, the main positive features of the suggested approach is its simplicity. All we need is just to take recurrence for global state probabilities in the form (Eq.3) and put it into relations (Eq.13) or it's modifications (Eq.14), (Eq.16). Another advantage is the possibility of increasing the efficiency of recursive algorithm by truncation of the state space used.

REFERENCES

[1]   Iversen, V.B. & Stepanov, S. N. The unified approach for teletraffic models to convert recursions for global state probabilities into stable form. Paper presented at IYTC 19, Beijing, China, august 29 - September 2, 2005. Participants proceedings, pp. 1559-1570.

[2]   Stepanov, S.N. & Iversen, V.B. & Kostrov, V. O. Optimised Dimensioning of Large Bandwidth Resources Under MPLS Technology. Proc. of St. Petersburg Regional International Teletraffic Seminar on Telecommunication Network and Teletraffic Theory. St. Petersburg, Russia. January 2002, pp. 50-63.

[3]   Iversen, V. B. Teletraffic Engineering and Network Planning. Chapter 10. Multi-dimensional Loss Systems. http.//www.com.dtu.dk/education/34340/material/telenook.pdf. COM, Technical University of Denmark. 2006. 350 pp.

[4]   Nilsson, A. & Perry, M. MultiRate Blocking Probabilities. numerically stable computations. ITC 15, Fifteenth International Teletraffic Congress, Washington, DC, USA, Elsevier, 1997, pp. 1359-1368.

[5]   Nilsson, A. & Perry, M. & Gersht, A. & Iversen, V.B. On Multi-Rate Erlang-B Computations. ITC 16, Sixteenth International Teletra_c Congress, Edinburgh, UK, June 1999. Elsevier 1999. pp. 1051-1060.

[6]   Berezner, S.A. & Krzesinski, A. E. An Efficient Stable Recursion to Compute Multi-service Blocking Probabilities. Performance Evaluation. Vol. 43 (2001). 151-164.

[7]   Fortet, R. & Grandjean, Ch. Congestion in a Loss System when Some Calls want Several Devices Simultaneously. Electrical Communications, Vol. 39 (1964). 4, 513- 526. Paper presented at ITC-4, Fourth International Teletraffic Congress, London, UK, 15-21 July 1964.

[8]   Kaufman, J. S. Blocking in a Shared Resource Environment. IEEE Transactions on Communications, Vol. COM-29 (1981). 10, 1474-1481.

[9]   Roberts, J. W. A Service System with Heterogeneous User Requirements - Applications to Multi-Service Telecommunication Systems. Pages 423{431 in Performance of Data Communication Systems and their Applications. G. Pujolle (editor), North-Holland Publ. Co. 1981.

[10] Delbrouck, L. E. N. On the Steady-State Distribution in a Service Facility carrying Mixtures of Traffic with Different Peakedness Factor and Capacity Requirements. IEEE Transactions on Communications, Vol. COM-31 (1983). 11, 1209-1211.

CHAPTER 8

A Network Management Framework for Emerging Telecommunications Networks

Augustine Samba

Department of Computer Science
Kent State University, Kent OH 44242, USA

Abstract. *Current Network Management (NM) procedures are concerned primarily with monitoring aspects. The architectures are centralized and based on static managed objects. They do not provide real-time control capabilities. Within this framework, effective network management depends on the coordination of monitoring operations across various Element Management Systems. The coordination is handled for the most part, by engineers/operators through manual procedures at designated Network Operations Center (NOC). The Subject Matter Experts at the NOC also use intuition and heuristic data to occasionally fine-tune the network parameters in order to maintain the advertised service-level objectives and control the traffic flow.*

Today's industry procedures, such as SNMP and TNM frameworks, presenta number of limitations for the complex and heterogeneous emerging telecommunication networks.

The next generation networks are expected to inter-connect different access network technologies and architectures in a multi-vendor environment. The access technologies such as optical Ethernet, dark fiber, wireless LAN and fixed wireless access systems will provide better and more resilient alternatives for multimedia traffic than existing DSL and Cable access networks. The core networks serving metropolitan regions will similarly migrate to increasingly heterogeneous technologies and architectures across different software platforms .More efficient technologies, such as Resilient Packet Rings, possibly Terabit and Gigabit Routers, OXC switching nodes and multiple service protocol platforms will provide the framework for transporting heterogeneous multimedia traffic over Dense Wavelength Division Multiplexing core networks.

A more efficient framework to facilitate network management and control in next generation networks is proposed. This framework employs a distributed architecture of autonomous and heterogeneous sensor entities. The architecture facilitates peer-to-peer networking under the supervision of a novel Integrated Network Management System (INMS). This approach provides automated decision making, rapid deployment of network management "service" functions; real-time monitoring of fault, configuration, accounting, performance and security management functions; real-time provisioning; and real-time NM control activation and removal. The novel framework provides an integrated view of end-to-end managed network entities

1. INTRODUCTION

Network management involves a set of activities and techniques that are required to plan, design, control, maintain and grow a network infrastructure and its associated services. These activities include monitoring the network and the ability to take prompt action to efficiently maintain the service-level objectives and to control the flow of traffic when necessary. Network management activities also include detection, identification, investigation and resolution of faulty network elements and transmission facilities. The overall objective is to provide efficient delivery of information between end-users and the various network elements.

In the current telecommunications environment, network monitoring is accomplished by logically connecting the network elements (e.g. switches, routers and servers) to remote Element Management Systems, which are under the control of one or more Network Management Systems (NMS). A high-level architecture of a typical network management environment is illustrated in Figure 1. The NMS is collocated with various Operations Support Systems in a Network Operations (NOC). Effective network management depends on the coordination of controls across the various Element Management Systems. These controls may include schedule changes, provisioning, fault and configuration management modifications. In today's environment, control coordinations are handled, for the most part, by NOC engineers/operators through manual procedures. The Subject Matter Experts (SMEs) at the NOC also use intuition and heuristic data to occasionally fine-tune the network parameters in order to maintain the advertised service-level objectives and control the traffic flow. The SMEs, therefore supplement the limited capabilities of NMS solutions in a rapidly changing and technology driven communications industry. Today's manual procedures present a number of setbacks for the complex and heterogeneous emerging networks. Specifically, the NMS capabilities for emerging telecommunication networks should include automated decision making, real-time provisioning, real-time schedule changes, real-time service adaptability, real-time fault, configuration and bandwidth management; and real-time traffic control capabilities.

Telecommunications networks continue to evolve to meet the increasingly diverse needs of end-users. Emerging Telecommunication networks will be more complex systems. These networks are expected to inter-connect different access network technologies and architectures in a multi-vendor environment. The access technologies such as optical Ethernet, dark fiber, wireless LAN and fixed wireless access systems will provide better and more resilient alternatives for multimedia traffic than existing DSL and Cable access networks. The core networks serving metropolitan regions will similarly migrate to increasingly heterogeneous technologies and architectures across different software platforms. More efficient technologies, such as Resilient Packet Rings, possibly Terabit and Gigabit Routers, OXC switching nodes and multiple service protocol platforms will provide the framework for transporting heterogeneous traffic over the core network.

The backbones interconnecting the Terabit routers in the core networks will similarly migrate to more resilient technologies such as DWDM. Within this framework, Service Developers will create heterogeneous service types that will exploit the emerging telecommunication network architectures and platforms across varying protocols.

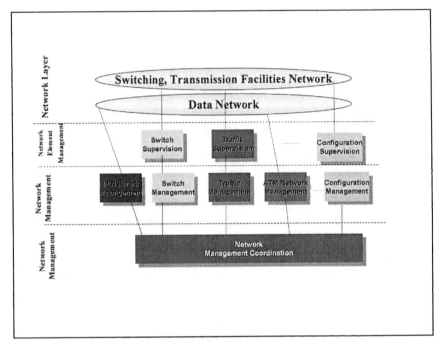

Fig. 1 Centralized Network Management

This paper describes a novel framework to facilitate network management for emerging telecommunications network. The framework is a distributed, peer-to-peer networking architecture under the supervision of one or more remote Integrated Network Management System (INMS). This approach facilitates automated decision making, rapid deployment of network management services, provisioning, schedule changes, fault management, configuration management, bandwidth management and multimedia traffic controls in real-time

Section 2 provides an overview of current industry practices for network management. The motivations for this work are outlined in Section 3.0. Section 4.0 describes the goals of the integrated network management model. A novel framework for network management in emerging telecommunications networks is described in section 5.0. The real-time network management and control architecture is described in section 6.0. A real-time network management scenario is described in section 7.0. The concluding remarks are summarized in section 8.0.

2. OVERVIEW OF NETWORK MANAGEMENT SYSTEMS - CURRENT INDUSTRY PRACTICES

The OSI reference model [1] classifies network management functions into five functional areas. The functional areas are illustrated in Figure 2 and summarized below.

- *Fault Management: The objective is to minimize and manage network element systems faults. This involves three basic tasks: detecting the fault, isolating the fault to specific component and finally correcting the fault. The underlying procedure for fault management includes the maintenance of error logs for fault detection, error detection algorithms and diagnostic testing.*

- *Configuration Management: The goal is to monitor and analyze network element configuration parameters, network topology and operational status of the network. This procedure involves the ability to change/adjust various configuration parameters in order to maintain the Service-level objectives*

- *Accounting Management: The goal is to provide a mechanism to measure the dominant network utilization parameters. The measurements are analyzed to identify usage patterns and determine cost structures based on resource utilization.*

- *Performance Management: The goal is to gather statistics on the operation of the network, maintain and analyze performance variables, determine thresholds values for the dominant performance variables, and optimize network operations by controlling the performance variables*

- *Security management: The goal is to create, modify, delete and control security services; provision security related information and report security-related events. The underlying procedure monitors strategic network element interfaces and control access to vulnerable network resources.*

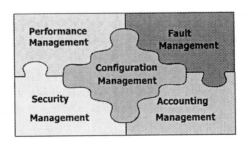

Fig. 2 Network Management Functions

The above network management functions are collectively referred to as "FCAPS".

Various techniques have been applied to facilitate the above network management capabilities in telecommunications and enterprise networks [2], [3], [4]. Today's NMS solutions for multi-vendor network environments, for the most part, are based on industry standards. The simple network management protocol (SNMP) framework is the dominant industry standard. The SNMP framework specifies the information structure and protocol requirements to perform network management in a distributed multi-vendor environment.

SNMP Management Framework In the distributed network management environment, a small set of nodes resides in Network Management Systems. Each NMS node is referred to as a Manager. The Managers communicate via the SNMP protocol with another set of remote nodes, referred to as Agents. The Agents reside in the managed Network Elements, and each Agent is pre-configured to respond directly to a Manager's requests or command.

The SNMP framework consists of three key elements:

i. *The standard Management Information Base (MIB):* Each NMS and managed network element has a fixed set of MIBs corresponding to the components of the protocol stack. The management information is stored as a collection of static objects in the MIB. Each MIB [1] object has a pre-determined type, size and access restrictions.

ii. ***The Structure of Management Information (SMI):*** The SMI provides a mechanism to name and organize objects. The SMI for SNMP is derived from the Abstract Syntax Notation One (ASN.1) [5]. The SMI defines the tree structure for the object identifiers

iii. ***The Simple Network Management Protocol (SNMP):*** specifies the protocol used between the Agent and the Manager. The protocol implements five messages. The Manager sends a request ("poll") to the Agent, and the Agent responds directly to the Manager's request [3].

3. MOTIVATIONS

The SNMP framework, despite its popularity, has a number of disadvantages [6] and [7]. Some of these are noted below.

The SNMP model assumes a static managed object. Every data item must be carefully pre-defined, including its type, size and access restrictions before it can be used in the MIB. The next generation networks will be highly complex and highly heterogeneous compared to today's network environments. Therefore the number managed objects in future generation networks will be relatively very high. Consequently, it would require tremendous time, effort and patience to accurately pre-define the wide variety of managed objects. The networks of the future will be very complex with heterogeneous (e.g., multimedia) service applications. Modeling these applications as static objects may lead to inaccurate representations.

To retrieve SNMP information, the Manager must first obtain (poll) all the values associated with the object(s), determine whether the retrieved values are of interest and then construct the information. The lack of direct filtering mechanism makes the information retrieval process cumbersome.

The polling of SNMP information over a WAN connection consumes a lot of bandwidth. The scenario may contribute to network traffic congestion due to the large volume of diverse data items in future communications networks.

The SNMP framework is concerned primarily with the monitoring aspects. The control mechanism is limited to the *"set"* request command.

Therefore SNMP framework does not provide real-time control capabilities.

The SNMP paradigm has a number of challenges in the context of emerging telecommunications networks: The paradigm is people intensive and expensive with regard to MIB design and deployment. It will be considerable slow due to the lack of an integrated provisioning capability. Record keeping is manual and non-integrated. Network monitoring will be complex due to the fragmented management environments; and hence the service restoration will be susceptible to high failure rates. Service restorations will also likely fail if the centralized Network Management system and/or the set of element management systems are overloaded. To effectively manage emerging telecommunications networks, there needs to be a paradigm shift from the centralized SNMP approach to a distributed and fully integrated, multi-vendor real-time network management and control framework.

4. THE GOALS OF THE INTEGRATED NETWORK MANAGEMENT MODEL

A decentralized and fully integrated network management model is proposed to effectively perform network management functions and control multimedia traffic in emerging telecommunications networks. The goals of the model are to facilitate:

- *Autonomous monitoring and control of heterogeneous network elements and bandwidth:* This is accomplished by embedding Autonomous Sensors Entities (ASEs) in managed network elements (NEs). Each ASE monitors its NE and bandwidth of associated transmission facilities without control or manual intervention from a network management operations center or a centralized network management system.

- *Real-time provisioning*: The goal is to provide rapid deployment of heterogeneous network management functions. This is accomplished by transforming the OSI reference model for network management functions to a set of heterogeneous network management services. The Integrated Network Management System (INMS) provisions customized services ("service elements") across the ASEs within the various network clusters.

- **Real-time end-to-end network management and control of heterogeneous nodes** is accomplished via *Peer-to-peer communication of autonomous sensor entities:* Each local ASE's behavior will evolve from one state to another based on the changing conditions of its network cluster. Each local ASE dynamically interacts with its neighbor and a unique Cluster ASE (CASE) based on its current state and the monitored conditions. The Integrated Network Management System (INMS) communicates with each of the CASEs in real-time; the INMS also communicates with Operations Support Systems Systems in near-real time.

- *Reliability.* An important aspect of network management is its effectiveness in ensuring Network reliability and network restoration. The INMS architecture supports two external network interfaces for communication with the embedded autonomous sensor elements. A dedicated Autonomous Sensor Network Interface (ANSI) allows the INMS to perform out-of-band network management functions over the Autonomous Sensor Network; while the Managed Network Interface (MNI) supports in-band network management capabilities over the "Access" and "Core" networks.

5. THE INTEGRATED NETWORK MANAGEMENT FRAMEWORK

Emerging telecommunications networks will provide latency-free communications, seamless and reliable delivery of heterogeneous multimedia services amongst a wide variety of multi-access end-user technologies and distributed servers. To accomplish these objectives, the networks will be logically partitioned into two key domains: The Access network domain will facilitate multi-technology, multi-access and higher-speed bandwidth over the last-mile. The Core network domain will be optical backbone to facilitate convergence of higher throughput data and voice networking, multi-access network technologies, heterogeneous protocols and services. Figure 3 illustrates a context diagram of the integrated network management framework for emerging telecommunications network. The network is logically partitioned into Access and Core network domains. Autonomous sensors are embedded into managed network elements.

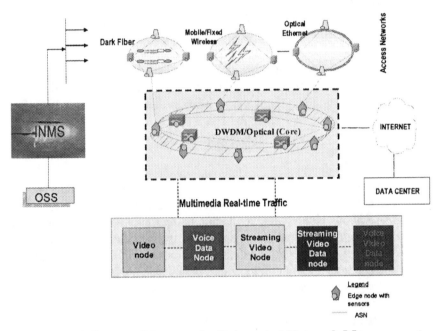

Fig. 3 Context Diagram for Integrated Network Management Framework

The Integrated Network Management System (INMS) communicates with the autonomous sensors using the access and core networks as the primary transport medium, and the Autonomous Sensor Network (ASN) as the secondary medium. The in-band network management is performed via the access and core networks. The out-of-band network management is performed via the ASN medium. The ASN is a packet network, designed to carry network management information amongst ASEs, CASE and INMS. The multimedia traffic, on the other hand, is carried exclusively over the Access and Core networks. The ASN traffic can also be transported over the Access and Core networks depending on the multimedia traffic load-levels. Figure 3 also illustrates Operations Support Systems for off-line processing.

Real-time network management and control capabilities are accomplished based on the shared intelligence of three independent systems: The Network Management-Service Creation System (NM-SC), the Integrated Network Management System (INMS) and the Autonomous

Sensor Entities (ASE and CASE). The functional capabilities of these systems are summarized below.

i. **The Network Management-Service Creation system (NM-SC)**

The purpose of the NM-SC is to create a set of heterogeneous NM Service Applications for FCAPS and Traffic Control. The FCAPS are based on the OSI Network Management reference model depicted in Figure 2. The NM-SC transforms the FCAPS into service elements templates based on the managed Device type and its local surroundings. The service element templates include an additional layer, beyond the OSI reference model, for Network Management Control capabilities. The service element templates are then installed on the INMS, where they are validated, verified and customized based on individual ASE/CASE requirements. Figure 4 illustrates the Integrated Network Management Service Element.

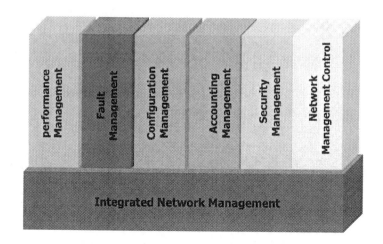

Fig. 4 Integrated Network Management Service Element

ii.   **The Integrated Network Management System (INMS)**

The basic functions of the INMS are as follows:

The INMS creates heterogeneous service elements by customizing the Network management Service Creation element templates based on the monitoring and control requirements of the embedded CASE/ASEs. These requirements are derived from the device types and its local surroundings.

The INMS partitions the Access and Core networks into logical network clusters respectively. Each cluster is characterized by a graph representing a set of nodes, edges and managed device types. The set of all graphs represents the CASE/ASE network topologies. The INMS maintains a golden copy of CASE/ASE network topologies

The INMS provisions executable versions of the service elements to appropriate CASE/ASE applications in real-time. The provisioning dynamically incorporates service elements into appropriate CASE/ASE applications. The INMS also maintains a golden copy of each provisioned service element.

The INMS receives audit or query requests from one or more Operations Support Systems to audit or query the status of remote managed nodes, via the embedded sensors.

The INMS communicates with the CASE and ASE in real-time

iii.   **The Autonomous Sensor Entities (ASE)**

An ASE's behavior is based on its state and its local surroundings. Each network cluster has at least one Cluster **ASE** referred to as CASE. The CASE coordinates the network management and control activities of ASEs within its network cluster.

The ASE communicates with neighbour ASEs and its network CASE in real-time. Additionally, each CASE communicates with the INMS and other CASE to facilitate real-time network management and control operations.

Both the CASE and ASE provide abstractions of hardware resources for measuring FCAPS metrics. The abstractions allow the autonomous sensor entities to detect Faults, network overload traffic, intra-node congestions, adjust configuration and access parameters; and to implement and remove network management controls.

6. REAL-TIME NETWORK MANAGEMENT & CONTROL ARCHITECTURE

The real-time network management and control framework is comprised of three autonomous systems. This section describes the software architecture of the three autonomous systems:

- Network Management-Service Creation system (NM-SC)
- Integrated Network Management System (INMS)
- Cluster / Autonomous Sensor Entities (CASE & ASE)

Figure 5 illustrates an overview of the integrated end-to-end network management and control system software architecture. The figure illustrates the logical association of three independent systems.

6.1 The Network Management Service Creation System (NM-SC) Architecture

The basic idea is to represent the OSI network management reference model for FCAPS as a set of generic service element templates [8], [9]. The service templates are then installed on the INMS as executables, where they are customized to individual ASE's requirements. The NM-SC is comprised of four logical layers:

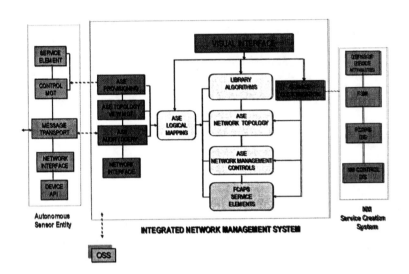

**Fig 5 Integrated View of End-to-End Network Management &
Control Systems**

Layer 1:   *The Network Management controls* provide extensions to
the OSI Network Management Reference model. This
layer specifies the procedures for activating and removing
network management controls. The controls allow
Automatic Sensor Entities (ASEs) to alter the flow of
traffic in the network in support of network management
objectives. The Network Management controls are
mapped to individual states defined by the FSM.
Programmable decision graphs [10] are employed to
activate and remove network management controls

Layer 2:   *Network Management function consist of five distinct
entities representing* Fault Management, Configuration
management, Accounting management, Performance
management and Security management respectively. Each
entity is defined based on a set of discrete states. Each
state represents a set of conditions and a finite set of
behavior (events) of the managed network element. The
operations of each Network Management entity will vary
depending on the state. Programmable Decision Graphs

[10] are used to control the execution of a network management function for each entity.

Layer 3:   **The Behavior** segment represents the dominant operations of the managed Network Element (assumed to be a digital system) as a set of State Machines. Finite State Machines (FSMs) define and emulate the transition of pertinent operations for a given Network Element (NE). This layer provides a software abstraction of the NE behavior.

Layer 4:   The **Attribute** field describes the characteristics of the autonomous sensor entities (CASE/ASE). This includes a unique ASE ID, network domain ID, Neighboring CASE/ASE Peer IDs (NID), Control and Network Management Parameters, Audit and Query types supported on the CASE/ASE and supervisory parameters for CASE

The layered-architecture for the Network Management Service Creation System is illustrated in Figure 6.

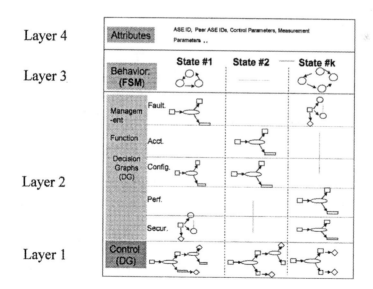

Fig. 6   The Network Management Service Creation Model

6.2   The Integrated Network Management System (INMS) Architecture

The INMS software architecture is comprised of four major components. The components provide the framework upon which platform independence, and real-time FCAPS network management and control are conceived.

The INMS software architecture is illustrated in Figure 7.0, and the major components are summarized below.

i.   **NM-SC Interface** decomposes into several subsystems, including Service Customization. This component allows the FCAPS service and control templates, created by the NM-SC, to be installed on the INMS.

ii.   **Visual Interface** provides support for administrative and service customization. A web browser or GUI can be used to customize the service elements to the requirements of each ASE's behavior. Service customization includes validation, verification, editing of decision graphs, specifications of attributes, exceptions and threshold values, administrative policies and constraints that are unique to a given ASE. The customized service elements are stored on the INMS and provisioned on appropriate ASEs and CASEs respectively.

iii.   **Network Element and Operations Support Systems Interface** decomposes into several subsystems, including Service Provisioning subsystem, View Management subsystem and the Audit-Query subsystem. This component facilitates real-time communications with embedded ASE and CASE; and offline communications with Operations Support Systems.

The Provisioning subsystem distributes the executable service elements to appropriate ASEs and CASEs. Service distribution includes the capability to insert new service elements, delete an existing service element or modify components of service elements on the ASEs and CASEs.

The View Management subsystem creates and manages integrated views of the end-to-end network. The View provides snapshots of the health of the network elements and network topology. The various levels of details are computed offline by the OSSs.

The Audit-Query subsystem manages scheduled and real-time audits/queries of the ASEs and CASEs. Audits and queries are initiated by either the Operations Support Systems or end-users via the Visual Interface subsystem.

iv. **The Host** consists of several subsystems, including CASE/ASE Network Topology, CASE/ASE Network Management Controls and FCAPS Service Elements.

The CASE/ASE Network Topology subsystem partitions the Core (and Access) Network Topologies into logical Clusters; and creates graph representations of each cluster. Each Cluster is comprised of a set of ASEs and typically one or two CASEs. Each graph represents the set of Nodes, Node types and Node Edges within the cluster. This subsystem maintains the golden copy of the logical network topologies

The CASE/ASE Network Management Control subsystem is the set of all executable network management controls defined for both the Access and Core Network topologies. This subsystem maintains the golden copy of all controls provisioned on the ASEs and CASEs; and Administrative policies provisioned on the CASEs.

The FCAPS Service Element subsystem is the set of executable customized services defined for Fault, Configuration, Accounting, Provisioning and Security management. This subsystem maintains the golden copy provisioned on the ASEs and CASEs

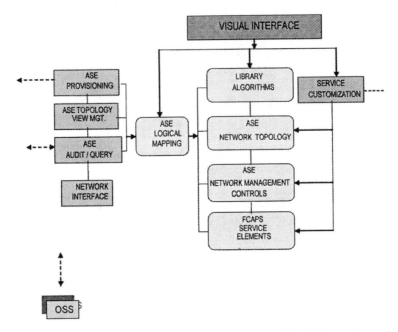

Fig. 7 Integrated Network Management Architecture

6.3   Autonomous Sensor Entity

The Autonomous Sensor Entity (ASE) is a middleware platform embedded in network elements including edge routers, core and access routers, switches, servers and multiplexors. The ASE hosts the customized network management service elements and provides runtime environment for autonomous network management monitoring and control functions. The Cluster Autonomous Sensor Entity (CASE) has the same architecture as the ASE. The CASE has all the functional capabilities of the ASE. Additionally, the CASE is provisioned with Supervisory functional capabilities for managing the ASEs in its network cluster.

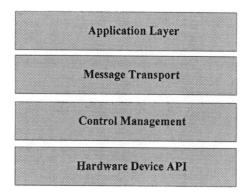

Fig. 8   Autonomous Sensor Entity

The ASE architecture is comprised of four key components as illustrated in Figure 8:

i.   **The Application layer** hosts the service elements provisioned by the INMS. The Application Layer provides the runtime environment to emulate designated hardware device operations and to execute Decision Graphs

ii. **The Message Transport layer** provides low-level messaging and I/O for peer-to-peer ASE, ASE -to- CASE, ASE-to-INMS, and CASE –to- INMS communications.

iii. **Control Management layer** is responsible for implementing and removing network management controls. The network control removal and activation is based on real-time performance data. When the CASE/ASE detects an internal stimulus, it performs an analysis of the performance data based on the administrative policies. The network management controls are then activated or removed depending on the result of the real-time analysis.

iv. **Low-Level Hardware Dependent API** provides an abstraction of hardware resources for measuring Fault, Configuration, Accounting, Performance and Security metrics.

The CASE/ASE stores Network Administrative polices in directories and uses lightweight directory access protocol (LDAP) to retrieve appropriate policies and interpret them prior to implementing network management controls.

7.   Real-time Network Management and Control Scenario

Figure 9 illustrates a scenario for activating automatic congestion control. In the scenario, the originating ASE (ASE-1) senses congestion within the machine (Switch, Edge node etc.).

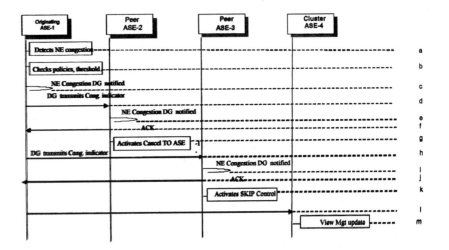

Fig. 9 Congestion Control Activation

a.   ASE -1 Performance DG detects congestion of its Network Element in state "K"

b.   ASE-1 Performance DG checks the Administrative policies, Congestion threshold values and constraints for state "K". The logical "AND"ing of these conditions specifies activation of controls.

c.   ASE-1 Performance DG notifies the Control DG responsible for Machine Congestion

d.   The ASE-1 Control DG transmits Automatic Congestion Control indicators to its peer ASE -2

e.   ASE-2 notifies its congestion DG, depending on its current state

f.   ASE-2 returns an ACK to ASE-1

g.   ASE-2 control DG activates a "CANCEL To"control for outgoing packets destined to ASE-1

h.   The ASE-1 Control DG transmits Automatic Congestion Control indicators to its peer ASE -3

i.   ASE-3 notifies its congestion DG, depending on its current state

j.   ASE-3 control DG activates a "SKIP" control for outgoing packets destined to ASE-1

k.   ASE-1 transmits a send-text message containing congestion Control indicators to CASE-4

l.   CASE-4 returns an ACK to ASE-1

m.   CASE-4 transmits View update message to the INMS

When the ASE-1 node Base Level Cycle takes normal time to complete assigned tasks, the ASE-1 Performance DG detects acceptable load levels for the node and notifies the Control DG to send control removal notification to its peers.

ASE-1 Control DG transmits control removal indicators to its peer (ASE-2, ASE-3) and the CASE-4 (or INMS if CASE-4 is unavailable).

ASE-2 and ASE-3 DGs subsequently remove the CANCEL TO and SKIP controls respectively. CASE-4 transmits a View update indicator to INMS, and the INMS updates the View for the network cluster containing ASE-1, ASE-2, ASE3 and CASE-4.

8. CONCLUDING REMARKS

This paper outlines a functional architecture for network management and control in emerging telecommunications networks. The proposed network management model is decentralized and provides an integrated view of the end-to-end managed network. The model facilitates autonomous monitoring and control of heterogeneous network elements and bandwidth; real-time provisioning for rapid deployment of network management functions; real-time end-to-end network management and control of heterogeneous nodes; and a framework that provides effective network reliability and network restoration. The model demonstrates how distributed network management and control capabilities are realized based on the shared intelligence of three autonomous systems.

The network management architecture is based on the vision that emerging telecommunications networks will be required to provide latency-free communications and reliable delivery of heterogeneous multimedia services across a wider variety of different network element technologies.

REFERENCES

[1]   International Organization for Standardization, Information Processing Systems - Open Systems Interconnection- Basic Reference Model-Part 4: Management Framework, ISO 7498-4-1989

[2]   H.G. Hegering, S. Abeck, and B. Neumair, *Integrated Management of Networked Systems: Concepts, Architectures, and Their Operational Application*, Morgan Kaufmann, 1999.

[3]   W. Stallings, *SNMP, SNMPv2, SNMPv3, and RMON 1 and 2*, 3rd edition, Addison-Wesley, 1999.

[4]   H.G. Hegering, S. Abeck, and B. Neumair, *Integrated Management of Networked Systems: Concepts, Architectures, and Their Operational Application*, Morgan Kaufmann, 1999.

[5]   International Organization of Standardization, "Information Technology - Open Systems Interconnection- Specification of Abstract Syntax Notation One (ASN.1)," ISO/IEC 8824, ANSI 1990

[6]   Thomas, Larry J. "The Distributed Management Choice." LAN Technology, April 1992:53-70

[7]   Y. Yemini, G. Goldszmidt and S. Yemini, "Network Management by Delegation", in Proceedings of the International Symposium on Integrated Network Management, pp. 95{107, May 1991

[8]   T. Nakano and T. Suda, "Adaptive and evolvable network services," in Proc. GECCO, Jun. 2004, pp. 151-162

[9]   Service Creation Development of TINA-Like Systems: The DOLMN Methodology, Ferdinando Lucidi et. al. Technology for ubiquitous telecom services, 5th International Conference on Intelligence in Services and Networks, ISN '98, Antwerp, Belgium, May 1998

[10] Statistics for Engineering and Information Science", M. Jordan, S.L. Lauritzen, J.F. Lawless, V. Nair, 2001 Springer-Verlag New York, Inc.

CHAPTER 9

Challenges of Tool Development Facing Rapidly Changing Market Demands

Gerta Köster

Siemens AG- Corporate Technology
Otto-Hahn-Ring 6, München

Abstract. *Industry research tends to be product oriented. Simulations are expected to mimic the behavior of a specific product, a product that is often in the design phase. The product's features, in turn, depend on ever changing market forecasts. At the same time, reliability is a must, since business decisions are based on simulation results.*

We look at the example of tool design for UMTS system level simulations to highlight the challenges of simulation tool development in the telecommunication industry and to discuss solution strategies.

1. INTRODUCTION

During the design and early introduction phase of UMTS (Universal Mobile Telecommunications System), simulations were a very important means to guide both, design and business decisions. Simulation tools were developed in parallel to the real systems, mimicking the existing parts and exploring alternative design options for those parts of the UMTS system that were not yet implemented.

Two types of simulations are common: link level simulations investigate the physical and system properties in a single cell. System level simulations treat the movement of mobile users through a network of radio cells. Results from link level simulations are a necessary input for the system level simulations. System level simulations are suitable to give capacity forecasts for different deployment scenarios.

We look at the example of tool development for system level simulations for UMTS. In section 2 of this paper we have a brief glance at the model. Then, in section 3, a selection of features, taken from a typical requirement list at one time, serves to highlight the conflicts of goals that the tool developers experience: At the one hand, they need to faithfully model a global image of standards and physical properties. On the other hand, they are supposed to react quickly to market demands, that is, to capture product specific and market driven features.

The tool must be extremely reliable because business decisions are based on the predictions derived from simulations. If, for example, capacity forecasts are wrong by a factor of 2, cost predictions will also be wrong by that factor. This means that either the company will loose business to competitors with lower offers or it will loose money if they sell their product for less then the production costs. Both are bad ideas. There is an obvious conflict of goals between reliability and a fast and flexible reaction to market driven questions.

In section 4 we discuss solution strategies. The platform approach adopted for our tool is described – including some difficulties encountered during the set-up phase. The idea of a developers' community is presented. Section 5 closes with a brief conclusion.

2. A GLANCE AT THE MODEL

Several hexagonal cells form a macro model of a mobile system according to the ETSI standardization document UMTS 30.03 [1]. A base station and antenna are situated at the center of each cell. See Figure 1. Mobile users move through the network according to a standardized model once again taken from UMTS 30.03 [1] to make results comparable across companies and research institutes.

The model tries to capture the most important physical features such as path loss, fading, interference. The user behavior and the characteristics of data transmission are modeled statistically. The model already contains many simplifications. This imposes a considerable restraint on the faithful capturing of physical properties and algorithmic features. For example, the speed of a mobile user modeled according to UMTS 30.03 is constant. This is neither true in reality, nor does it allow to investigate algorithms that deal with varying speed.

Figure 2 gives an overview over the functionality of a system level simulation tool. A multitude of interconnected modules is necessary to run a single simulation.

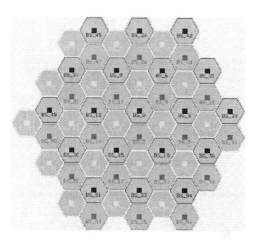

Fig. 1 Macro model according to UMTS 30.03. mobile users move through a network of cells. a base station and antenna is situated at the center of each cell.

The model captures physical properties and mechanisms that form a part of layer 1 and layer 2 of the OSI layer model as described, for example, in [2].

Very important mechanisms are: interference calculations, power control, interleaving, antenna diversity and the properties of smart antennas (layer 1), or scheduling algorithms and medium access control (layer 2). Also, radio resource management must be covered with admission control, handover, DCA (dynamic channel allocation), intra cell handover. Thirdly, the environment is modeled with mobility, propagation properties, and the base station deployment. A traffic generator creates a system load with speech users, packet users or mixed services. Discontinuous transmission must be taken into account. Finally, the software interface deals with input parameters and generates meaningful and readable output.

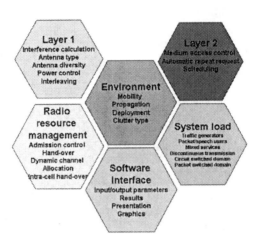

Fig. 2 Functionality of the system level simulation tool. a multitude of inter working modules are necessary to run each simulation.

3.   CONFLICT OF GOALS

The following list of features is taken of a typical requirement list, at some time about 2 years ago, to highlight the difficulties experienced during tool development.

- Adjacent channel interference: Energy spills from one frequency band to a neighboring band during transmission and causes interference. This basic physical fact applies to all UMTS varieties (FDD – frequency division duplex, TDD – time division duplex), indeed to all radio technologies, and to all systems. As such its occurrence is independent of the manufacturer. The impact, however, depends strongly on the implementation – such as the quality of filters used or the width of guard bands that are employed. In short, this is a basic physical feature with additional implementation aspects.
- Variation of update time for power control: Power control algorithms regulate the transmission power of a radio signal

according to the link quality. If, for example, a mobile user is far away from the base station more power is needed to transmit an intelligible signal than close to the base station antenna. Shadowing and interference play a similar role. Power control strongly depends on the specific implementation which usually is based on company owned patents. Thus simulating power control means to investigate operator dependent system parameters.

- DCA (dynamic channel allocation) with directional information: The allocation of mobile users to channels varies with time according to the interference situation. Directional information may help to assess the interference situation. DCA was considered, for a certain time, as a remedy to many capacity problems. Unfortunately, not all of these expectations can be met because the mechanism only works if there are interference "free" channels left to which users can be reallocated. DCA takes into account a mix of physical, statistical and system features. As such it represents, at one time, a research topic and an urgent request from the sales department.

- Avoid double counting of beam forming gain of smart antennas: Smart antennas form a beam in which the energy is concentrated. This beam is modeled on both, the system level and the link level, to correctly assess the interference in both simulations. However, feeding the link level results into the system level experiments means that the beam forming gain is considered twice. This would beautify the end results. To correctly simulate the radio system this logical bug must be eliminated.

From the list above, we see that the basic model with its physical and statistical properties is developed in parallel with algorithms and features that depend on the current market interest.

Also, there is a conflict of so-called 'urgent' features with 'important' features. Urgent features are driven by the hectic time frame of a project. A customer demands a special functionality or the sales department may see the need to offer certain system properties. Time is pressing and there is a strong wish to simply get the tool to work so that predictions can be made regarding the latest feature before the feature is actually implemented in the real system. Often simulation is the only way to quantitatively judge the impact of a new feature - besides the engineer's 'gut feeling'.

On the other hand, we need a very stable tool that produces reliable results to ensure that these results are indeed quantitatively meaningful. Simulation results are used as a basis to decide on features and even on the deployment of whole systems. Simulation data is fed into capacity and price calculations. Sometimes, simulation results are presented to the customer and there may be a risk of contract penalties. In view of this, clearly, reliability is extremely important. Experiments must yield the same correct result when repeated or performed with later tool versions. However, it requires time to achieve reliability. Thus, unfortunately, urgent features have a tendency to win over important features.

4. SOLUTION STRATEGIES

How to tackle such a challenge? How to make sure that we indeed save or make money through simulation aided development?

In this instance, the idea was to separate the development of 'urgent' features from the development of 'important' features, that is, to make flexible feature implementation independent of the important basic parts of the tool.

The physical properties and general system properties are captured in a platform common to all users and flavors of the tool. Version updates for this 'kernel' are programmed by a dedicated 'platform team' and are released every 6 months independent of market requirements.

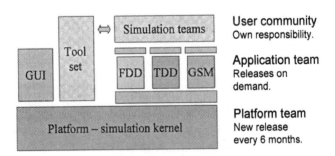

Fig. 3 Platform Approach to Tool Development.

Another team of developers provides simulation modules that are adapted to system specifics such as FDD, TDD or GSM. Here versions can be released according to demand and the users who conduct experiments can prioritize their requirements.

Interfaces on top of the system specific application modules can be accessed with the help of a tool set. This enables the users to write their own codes and to conduct rapid experiments with emerging features without influencing the development and stability of the tool itself.

In our case, this approach proved to be successful. However, we encountered a number of difficulties during the set-up phase. We briefly state some of the major hitches:

- The complexity of the system is reflected in a multitude of input and output parameters. These parameters are very difficult to control for a user who is not involved in the development.
- One additional obstacle of today's globally distributed team work is that the parameter sets are shifted between locations. Also, version updates must be kept aligned between locations. This makes experiments difficult to reproduce. Faithful reproduction, however, is a must to check modeling and coding and thus to build confidence in the simulation results.
- Also, interfaces are subject to change when one simultaneously develops a tool and conducts experiments with it.

Two very simple insights can be obtained from these difficulties:

- On the one hand, we need to understand the real system to code effectively, that is, to correctly capture the system features. Otherwise we risk a significant gap between model and reality.
- On the other hand, we need to understand the tool very well to be able to code correctly and simulate correctly, that is, not to introduce bugs into the tool and the experiment.

Evidently, very close cooperation between the simulation teams and the programming teams is called for. In our case, user features were coded by the platform and application programming teams for a considerable stretch of time.

The tool developers were even involved in the first simulation experiments to enhance their understanding of the system and, at the same time, to enhance the simulators' understanding of the tool.

Notwithstanding the success of the approach, we would like to go one step beyond: One of the bigger shortcomings of the simulation tools developed for UMTS may be – despite the use of a common physical and statistical model [1] – their lack of comparability. Tools differ widely across companies and research institutes so that comparisons and commenting of results are still very difficult. E.g. can a result derived from a snapshot simulator be compared to the output from a truly dynamic simulation tool? And which one is more correct if aspects such as the faithful reproduction of the environment are taken into account?

Here, a commonly available tool controlled by a neutral body could be of great use. Following the example of open source tool development, a community of users could access the tool and add their own code to it according to their need. Useful applications could be made available – after thorough testing through the platform owner – to the community.

5.   CONCLUSION

In this paper we discussed the difficulties for tool development arising from constant change in feature requirements derived from market demands. We suggested separating the development of urgent market features from the development of the tool basis in order to maintain reliability and stability. This seems best done by providing a platform on top of which users can add – possibly short living code – at their own need.

This platform approach proved successful in developing tools for UMTS system level simulations. However, comparisons between companies and institutions remain difficult as long as tools differ widely. In view of this deficiency, we brought up the idea of a user community. The programming activities of such a community would have to be controlled by a neutral body, such as a university, to ensure the reliability and stability of a common tool.

REFERENCES

[1] UMTS 30.03, *Tech. Rep. ETSI TR 101 112 v.3.2.0*, ETSI SMG2, April 1998.
[2] A. Tanenbaum, *Computer Networks*, Prentice Hall, [4]2000.
[3] B. Hampl, *Siemens' MoRSE User Manual*, 2003.
[4] B. Sklar, *Digital Communications*, Prentice Hall, [2]2001.

CHAPTER 10

Packaging Simulation Results With CostGlue [*]

Matevz Pustišek[1], Dragan Savić[1], Francesco Potortì[2]

[1] University of Ljubljana (SI),
[2] ISTI-CNR, Pisa (IT).

Abstract. *Researchers performing simulations in the field of computer telecommunications are often faced with the time-consuming task of converting huge quantities of data to and from different formats. We examine some of the requirements of the telecommunications simulation community and propose an architecture for a general purpose archiver and converter for big quantities of simulation data to be released as free software.*

1. INTRODUCTION

We describe the motivation, design issues and the approach to the implementation of a low-overhead software package that can import simulation or measurement results into a common data structure, launch post-processing applications on the imported data and store data or export it into various output formats.

Modern simulation packages and statistical tools use different, even proprietary formats for the results. This can be a major obstacle to the

[*] This work was funded by the European Commission under the COST 285 Action, by the CNR/MIUR program "Legge 449/97" (project IS-Manet) and by the Ministry of Higher Education, Science and Technology of the Republic of Slovenia (program P2-0246).

efficient exchange of scientific data. Moreover, if we consider issues like storage and data management, documentation and meta description, to handle simulation results becomes apparent. This topic has been addressed [1] in the framework of the European COST 285 Action "Modelling and Simulation Tools for Research in Emerging Multi-service Telecommunications" (See Chapter 1), a forum where researchers from all around Europe periodically meet to address issues related to simulation of communications systems. The point made was that apparently no general purpose tools exist for exchanging big quantities of simulation data coming from different sources in different formats. Not only the need of a common format for exchanging data was highlighted, but also the need of feeding this data to different tools for postprocessing them, each requiring a different input format.

To better understand the scope of different requirements we define a reference model that encompasses data creation, flows and processing in the analysis of the telecommunications systems by simulation. The main functional parts composing such a model (simulators, data collectors, graphing tools, statistical tools) are covered by the many existing tools that are used by the research community; we focus on the input data in form of simulation, like ns-2 [2] traces, or measurements, like Tcpdump [3] traces. Raw data can be post-processed (e.g. calculating average packet delay) and the results stored separately from the raw data or complementing it, so that further analysis is possible based on both raw and preprocessed data Finally, the results can be exported in various more or less widespread output formats, like ASCII or XML [4].

Data storage is based on the HDF5 [5] data format, which was selected after an analysis of the available options. HDF5 has been successfully applied in several scientific projects; it enables efficient data storage and lookup. Among the features most relevant to our purpose, it provides support for extremely large quantities of data, for meta descriptors and for embedded compression. A set of programming libraries is available in C and Python, which simplifies software development based on HDF5.

The proposed CostGlue software architecture is modular, to make it possible to include the future development contributions from other research communities. It is composed of three building blocks. The core is a Python application called CoreGlue, which provides HDF5 database connection, a simple command line, an HTML based interface and basic functions for import, filter and export. It also controls and executes other parts of the application. Specific functions, such as conversion from specific data format and calculations, are implemented as sets of self-descriptive loadable modules. A graphical user interface is envisaged via the use of a web browser, allowing user friendly and efficient remote access to the application.

CostGlue will be released as free software. The next sections describe into deeper detail the proposed architecture of CostGlue and particularly of its core, CoreGlue.

2.   A MODEL FOR THE SIMULATION PROCESS

In order to obtain a general and systematic overview of data creation, flow and processing, we define a reference model for the simulation process, which is depicted in Figure 1. The model provides a layered decomposition of main functions that are usually encountered when using simulation as a research method. There are three layers in the model.

The *source layer* provides raw simulation output, describing a simulation run into the smallest detail. Usually one or more records for every event during the simulation run are created at the source layer. Structure and volume of the data at this point is entirely dependent on the simulator. Most frequently it is in the form of large tabular traces, in ASCII or binary format. An optional source recoding sublayer handles the raw source data: its main purpose is to convert between different formats (e.g. from ASCII to binary or vice versa), data compression (e.g. Gzip [6], Bzip2 [7]) or removing private information, e.g. real IP addresses, from simulation traces. The source layer supports both raw simulation data and real measurements data. In fact, during our discussions, we found out that nearly the same model can be applied to the analysis of real network traffic traces. In this case the raw data is not a result of a simulation, but for example, data traces captured in a network link. Apart from the different tool (traffic capture tool, like Tcpdump or Ethereal [8] instead of a simulator) that generates raw data, all the functions of the upper layers remain the same in both cases.

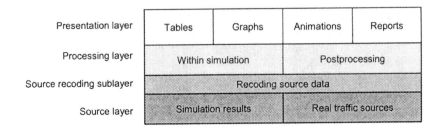

Presentation layer	Tables	Graphs	Animations	Reports
Processing layer	Within simulation		Postprocessing	
Source recoding sublayer	Recoding source data			
Source layer	Simulation results		Real traffic sources	

Fig. 1 Reference Model of a Simulation Process

The *processing layer* is responsible for simulation data analysis. At this level cumulative results can be derived from raw data (e.g. mean packet delay is calculated) or statistical confidence of the results determined (and consequently additional simulation runs conducted). An important characteristic of the data processing layer is that the amount of data received from the source layer is usually much larger that the amount of results of post-processing.

The *presentation layer* is the final stage where the results are organized in a form useful for communicating the most important findings with other interested practitioners. In case of simple tabular printouts of the results, this layer is void or only does trivial modification of the data (e.g. changes in number formats, column spacing). But frequently data is shown in 2- or 3-dimensional graphs (e.g. being part of scientific reports or research papers, web pages) or even presented in animated form (e.g. nam - Network AniMator [9]). At this layer the predominant requirement is flexibility of presentation and a possibility to create new or modified presentation objects from new or changed simulation results without reformatting.

We can map the functionality of particular tools used in simulation to the layers of our model. Usually, a single tool provides more than one functional layer or even all of them. In the most favorable situation, it would encompass all the functions needed and implement them adequately to meet all the researcher's needs. In practice this occurs very rarely and there is usually a set of complementing tools applied to cover the required scope of functions within the model. The selection of tools is made on arbitrary conditions, including their capabilities and performance, researchers' past experience with a particular tool or availability of tools. In case of simulated results, raw data can be generated by discrete event simulators (e.g. ns-2, Opnet [10]). Raw data can be captured in real networks, with sniffers, such as Tcpdump or Etheral. Source recoding can be done with small dedicated tools (e.g. Gzip, Tcpdpriv [11]) or different proprietary shell scripts. Data analysis can be performed with generic tools for mathematical computation (e.g. Octave [12], Matlab [13], Mathematica [14], Excel [15]), special statistical tools (SPSS [16], R [17]) or proprietary and dedicated programs or scripts. Often, the simulation package provides the functionality for data processing and analysis and it is up to the researcher to decide whether this is adequate or if an additional more powerful and flexible tool should be used. Other than in dedicated graphing or animation tools, presentation ability can be provided in generic mathematical tools and sometimes even simulators.

3.   FORMATS FOR EXCHANGING DATA

CostGlue acts as a central repository for data generated by various different simulation programs and as a converter both from several output formats and to several input formats. Therefore, it is important to know what programs and formats are generally used by telecommunications systems practitioners. To this end, we used the information that we got from the COST 285 participants, representatives from more than ten European nations, about the kind of tools they use for their simulation work. From this sample, we learned that no single simulation tool has a dominant position, but there is a great variety of used tools.

Among those that use a predefined format output, the most used appears to be the network simulator ns-2. Among the generic tools for mathematical computation that are used to run simulations, Matlab appears to be used by many. A large part of the simulators is composed by standard scripting or programming languages and generally by ad hoc simulators. On the side of tools used for postprocessing or graphing, there is an even greater variety.

These observations, while limited in scope, show that some sort of tabular ASCII format is of generally common use, and thus being able to read and write ASCII tabular data is certainly a requisite for our proposed archiver and converter. But the variety of tools used also calls for a general way of reading and writing many formats: that's why we consider the modular architecture of CostGlue a necessary feature for the tool to be useful at all.

Another interesting point is that simulation data and measurement data have a lot in common, and a tool useful for one can be useful for the other. However, measurements are often output in particular formats, and an input converter is very frequently needed. An interesting feature that can be made part of CoreGlue is the ability to give a similar treatment to data coming either from measurement or from simulation of the same environment, and archive them in the same format. This is the reason why the first prototype of the CostGlue will include the ability read ns-2 data and Tcpdump data, store them into a common format and write in both formats. This capability would make it easy to use the many tools available that are able to analyze and graph data obtained by both ns-2 and Tcpdump.

All the above discussion leads to a scenario where a simulation tool is run several times, each time producing tabular data, that is, data that can be conveniently stored into a two-dimensional structure having relatively few columns and a possibly huge number of rows. What about data that cannot be naturally converted to a two-dimensional format? In this case, the inner structure of the archived data needs to be different. One of the challenges of the proposed tool is to being efficient in the most common case of collections of tabular data, but still be useful in the case of non-tabular data.

4. THE DATABASE

A common file format solves several problems regarding the exchange of the simulation data. Therefore we made a thorough analysis of erent data formats and their corresponding libraries for data manipulation. Among many we have focused on the following set of data formats: HDF4, HDF5, netCDF, ODB, FITS and OpenDX [18]. Beside these, we also considered using plain text formats, XML and SQL databases. The results of the analysis makes it clear that the HDF5 file format is the most suitable for this task since it meets all the requirements of data organization e.g., separation of raw data and metadata and different requirements of contemporary computer system architectures, such as managing big quantities of data, offering a general data model, supporting complex data structures, portability among different computer platforms, parallel data access and processing, diversity of physical file storage   media, etc.

4.1   The Database Structure

Summarizing what we said above, most simulation data in the computer communications area are collections of tables of numeric data: each simulation run generates a table of data having few columns and a possible huge number of rows. Each table is associated with certain parameters that are specific for a simulation run that generated it, and is uniquely identified by the values of those parameters. We are interested in defining a database structure that is able to efficiently accommodate this type of data. To meet these requirements we have chosen to build a database in HDF5.

HDF5 is a data format and an associated software library consisting of two primary objects: dataset and group. A dataset represents a multidimensional array of data elements, which can hold different types of data. The data stored in datasets can be either homogenous (only one data type used within single dataset – simple dataset) or compound (different number of data types within one dataset – compound dataset). Since tabular data collected from certain simulators often contains data with different types (e.g. integer, float, char), we use compound datasets to accommodate the nature of simulation outputs. An HDF5 group is a structure containing zero or more HDF5 objects. By using two primary HDF5 objects, data can be organized hierarchically by means of a tree structure where an arbitrary number of HDF5 objects are derived from the main "root" group. Groups and datasets have a logical counterpart in directories and files in a hierarchical file system and, similarly to a file, one can refer to an object in an HDF5 file by its full path name.

To meet the requirements of effective data storage, especially those that are critical to management, understanding and reuse of scientific data, each HDF5 object may have associated metadata stored in the HDF5 file – referred as archive – in a simple attributes form. Attributes usually represent a small dataset connected to a certain group or a dataset. Their purpose is to describe the nature and/or the intended usage of the object they are attached to.

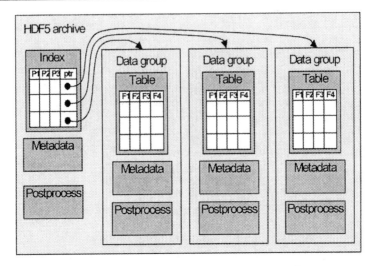

Fig. 2 The Logical Structure of the Database

In the design of the database structure our goal was a flexible representation of the stored simulation data by using one multidimensional array, where a user can easily extract a desirable portion of the simulation data. Even though HDF5 supports multidimensional arrays, it is not efficient to store huge amount of data in just one array. Furthermore, due to the HDF5 primary aspect of use, which is to have data organized hierarchically, storing everything inside a single multidimensional array means losing on the side of flexibility. We then introduced an indexing table which maps the logical view of a multidimensional matrix into an HDF5 hierarchical structure as detailed in the following.

Figure 3 presents a detailed overview of the proposed database structure where the *indexing table* represents the logical part and all other groups and datasets represent actual data where the raw simulation data, metadata and postprocessing data is stored. The whole database is treated as one archive containing a "root" group from which all other groups and datasets form a two-level tree.

An immediate extension to having tables of numeric data is having tables of fixed-length data, which can be flags, numbers or strings. The PyTables library [19] allows efficient manipulation of 2-dimensional HDF5 compound datasets from Python referred to as *tables* from now on. Each table, together with metadata and postprocess data, is attached to a *data group*, which will usually holds the data produced during a single simulation run. Data groups are indexed by vectors of *parameters*. An *index* is a 2-dimensional array, referred above as the indexing table, where the parameters relative to each data group are stored: each column corresponds to a different parameter, and each row contains the values of the parameter relative to a data group. So an index is used as the data structure for accessing a data group using an array of parameter values relative to that data group as the key. To each data group a table is attached, where each row is filled with the values of the *fields*, each field corresponding to a column of the table.

The overall structure is then a collection of 2-dimensional tables indexed by arrays of P parameters. Overall, this can be logically seen as a matrix with $P+2$ dimensions, where the first P dimensions are sparse and the last 2 dimensions are dense. We define the first P indices as parameters that identify a data group. As for the last two indices, the first index represents the field in the data group's table, and the second index the row number of the data group's table.

Let us describe how the results of an example real-world simulation can be stored in the described structure. We are simulating the behavior of packet switches in ns-2; each run is characterized by several parameters, such as architecture type, buffer size, number of I/O ports and traffic load. Each simulation run differs in at least one of these parameters. When storing the results of one simulation run in the archive, we populate a new row in the indexing table: values in each row uniquely identify a simulation run. Each row contains the full path to a data group, containing the table with the results of the simulation run, metadata and processing data (see Figure 3). Metadata stores information about the type of scripts used to generate that simulation run, type of network topology, traffic patterns, etc. Post-processing data include packet loss probability, maximum, minimum and average packet delay. Metadata and processing data are also associated with the whole archive, and contain information relative to the whole set of simulation runs.

The CoreGlue manages the index and the database structure, including the tables. Modules are responsible for metadata and postprocess contents, both for the single data groups and for the whole archive. The CoreGlue and the modules together constitute the whole application, which is named CostGlue.

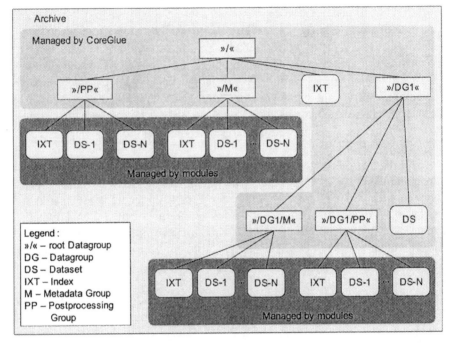

Fig. 3 HDF5 Database Structure

5.   COSTGLUE ARCHITECTURE

The CostGlue is being written in Python [20]. This language was chosen because of anecdotal evidence of its efficiency both in memory usage and processing power and for its programmability ease, due to automatic garbage collection and many native functions and types. Portability among operating systems is excellent and library availability for many tasks, especially mathematical ones, is rich. With respect to its main competitor, Java, Python has a generally smaller memory footprint and, being the implementation completely free, does not suffer being controlled by a single private entity.

The architecture of the CostGlue application can be seen in Figure 4. The CoreGlue connects other parts of the application by performing the following tasks: it reads module descriptions, executes modules, passes parameters, reads module execution results, handles the HDF5 database, provides a CLI (Command Line Interface), provides an HTML interface, generates XML for additional user interface and takes care of exporting data in various formats (CSV, TXT ...).

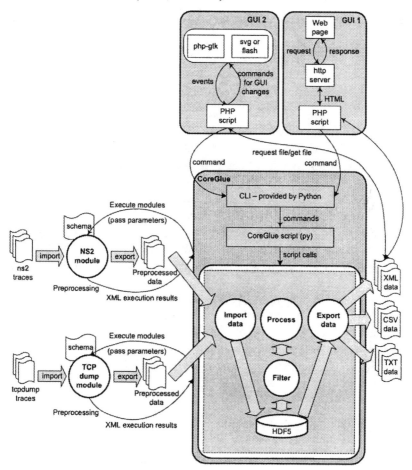

Fig. 4 CostGlue Software Architecture

5.1   Overview Of The Module API

Python, like many modern languages, supports dynamic module loading. We exploit this possibility by providing a module library with a restricted pool of modules, and leaving the possibility open to other parties to write other modules. Modules are Python self-describing programs that reside in a fixed place. The CoreGlue can look for available modules and query them one by one in order to know their capabilities; this can be used for example for building a menu for a GUI (graphical user interface). Modules are able to describe the parameters they need and the type of output they produce. The needed parameters can then be provided to the module by the CoreGlue, with or without input from the user; in case input is required, CoreGlue can check that the parameters provided are consistent with the constraints defined by the modules. A GUI can also use this information to present the user with choices for the input parameters.

Modules interact with the CoreGlue through a well-defined API which contains a minimum of classes and methods for interacting with the database. The methods allow a module to manipulate the data group index in order to add or remove data groups, and to manipulate the data groups, in order to add or remove data, metadata and postprocess data to it.

The API also includes methods for accessing data with *selectors* written with the index notation that is widely used in matrix computations programs such as Matlab or Octave. In this notation, each of the $P+2$ indices can be "1", indicating the smallest index; "end", indicating the highest one; "$n{:}m$", indicating all the indices between n and m included; "$:$", indicating the whole range from smallest to highest, "$n{:}s{:}m$", indicating the range from n to m in steps of s; an array like "[1 5 6 8]" indicating the selected indices. Using the Matlab index notation one can take complex orthogonal slices of the multidimensional matrix composed of all the data in the database; in fact, the database structure can be seen as a sparse matrix with $P+2$ dimensions, and being able to take a slice of this matrix could prove a powerful feature of CostGlue.

5.2   The Command Line And The HTML GUI

The CoreGlue always needs to load a module to do anything useful. When invoked by a command line, its first argument is the module name, followed by the parameters needed by the module. When invoked with the `--html` option, the CoreGlue acts as an HTTP server, providing a graphical user interface. Through this interface, the user can look at the list of available modules and, for each of them, look at a description, at the input they require and at the output they provide. The CoreGlue provides an interface for the input of module parameters, complete with checking, thanks to the information that it reads from the modules themselves. In the simplest case, this is analogous to calling the module on the CoreGlue command line, but more convenient for interactive use. A module can also provide a graphical interface or graphical output by itself.

6. CONCLUSIONS

The purpose of the conversion and storage tool whose design we described in this paper is to facilitate the exchange and management of simulation data among researchers, and to ease the task of using different simulation, measurement, data processing and visualization tools. If this design will develop into a usable package, then it will also be useful outside of the academic community and will hopefully be enhanced by other simulation practitioners thanks to its modular structure. Once its usefulness will be proved, some extensions will need to be provided, such as being able to accommodate non-tabular data.

We believe that software developed as part of research activity should be released with a free software license, because research results should be made available for use by anyone, for any purpose and be freely modifiable, in order to further knowledge and usefulness [21]. The choice of license will be among the MIT X license, the GNU LGPL and the GNU GPL licenses, which we think the best serve the purpose of free research software [22].

REFERENCES

[1]   Arnold Bragg, "Observation and Thoughts on the Possible Approaches for Ad-dressing the Tasks Specified in the COST 285 Work-Plan", COST 285 TD/285/03/15,
www. COST285 .itu.edu.tr/ tempodoc/
[2]   "The Network Simulator - ns-2", http://www.isi.edu/nsnam/ns/
[3]   "Tcpdump public repository", http://www.tcpdump.org
[4]   "Extensible Markup Language (XML)", http://www.w3.org/XML
[5]   "HDF5 Home Page", http://hdf.ncsa.uiuc.edu/HDF5/
[6]   "The Gzip home page", http://www.gzip.org/
[7]   "The Bzip2 home page", http://www.bzip.org/
[8]   "Ethereal: A network protocol analyzer", http://www.ethereal.com/
[9]   "Nam: Network Animator", http://www.isi.edu/nsnam/nam/
[10]   "Opnet modeler", http://www.opnet. com/products/ brochures/ Modeler.pdf
[11]   "Tcpdpriv, A tool for anonymising tcp traces", http://www.openbsd.org/3.5_packages/vax/tcpdpriv-1.1.10.tgz-long.html and http://fly.isti.cnr.it/software/tcpdpriv/
[12]   "Octave home page", http://www.octave.org/

[13] "The MathWorks - MATLAB and Simulink for Technical Computing", http://www.mathworks.com/

[14] "Mathematica - Wolfram Research, Inc.", http://www.wolfram.com/

[15] "Excel 2003 Home Page", http://office.microsoft.com/en-us/FX010858001033.aspx

[16] "SPSS home page", http://www.spss.com/

[17] "The R Project for Statistical Computing", http://www.r-project.org/

[18] "HDF5 Wins 2002 R&D 100 Award", http://hdf.ncsa.uiuc.edu/HDF5/RD100-2002/All_About_HDF5.pdf

[19] "PyTables – Welcome Page", http://pytables.sourceforge. net/html/ WelcomePage.html

[20] "Python Programming Language", http://www.python.org/

[21] Francesco Potortì, "Free software and research", short paper, proceedings of the International Conference on Open Source Systems, Genova (IT), pp. 270-271, July 2005.

[22] David Wheeler, "Make Your Open Source Software GPL-Compatible Or Else.", web essay at http://www.dwheeler.com/essays/gpl-compatible.html, revised February 2005.

CHAPTER 11

Modeling Grids in (Near) Real Time

Arnold Bragg[1], Harry Perros[2], Mike Devetsikiotis[3], Ilia Baldine[1],
Dan Stevenson[1]

[1] Center for Advanced Network Research
RTI International, Inc. Box 12194
Research Triangle Park, NC 27709 USA

[2] Computer Science
North Carolina State University Raleigh,
NC 27695 USA

[3] Electrical & Computer Engineering
North Carolina State University Raleigh,
NC 27695 USA

Abstract. *A Grid is a family of technologies for dynamically and opportunistically provisioning computing power from a pool of resources. Some experts believe that Grids are the next stage in the evolution of distributed systems. Enterprise-level Grids are beginning to be deployed, and researchers are investigating a number of novel Grid applications and services in Grid testbeds. Production-grade commercial Grids are expected to be large and widely-distributed systems.*

Grids are enormously complex computing systems. They have large numbers of geographically dispersed resources at their disposal, and they lack some characteristics of earlier computing systems that simplified the analysis of those systems – e.g., homogeneous and closely-coupled components, low-latency communications paths, global state maintenance, and deterministic controls. Performance analysts have very little understanding of the dynamic behavior of Grid components at different layers and at different time scales, and of the complex interactions among Grid components – functionally, spatially, and temporally.

We describe an architecture for a scalable, modular, configurable, plug-and-play, "Grid-in-a-lab" tool for modeling Grids, and for analyzing the performance of applications and middleware services offered on Grids. The tool addresses all of the components that are relevant to Grid performance – hardware, middleware, infrastructure, services, applications, traffic – and the complex interrelationships among the components. It uses a novel combination of modeling techniques – e.g., analytical models with closed form expressions, discrete event simulation, emulators for protocol stacks and resource scheduling, rare event simulation methods, and measurement-based optimization. The tool's components are modular and configurable so as to provide different degrees of fidelity over a range of time scales, and to provide insight about Grid dynamics and interactions among various components, applications, and services.

1. INTRODUCTION

1.1 Definitions – Grids, and Applications and Services Running on Grids

A *Grid* is a family of technologies for dynamically and opportunistically provisioning computing power from a pool of resources. *Resources* include processors, memory, file and data storage, cache, depots, databases and data warehouses, and a network infrastructure that interconnects the resources. Resources may be geographically dispersed, and may vary in capability, capacity, and availability. *Provisioning* includes mechanisms for locating, authorizing, assembling, scheduling, and accounting for resources and their usage [76, 99, 100, 119, 126, 127].

Grid-aware *applications* share the resource pool; e.g., [1, 3, 9, 15, 24, 27, 35, 36, 38, 44, 46, 58, 59, 79, 85, 86, 94, 95, 96, 104, 118, 123, 132, 133, 135, 138, 143, 144, 145, 148, 151, 163, 181, 209, 214]. Grid applications can be characterized in a number of ways: *computationally intensive* with high CPU, memory, and/or performance requirements (e.g., distributed supercomputing for experimental sciences); *data intensive* with high input/output and/or communications bandwidth requirements (e.g., data mining); *throughput intensive* with large numbers of idle resources used for quasi-independent scatter/gather tasks (e.g., cryptography); *collaborative* with interactive steering and/or multicast requirements (e.g., distributed interactive simulations); *time sensitive* with stringent latency and jitter requirements (e.g., visualization, virtual reality); or *time bounded* ("finish no later than") which may be more sensitive to cost than to performance (e.g., batch processing) [95].

Most Grid-aware applications have resource requirements that vary during the application's lifetime. Resource availability will also vary in response to competing demands from other applications. Grid *middleware* is used to bind applications to resources by adaptively mapping resource demand to resource availability during execution [142]. Middleware also provides *services*[1] for data and replica management, data/metadata access [20], data transfer, fault detection [121], job management, scheduling, monitoring, communications (bandwidth, message passing, etc.), reliable transport, resource discovery/inquiry, service level agreements, virtualization, etc. (For examples of middleware services, see [2, 4, 7, 10, 19, 35, 42, 45, 47, 52, 54, 57, 60, 78, 80, 81, 94, 107, 122, 134, 141, 154, 178, 186, 192, 196, 197, 210, 223, 224]).

1.2 Why Grids are Important

Some experts believe that Grids are the next stage in the evolution of distributed systems – following batch and timeshared systems, teletraffic systems, multiplexed communications networks, multiprocessor and parallel systems, and tightly-coupled distributed computing platforms. Enterprise-level Grids are beginning to be deployed, and researchers are investigating a number of novel Grid applications and services in multi-service research testbeds [114, 115, 116]. Production-grade commercial Grids are expected to be large and widely-distributed systems [98, 130, 139, 140, 147, 152], and Grid deployments are expected to grow exponentially at least through 2010 [43, 87, 112].

Grids are the focus of several wide-ranging European Union (EU) initiatives [68, 69, 70, 71, 72, 73]. One initiative maintains that "[a] view of the future is closely related to developments in the Grid, seen by many as the future of the Internet and said to be just as significant. It will allow

---

[1] Not to be confused with lower-level "Grid Services", a family of platform- and language-independent technologies used to create client/server applications for loosely-coupled distributed computing systems.

users, both business and domestic, to collaborate and share computational and storage resources, treating them as utilities in a wide range of applications" [73, p. 53.]. Several notable EU Grid initiatives are summarized in Table 1[2].

Grids are also focus areas in at least three EU COST Actions – Action 282, *Knowledge Exploration in Science and Technology* [3]; Action 283, *Computational and Information Infrastructure in the Astronomical Data Grid* [4]; and Action 291, *Towards Digital Optical Networks* [5].

1.3 Challenges

There are at least two major challenges for modeling, simulating, and analyzing Grids – (1) the enormous *complexity* that Grids introduce, and (2) the *limitations* of today's methods and tools for analyzing Grid performance and dynamics.

The evolution of methods and techniques for the performance analysis of complex computing systems usually passes through three distinct stages. The first is observational, and focuses on instrumentation, metrics, measurements, and simple empirical representations. Performance frameworks, analytical and discrete-event simulation models, and rudimentary theory emerge during the second stage. Mature models with closed-form expressions, realistic stochastic models, and formal theory appear during the third stage.

Grid performance analysis is poised to begin its second stage. Unfortunately, Grids lack many of the simplifying characteristics of earlier distributed systems – e.g., homogeneity, closely-coupled components, low latency, quasi-global state maintenance, and closed-loop controls. Performance analysts will require a better understanding of the dynamic behavior of Grid components at various protocol layers and at different time scales; and insight into the complex interactions among disparate Grid components – functionally, spatially and temporally.

---

[2] http://www.gridstart.org/index.shtml .
[3] http://www.mpa-garching.mpg.de/~opmolsrv/COST282/ .
[4] http://www.cordis.lu/cost/src/283_indivpage.htm .
[5] http://cost.cordis.lu/src/action_detail.cfm?action=291 .

Table 1. Key EU Grid Initiatives.

Initiative	URL	Focus Area(s)
CrossGrid	www.crossgrid.org	Techniques for large-scale Grid-enabled real-time simulations and visualizations, and interactive compute- and data-intensive applications
DAMIEN	www.hlrs.de/organi zation/pds/ projects/damien	Building blocks for a middleware environment for distributed simulation and visualization, and software supporting the Grid infrastructure
DataGrid	eu-datagrid.web. cern.ch/eu-datagrid	Techniques supporting processing, data storage, and infrastructure for scientific research and applications in high energy physics, environmental science, bioinformatics
DataTAG	datatag.web.cern.ch /datatag	Research and technology development for a transatlantic Grid to support reliable and high-speed collaboration across widely distributed networks
EGSO	www.mssl.ucl.ac. uk/grid/egso	European Grid of Solar Observatories' virtual archive of solar data centers as a Data Grid
EUROGRID	www.eurogrid.org	Develop core Grid software components
GRIA	www.gria.org	Application-driven Grid testbed focused on outsourcing of secure and reliable computational services
GridLab	gridlab.org	Develop software able to fully exploit dynamic Grid resources
GRIP	www.grid-interoperability.org	Interoperability of leading Grid software packages for remote access to supercomputer resources and simulation code

1.3.1   Complexity of Grids

Grids are complex for a number of reasons. Grid *resources* are heterogeneous, and may be distributed over a wide area. Resource assignments are opportunistic, and resource owners may retain preemptive priority over the shared use of their resources. Resource requirements and availability vary unpredictably during an application's lifetime.

Grid *applications* may have stringent service requirements. Intelligent pre-fetching and depoting of data may be required to mitigate the effects of delay and jitter for real-time applications (e.g., visualization, tele-immersion, virtual reality).

The Grid network *infrastructure* must support a large number of different traffic types – long-lived static connections, dynamically provisioned sessions with lifetimes of a few tens of seconds, flows between source-destination pairs, bursts of a few tens of packets, and irregular single packet transfers. Service quality requirements can range from best effort to circuit emulation [88, 183]. Wide-area Grids may have hundreds of millions of packets in circulation at a time.

Grid *resource scheduling* and other middleware services are more efficient when coupled with a global information base, but the message traffic (request/response/update) required to main consistent state is very sensitive to latency and loss. Centralized Grid services tend to be inefficient and prone to single points of failure; distributed Grid services tend to be very difficult to synchronize.

Fast provisioning of network resources is a fundamental requirement in some Grids. The Global Grid Forum has identified requirements for Grid network resources; viz., calling for a scalable, flexible, and rapidly reconfigurable network infrastructure; bandwidth on demand between arbitrary endpoints; and application provisioning and control of bandwidth [113]. The high energy physics community uses rapidly provisioned "lambda Grid" network infrastructures for Terabyte file transfers. Visualization applications usually require data to be speculatively pre-fetched using a Grid of high bandwidth connections, and stored in close proximity to the endpoint so that the latency seen by the application is quite small [8, 53]. However, dynamically-provisioned high capacity networks are difficult to model and analyze due to: widely disparate traffic sources; controls (e.g., TCP rate controls, optical wavelength provisioning controls); complex network elements (e.g., routers with active queue management schemes, switches with preemptive blocking); cross-layer protocol effects; enormous asymmetries in bandwidth requirements and session lengths; and confounding effects caused by interactions among these factors.

Hence Grids, and the applications and services they support, are enormously difficult to model and analyze. Measurements on research Grids indicate that the confounding effects caused by interactions among the various components can also have a profound impact on performance [62, 172, 185, 223, 224].

1.3.2   Limitations of Today's Methods and Tools

Large-scale Grids cannot be realistically analyzed with a single performance model, method or tool. *Discrete event simulators* (DES) may be appropriate for some Grid components[6]; however, a monolithic DES cannot scale to the numbers of events that a wide-area heterogeneous Grid will generate, even if sophisticated parallel or distributed extensions are used (e.g., [105, 106]). Realistically simulating a single job submitted to a wide-area Grid may spawn hundreds of thousands of events.

Analytical techniques (e.g., queuing theory, Markov processes) are appropriate for modeling some Grid components at an abstract level, but obviously cannot model a dynamic, stochastic, multi-component Grid in great detail.

Hybrid simulation/analytic techniques are effective for modeling large and complex systems where one or more subsystems are represented in the simulation model using queuing-based analytic expressions. However, crafting a hybrid model requires expert knowledge, fine tuning, and a precise balance between model fidelity and the resources available to the modeling tool (e.g., CPU cycles, simulator memory, simulation run time). For example, background traffic behavior in Grids with high-capacity, dynamically provisioned, multi-channel optical links is an excellent candidate for hybrid techniques, but such traffic has not been characterized (to our knowledge).

---

[6]   E.g., *GridSim* is a simulation toolkit for modeling distributed resources, resource allocations, and application scheduling that utilizes parallel and distributed cluster computing systems for execution [37, 157, 198, 199, 200].

Emulation is an effective technique for some types of Grid components (e.g., network protocols). Emulators run (nearly) unmodified protocol stacks on a few tens of 'instances' – commodity single-board computers, or processes executing on multi-tasking hosts, or both – in (near) real time. Emulation is not suitable for every Grid component, but may be appropriate for Grid schedulers, and for modeling some network infrastructure and Grid middleware services.

Traffic behavior may not even be *measurable* in Grids with high-capacity optical links. Optical monitoring equipment may be prohibitively expensive, there may be tens or hundreds of rapidly-provisioned/released optical channels sharing a single link, and the sheer volume of data transiting the link may require costly sampling, storage and analysis tools.

Clearly, a combination of modeling methods and techniques are required to analyze Grids.

1.4   Relevance to EU COST Action 285

Modeling, simulating, and analyzing Grids is relevant to COST Action 285 [74] (See Chapter 1) for several reasons. It addresses three COST Action tasks – task 1.1 (*techniques for reducing simulation time*), task 1.3 (*multimedia traffic behavior at different time scales*), and task 2.2 (*multilayer traffic modeling*) as applied to Grid infrastructures.

The family of large and complex networking infrastructures deployed in Grids is clearly an emerging multi-service telecommunications network backbone technology. Modeling this network technology requires a combination of methods to reduce simulation time (discrete event simulation, analytical models with closed-form expressions, efficient sampling techniques, measurement-based optimization, emulation, hybrid techniques, etc.) as these networks have aggregate traffic volumes 4-5 orders of magnitude larger than what today's discrete event and hybrid simulators typically handle.

New modeling and simulation methods and tools are also required to capture the behavior of Grid resources, applications, and services at various protocol layers and time scales, and to decipher the complex functional, spatial, and temporal interactions among the components.

Grid dynamics and operational behaviors are not understood, which is a "real" problem facing Grid infrastructure providers. In the near term, configurable and scalable methods and tools for exploratory analysis and experimentation will be required to provide network operators and network service providers with insight about how Grids actually work [28].

2.   REFERENCE MODEL FOR MODELING GRIDS

The reference model has six structural dimensions: Grid applications; tasks; Grid services; Grid resources and resource requirements; performance requirements; and performance metrics (Figure 1).

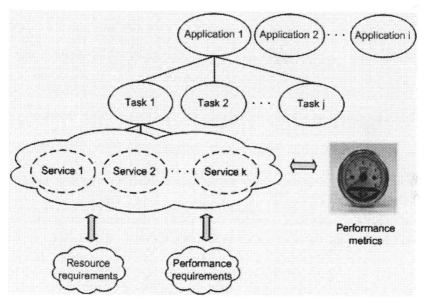

Fig. 1 Reference Model with Six Structural Dimensions.

2.1 Applications

Grid *applications* are jobs submitted to the Grid for processing. As noted, applications may be characterized in a number of ways (e.g., computationally intensive, collaborative, time sensitive, etc. [17, 94]).

2.2 Tasks

A Grid application comprises a set of *tasks*; e.g., a tele-immersion application may spawn tasks for control, data transport (text, video), tracking, database transactions, simulation, haptics[7], and resource rendering [63]. Each task has resource and performance requirements. One may also view tasks as a way to characterize the various programming paradigms used in Grids and distributed systems; e.g., shared memory, message passing, remote procedure calls, remote input/output, intelligent agents, batch, interactive, client/server, scatter/gather [94, 99]. The number of tasks spawned by a single Grid application may be quite large.

2.3 Services

Each task uses some number of Grid middleware *services* to map its application's resource and performance requirements to Grid resources. As noted, services include data management, data access, data transfer, fault detection, job management, scheduling, monitoring, communications, transport, resource management, service level agreements, etc. The number of Grid services required by a single task may be quite large [84].

2.4 Resources and Resource Requirements

Services require *resources*. Grid resources are geographically dispersed, and may vary widely in capability, capacity, and availability. Resources and resource components comprise the following:

- *Computing infrastructure*: devices (e.g., supercomputing 'supernodes', vectors, highly/massively parallel with COTS processors, clusters, arrays of desktop machines); computational types (e.g., single or multithreaded, distributed [94]); input/output (e.g., local, parallel, distributed file systems [94]); memory (e.g., shared, unshared, partitioned [56, 89, 91, 92, 93, 201]); operating systems; virtual machines; etc. [77, 101, 102, 103, 217].
- *Network infrastructure*: interconnection technology (e.g., Infiniband); LAN technology (e.g., Gigabit Ethernet); WAN technology (e.g., SONET, WDM); protocols (e.g., streaming, transport; reliable, unreliable; real time, non-real time; connection oriented, connectionless; reservation-based); resource provisioning methodology; data plane characteristics (e.g., payload, framing,

---

[7] The science that deals with the sense of touch.

formatting); control plane characteristics (e.g., signaling, management, in-/out-of-band); unicast, multicast, broadcast services; other characteristics (e.g., bandwidth); etc. [117, 120, 189, 190, 191].

- *Storage infrastructure*: local disks, tapes, mass storage systems; cache (higher speed, "virtual disk" type devices, spools), staging storage (low and high speed); data repositories, distributed file systems (e.g., NFS, AFS, CXFS); storage area networks; depots for pre-fetched data; mirrored/replicated sites; replication technologies [168, 170, 171, 193, 194]; etc. [82, 83, 90, 146, 184, 205, 206, 207].
- *Input and output devices*: sensors, scientific instruments; performance monitoring equipment; display walls; etc.
- *Data resources*: databases, database management systems, data warehouses; repositories for metadata and derived data; data catalogs/directories and services; data libraries; etc.
- *Resource management*: creation, discovery, search engines; brokering, negotiation, preemption; data migration/staging, resource redistribution; scheduling; state maintenance, resource databases with update capability; local services, global services; performance tuning (open loop, closed loop); etc. [6, 11, 55, 97, 158].
- *Fault management*: fault recovery (e.g., checkpointing and restart); fault avoidance (e.g., redundant processing); etc.
- *Communications*: provisioned circuits; short-lived connections; flows; bursts; messages; packets; etc. [66, 166, 220, 221, 222, 226, 227].
- *Composite elements* [50]: local Grids; high performance virtual machines (HPVMs); SANs; characterization (capacity, aggregate compute/communications/storage performance, availability); etc. [5].

2.5 Performance Requirements

Applications, tasks, and services have *performance requirements*. These include minimum/peak/mean throughput, "goodput"; response time, bandwidth, rate; latency; jitter (delay variation); error rate; loss rate; reliability, etc. Services may have absolute guarantees, relative guarantees, or no guarantees (e.g., best effort) [149, 176, 177, 182, 195, 208, 211, 215, 216, 225].

2.6 Performance Metrics

Performance metrics can be application-, service-, or task-focused (e.g., reflecting application service requirements like response time and fairness); or infrastructure-focused (e.g., that optimize resource utilization, efficiency, and/or reliability). Higher-level performance analyses include identification of bottlenecks; root cause analyses; diagnostics; tuning and adaptation, etc. [173].

2.7 Attributes

Each component in each dimension has a number of *attributes*; e.g., applications have "home" locations, and tasks/services/resources can be characterized in various ways.

2.8 Other

We do not consider authentication, authorization, security, accounting, inter-domain policy, credential management, access control, portals, assurance mechanisms, or audit functions in this reference model. These are important components, but they have far less of an impact on Grid performance than others.

3. "GRID-IN-A-LAB" TOOL FOR MODELING GRIDS

A goal of our work is to define an architecture for a modular, configurable, scalable, plug-and-play tool for measuring, modeling, analyzing, evaluating, and predicting Grid performance, and for exploratory analysis and experimentation over simulated Grid infrastructures. We briefly describe the design concepts, components, and several implementation issues in this Section.

3.1 Design Concepts

The tool imitates the operation of all of the Grid components that are relevant to Grid performance; viz., hardware, middleware, infrastructure, services, applications, and traffic. Some components have sub-components or other layers of abstraction [12, 13].

The tool supports a large number of modeling technologies: viz., analytical models, discrete event simulation, hybrids with simulation and analytical models, measurement-based methods, emulation code, transfer functions, simple mathematical expressions, finite state machines, neural nets, empirical distributions, etc.

A tool module that represents a particular Grid component – e.g., a multi-scheduler – can be implemented in many different ways – as a finite state machine, as a suite of conservation laws, or as an actual Grid scheduler executing on commodity hardware in real time. The choice depends on the analyst's requirements for model fidelity, time scale, and execution (or "wall clock") time available for simulation runs. Tool modules are configurable so as to provide different degrees of fidelity over a range of time scales.

The tool is <u>not</u> a monolithic discrete event simulator (DES), although a designer may choose to use DES in the implementation of some of the module(s). DES simply cannot support the sheer volume of stochastic events required to model high performance Grids.

The tool, when completed, will be made available to the research community. We believe this will significantly increase the community's understanding of how Grids operate, how to design and build better Grid applications and middleware services, and how to achieve considerably better and more predictable performance of Grid applications and services.

3.2 Components

The reference model is a blueprint for tool design and development. A simple tool requires at least seven components (Figure 2): a class-based application generator; a Grid scheduler; a set of models that represent the state of resources at nodes and on links; a representation of the Grid's network infrastructure; a repository for state information about the Grid topology and its resources; a set of models that generate and inject impairments; and a supervisory kernel (not shown in Figure 2).

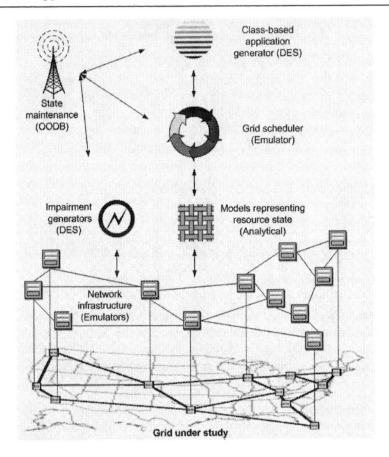

Fig. 2 Reference Model for a Simple "Grid-In-A-Lab" Tool.

3.2.1 Class-Based Application Generator

Each class of applications has its own arrival rate and interarrival distribution. Each application instance generates a workflow of tasks [22], and each task has a service time distribution (based on its class), a partial ordering (relative to sibling tasks), resource requirements, and performance requirements. Applications may also have global performance requirements (e.g., "complete all tasks by ...").

The class-based application generator is similar in many respects to job arrival models used for decades in simulations of batch computer systems, and can be implemented as a discrete event simulation model that executes in (near) real time. For example, a model might generate application arrivals based on the rate at which class r (r = 1, 2, ..., R) jobs arrive, and the burstiness of the arrival rate (e.g., estimated by the coefficient of variation). For each class r, a work flow profile is generated that includes characteristics such as the sequence of execution of different processes and input/output requests, which job tasks may be executed in parallel, the computational resource requirements per task, and file requirements for the input/output operations. A first-generation generator will be based on workflow traces from an operational research and educational Grid [159, 160, 161]. The output of the generator is passed to the *Grid scheduler*.

3.2.2   Grid Scheduler

The Grid scheduler determines in real-time where and how each application task will be executed [110, 169, 188, 218]. The scheduler has some knowledge of Grid resources – resource location, network topology, current resource utilization and job queue lengths at each node, networking performance measures such as delay and packet loss rates between nodes, etc.

The tool's scheduler can be implemented via emulation [75, 162, 179, 180, 203, 204, 213, 219]; i.e., by executing an off-the-shelf Grid scheduler on a commodity PC, or as a task on a multitasking platform in (near) real time. Being able to run a batch, application, or dynamic Grid scheduler (e.g., NQE, AppLeS, MARS) in real time is an important feature. The emulator maps resource information (type, number, location) to application requirements, and supports dynamic and transient effects such as preemption, contention, negotiation, and faults [14, 18, 21, 41, 91, 109, 155, 187, 212]. The output of this model is translated to arrival rates, routing probabilities, CPU and input/output service times, and other metrics used in various other models.

Performance measures are provided by the *resource state models* and the *network infrastructure models* (discussed below).

3.2.3   Resource State Models

A family of models is used to represent the state of resources at each node and on interconnecting networking links – e.g., queuing networks with product form solutions can be used to associate groups of similar tasks with a chain in the queuing network, and to depict the path that a task follows through the Grid's resource space. A queuing network model can also be used to depict the resources at each node, and the node's interconnecting networking links. A customer represents a particular task. The execution sequence of each task is represented in the queuing model by the routing probabilities and service times at each queuing node.

These resource state models are valuable because they provide performance estimates without the run time requirements and scalability ceiling of discrete event simulation. However, these types of analytical models are limited to certain types of flows, and do not support some events that arise in the execution of processes (e.g., fork/join operations, client/server interactions, interleaved execution of many processes onto one CPU, feedback loops).

E.g., a process may spawn a number of processes or input/output requests that are executed in parallel at different nodes, and the parent process must wait until all the child processes or requests are completed before it can continue its execution. The time it takes to execute a process is a function of the current workload and the scheduling algorithm. However, these models are able to capture essential features of computing resources (vectors, clusters, arrays of desktop machines), computational types (single, multithreaded, distributed), input/output systems (local, parallel, distributed), memory (shared, unshared, partitioned), etc. [16, 29, 39, 40, 48, 51, 61, 64, 108, 111, 128, 131, 136, 156, 165, 166, 167, 174, 175, 202].

We also use *hybrid techniques* that combine analytical models with very fast discrete event simulation. The analytical components are derived by studying sub-systems in isolation [23, 150]. Hybrid models have a dual, time-varying character:

```
hybridModel = α(t) × analyticalComponent + [1-α(t)] ×
                        simulationComponent
```

where the choice of α depends on a number of factors. Model fidelity, computational memory, and simulation run times vary inversely with α but decay at different rates. Extensions of Kalashnikov's method, a self-adapting, hybrid technique that divides the problem space into analytical and simulation components, can also be used[8].

---

[8]   These extensions consist of analytical expressions, transform-expand-sample processes as event generators, likelihood ratio method sensitivity estimates for transformation and abstraction, and piecewise linear aggregates as the formalism for the simulation component [67, 129, 137, 164].

3.2.4   Network Infrastructure Models

High capacity optical backbone networks used in Grids typically have two components: resources for data traffic (*data plane*), and a small set of resources for control traffic (*control plane*). The control plane carries messages that configure network elements, and that convey routing and network management messages. A data plane path may be configured as a provisioned circuit dedicated to a specific application, or as a tunnel that is shared by applications with the same destination, or for a short-lived' on-the-fly' data burst that may carry only a few tens of packets. Data paths can be provisioned in milliseconds in some Grids, and may have lifetimes from a few seconds to days.

Nearly all relevant information about the data plane is contained in the signaling messages that transit the control plane – e.g., source/destination addresses, routing and forwarding information, QoS requirements, link characteristics, exceptions, etc. Once a data path is provisioned (for a long-lived static connection, or a dynamically provisioned session with a lifetime of a few tens of seconds [113], or a short-lived flow or burst), then the control plane carries relatively little additional information about that data path. So one can model a high capacity optical backbone Grid network in large part simply by modeling its control plane.

Performance measures of interest that can be derived from signaling messages in the control plane include blocking and preemption on the data channels; queuing delay and delay variation (jitter), loss, and error events on the signaling channels; data channel characteristics such as transmission counts, transmission durations, transmission arrival and interarrival processes and distributions; short- and long-range dependencies in each realm; confounding effects arising from interactions between the data and signaling realms; topological information (nodes, links, faults) from forwarding and routing messages carried on the control channels; and relevant data about network impairments.

The Grid's network infrastructure can be implemented by using emulators running (unmodified) network control plane protocols to establish connections (flows, tunnels, etc.) between the nodes containing applications and resources. This allows one to represent almost any network architecture and multiplexing scheme (circuit, burst, flow, packet) by running protocol stacks on commodity devices in (near) real time. We use control plane messages to deduce the state of the Grid network infrastructure's links and devices and to compute performance measures from control traffic, and we use additional queuing network models to estimate the end-to-end delay between a source/destination pair. Emulators are also used for components related to routed traffic, congestion, latency, loss, and error using inexpensive, single-board commodity computers and computing clusters.

Measurement-based *flow methods* are used to improve run time performance and reduce memory requirements by dividing Grid flows into background and explicit components. Background flows consists of a baseline load which is static per link, while aggregate flows are defined between end-to-end pairs and based on traffic matrices or imported traffic traces (e.g., RMON-2) [124, 125]. These provide coarse estimates of mean link utilization and delay, capture important effects, and introduce fine-grained variability and stochastic behavior without the resource requirements of discrete event simulation.

3.2.5   Repository for State Information

A repository is required for the tool to maintain state information about the Grid topology and its resources. This can be implemented in any number of ways – e.g., as an object oriented database with multiple indices for fast access and update.

3.2.6   Models that Generate and Inject Impairments

Network impairments affect loss and error rates in Grids, and arise from a number of dynamic and transient phenomena [49]. We use a family of models to generate and inject impairments – e.g., transient traffic phenomena, stochastic network impairments, random and correlated loss and error events, resource failures, service faults, pre-empted tasks. Implementations can be based on flow methods that use empirical results derived from measurements on operational research, educational, or enterprise Grids.

A fast hybrid simulator can be used to generate and inject traffic from other models; to represent transient traffic behaviors; to generate and inject stochastic impairments; and to maintain other quasi-global information. Implementations are based on results from high performance networks that are used as Grid backbones; e.g., [30, 31, 32, 33, 34, 25, 26 65, 153]. Simple source models are sufficient to obtain first-order estimates for verification and testing, and to derive gross performance bounds. We shall also use measurements from an operational research and educational Grid to derive more realistic models.

3.2.7 Supervisory Kernel

A software supervisory kernel (not shown in Figure 2) is required to manage and interconnect the tool's components. All inter-component communication is managed by this kernel.

3.3 Implementation Issues

Three important implementation issues are granularity, performance measures, and validation. Each is briefly described in this Section.

3.3.1 Granularity

The tool must support various *granularities* in detail and fidelity for each functional component and hierarchical layer, and must permit analyses at different time scales and at different levels of abstraction.

One way to support different granularities in function, space, and time is to design each of the tool's components to be both configurable and viz., analytical models, discrete event simulation, hybrids with simulation and analytical models, measurement-based methods, emulation code, transfer functions, simple mathematical expressions, neural nets, empirical distributions, etc.

For example, the module that implements the Grid scheduling component of Figure 2 can be implemented in different ways – as a finite state machine, as a suite of conservation laws, or as an actual Grid multi-scheduler executing on commodity hardware. The choice depends on the analyst's requirements for model fidelity, time scale, and execution or "wall clock" time.

3.3.2 Performance Measures

Performance metrics can be application- or task-focused (e.g., response time and fairness), or infrastructure-focused (e.g., resource utilization and efficiency, reliability). A list of minimum requirements includes: *infrastructure performance* (e.g., transmission volume, transmission minimum/peak/mean throughput, transmission goodput, transmission response time, transmission latency and delay variation, transmission error rate, loss rate, reliability metrics, blocking and preemption events); *resource performance* (e.g., arrival and interarrival statistics, bottleneck identification, root cause analyses, utilization, efficiency, service time, waiting time, idle time); and *application and task performance* (e.g., time and resources required, fairness, efficiency, throughput, goodput).

3.3.3   Validation

We shall use measurements from an operational research and educational Grid to develop and validate the tool modules [159, 160, 161]. The Grid is one of a few large-scale operational research and educational computing Grids in the United States. We shall model its behavior, and compare simulated results to actual measurements from the Grid.

Laboratory experiments will provide insight about the problem space, and the tool's performance and scalability. Measurements include delay and delay variation (jitter), loss, recoverable and unrecoverable error, blocking probabilities, resource utilization, and throughput. Measurements will focus on empirical distributions and stochastic effects, including short range and long range autocorrelation. We shall capture workflows using RMON-2 traffic traces, and use them to generate estimates of mean link utilization and delay at the edge and in the core.

We shall focus on Grid edge and Grid core independently to avoid some scalability problems related to the size of the simulated Grid infrastructure and the abstraction paradigms used (e.g., packets and message flows). We shall focus on the performance of data going through the on-line measurement and scheduling at the Grid edge at the *packet* level of abstraction, since packet-level traffic characteristics are closely coupled with the performance of these components. We shall focus on the dynamic behavior of the Grid's core at the flow level since pertinent performance metrics can be captured at this level.

4.   IMPACT ON (NEAR) REAL-TIME MODELING OF GRIDS

We believe this approach to tool design has a number of advantages and novel features, and will have a significant impact on (near) real-time modeling of Grids.

4.1 Insight

We believe the tool will provide insight into enormously complex problems in Grids in (near) real time. Grids with dynamic resource provisioning are impossible to analyze by direct measurements, and discrete event simulations may require days of computational time for each minute of simulated time. The tool will be used for exploratory 'what-if' analysis and experimentation to gain insight about Grid dynamics, and about the interactions among various Grid components and functional areas.

4.2 Extended Run Times

We believe the tool will be capable of running at (nearly) the same rate as in the modeled Grid, so very long (wall clock) run times can be supported. Long run times may generate rare events not seen with short simulation times. E.g., some scheduling algorithms fragment the resource state space in the same way that multitasking operating systems fragment memory. This phenomenon may appear only after several million job tasks have completed.

4.3 High Fidelity

We believe the tool will provide high fidelity by emulating unmodified network protocols, Grid schedulers, middleware, etc. It captures the behaviors of a number of influential components – resource allocation and blocking, delay, delay variation, loss, errors, congestion, and network routing.

4.4 Reproducibility

A major advantage of any tool that operates in (near) real time is its ability to perform a sufficiently large number of replicated measures to guarantee statistical validity. The tool generates results that are 100% reproducible, which is important to analyses that focus on influential factors and parameters, the sensitivity of those factors and parameters, interactions among factors, and the confounding effects of those interactions.

4.5 Insight

We believe the tool will support all operationally deployed Grid infrastructures. The Grid topologies that can be modeled are arbitrary, and are specified by the performance analyst.

4.6 Scalability

We believe the tool will be scalable over a range of Grid topologies, diameters, sizes, and network control plane technologies (e.g., the IETF's GMPLS). It does not require tools to be modified (repartitioned or remapped) to accommodate larger problems.

The tool itself is also scalable. Small Grids can be analyzed on a multitasking Linux platform with a single CPU. Larger Grids, or Grids with more elaborate modules, may require a multiprocessor platform, or a computing cluster, or a constellation of commodity PCs sharing a local area network.

4.7 Modularity

The architecture is modular, so that different models, schedulers, protocols, etc. can be interchanged in a plug-and-play fashion. The software kernel can be expanded to include GUI and other presentation features. One can vary the tool's functional and temporal granularity and level of abstraction by changing modules.

4.8 Data-Driven

The tool supports injection of traffic, impairments, and other behaviors collected from operational Grids.

5. CONCLUSIONS

We believe the architecture described herein will provide performance analysts with analytical methods and tools capable of capturing the dynamic behavior of Grid components at various protocol layers and time scales.

The framework is the first of its kind for modeling Grids, and provides a much-needed structural component for measuring, analyzing, and predicting the performance of Grid applications and services. It addresses all of the components that are relevant to Grid performance – hardware, middleware, infrastructure, services, applications, traffic – and the interrelationships among components. It supports different degrees of detail and fidelity in each component and hierarchical layer, and analyses at different time scales.

Tool modules are configurable and 'pluggable' so that one can assess the impact of a change on the performance of the entire system. A module may consist of a mathematical expression, analytical model, finite state machine representation, transfer function, emulation code, empirical distribution, or other abstraction.

REFERENCES

[1]   Agrawal S., "NetSolve: Past, Present, and Future; A Look at a Grid Enabled Server," in Berman F. et al. (eds.), *Grid Computing: Making the Global Infrastructure a Reality*, Wiley, March 2003.

[2]   Allcock, B. et al., "Protocols and Services for Distributed Data-Intensive Science." *ACAT 2000 Proceedings*, pp. 161-163, 2000.

[3]   Allcock B. et al., "High-Performance Remote Access to Climate Simulation Data: A Challenge Problem for Data Grid Technologies." *Proc. SC 2001*, November 2001.

[4]   Allcock, B. et al., "Secure, Efficient Data Transport and Replica Management for High-Performance Data-Intensive Computing." *Proc. IEEE Mass Storage Conf.*, 2001.

[5]   Allen G. et al., "Supporting Efficient Execution in Heterogeneous Distributed Computing Environments with Cactus and Globus." *Proc. SC 2001*, November 2001.

[6]   Allen G. et al., "The Cactus Worm: Experiments with Dynamic Resource Selection and Allocation in a Grid Environment." *Intl. J. High-Performance Computing Applications*, Vol. 15(4) 2001.

[7]   Allcock, B. et al., "Data Management and Transfer in High Performance Computational Grid Environments." *Parallel Computing Journal*, Vol. 28(5), May 2002.

[8]   Allcock W. et al., "Grid Mapper: A Tool for Visualizing the Behavior of Large-Scale Distributed Systems." *Proc. 11th IEEE Intl. Symp. on High Performance Distributed Computing (HPDC-11)*, July 2002.

[9]   Allen G., "Classifying and Enabling Grid Applications," in Berman F. et al. (eds.), *Grid Computing: Making the Global Infrastructure a Reality*, Wiley, March 2003.

[10]   Allcock, W. et al., "Reliable Data Transport: A Critical Service for the Grid." *Building Service Based Grids Workshop*, Global Grid Forum 11, June 2004.

[11]   Angulo, D. et al., "Design and Evaluation of a Resource Selection Framework for Grid Applications." *Proc. IEEE Intl. Symp. on High Performance Distributed Computing (HPDC-11)*, Edinburgh, July 2002.

[12]   Angulo, D. et al., "Toward a Framework for Preparing and Executing Adaptive Grid Programs." *IPDPS*, 2002.

[13]   Atkinson M., "Towards a Grid Architecture Roadmap", in Berman F. et al. (eds.), *Grid Computing: Making the Global Infrastructure a Reality*, Wiley, March 2003.

[14] Au, P. et al., "Coordinating heterogeneous parallel computation," *Proc. EuroPar Conf.* Berlin, 1996.

[15] Baxevanidis K. et al., "Grids and Research Networks as Drivers and Enablers of Future Internet Architectures." *Computer Networks*, 2002.

[16] Baskett F. et al., "Open, closed and mixed networks of queues with different classes of customers", *J. ACM*, 22:248-260, 1975.

[17] Berman F. et al., "The GrADS Project: Software Support for High-Level Grid Application Development." *Intl. J. High-Performance Computing Applications*, 15(4), 2002.

[18] Berman, F. and R. Wolski, "The AppLeS Project: a status report." *Proc NEC Symp. On Metacomputing*, 1997.

[19] Benger W. et al., "Numerical Relativity in a Distributed Environment." *Ninth SIAM Conf. on Parallel Processing for Scientific Computing*, April 1999.

[20] Bester, J. et al., "A Data Movement and Access Service for Wide Area Computing Systems." *Sixth Workshop on I/O in Parallel and Distributed Systems*, May 1999.

[21] Berman, F., "High-Performance Schedulers", in Foster and Kesselman (eds.), *The Grid: Blueprint for a New Computing Infrastructure*, Morgan-Kaufman, San Francisco, 1999.

[22] Blythe, J. et al., "Planning for Workflow Construction and Maintenance on the Grid." *ICAPS 2003 Workshop on Planning for Web Services*, 2003.

[23] Bolla, R. et al., "Hybrid analytical/simulation mechanism for access control and on-line bandwidth optimization in diffserv environments", *Proc. PlanetIP Workshop*, Courmayer Italy, January 2002.

[24] Bourne P. and K. Baldridge, "The New Biology and the Grid," in Berman F. et al. (eds.), *Grid Computing: Making the Global Infrastructure a Reality*, Wiley, March 2003.

[25] Bragg A. "Simulations with self-similar traffic models." In Ince A. (ed.). *Modeling and Simulation Environments for Satellite and Terrestrial Communication Networks*. September 2000. Kluwer.

[26] Bragg A., et al. "Improving the performance of IP traffic over ATM networks". *Proc. World Telecommunications Congress XVII*. Birmingham UK. May 2000.

[27] Brady, M. et al., "eDiamond: a Grid-enabled federated database of annotated mammograms," in Berman F. et al. (eds.), *Grid Computing: Making the Global Infrastructure a Reality*, Wiley, March 2003.

[28]   Bragg A., H. Perros and M. Devetsiokis. "An Architectural Framework and Tool for Modeling Grids". *Proc. SCS Spring Simulation Multiconf (ATS)*. San Diego CA. April 2005.

[29]   Bruell, S., *Computational Algorithms for Closed Queuing Networks*, Elsevier, 1980.

[30]   Bragg A. and W. Chou. "Traffic analysis for high-speed networks." *Proc. IEEE Globecom '91*. Phoenix AZ. December 1991.

[31]   Bragg A. and W. Chou. "Analytic models and characteristics of video traffic in high-speed networks." *Proc. IEEE Second Intl. MASCOTS '94*. January 1994.

[32]   Bragg A. and W. Chou. "The locality and transitional behavior of ARIMA network traffic models." *Proc. IEEE Third ICCCN '94*. San Francisco CA. September 1994.

[33]   Bragg A. and W. Chou. "Real-time forecasting of bandwidth demand in high-speed communications networks." *Proc. 29th Conf. on Information Sciences and Systems (CISS '95)*. Baltimore MD. March 1995.

[34]   Bragg A. and W. Chou. "Real-time computation of empirical autocorrelation, and detection of non-stationary traffic conditions in high-speed networks." *Proc. IEEE Fourth ICCCN '95*. September 1995.

[35]   Bresnahan, J. et al., "Communication Services for Advanced Network Applications." *Proc. Intl. Conf. on Parallel and Distributed Processing Techniques and Applications 1999*, Las Vegas, June 1999.

[36]   Brown M. et al., "The Intl. Grid (iGrid): Empowering Global Research Community Networking Using High Performance Intl. Internet Services." *iGrid*, April 1999.

[37]   Buyya R. and M. Murshed, "GridSim: A Toolkit for the Modeling and Simulation of Distributed Resource Management and Scheduling for Grid Computing," *J. Concur. and Computation (CCPE)*, 14(13), 2002.

[38]   Bunn J., "Data Intensive Grids for High Energy Physics," in Berman F. et al. (eds.), *Grid Computing: Making the Global Infrastructure a Reality*, Wiley, March 2003.

[39]   Buzen, J. "Fundamental Operational Laws of Computer System Performance," *Acta Informatica* 7,2, 1976.

[40]   Buzen, J., "Operational Derivations of some Basic Results in Queuing Theory", *Proc. Int. CMG Conference*, pp. 356-361, 1982.

[41] Budenske, J. et al., "On-line use of off-line derived mappings etc." *Proc. Heterogeneous Computing Workshop*, IEEE Computer Society, 1997.

[42] Cai, M. et al., "A Peer-to-Peer Replica Location Service Based on A Distributed Hash Table." *Proc. SC2004 Conf. (SC2004)*, November 2004.

[43] Canarie, http://www.canarie.ca/canet4/ .

[44] Chervenak, A. et al., "The Data Grid: Towards an Architecture for the Distributed Management and Analysis of Large Scientific Datasets." *J. Network and Computer Applications*, 23:187-200, 2001.

[45] Chervenak, A. et al., "A Framework for Constructing Scalable Replica Location Services." *Proc. Supercomputing 2002 (SC2002)*, November 2002.

[46] Chien A., "Architecture of an Commercial Enterprise Desktop Grid: The Entropia System," in Berman F. et al. (eds.), *Grid Computing: Making the Global Infrastructure a Reality*, Wiley, March 2003.

[47] Chervenak, A. et al., "Performance and Scalability of a Replica Location Service." *Proc. Intl. IEEE Symp. on High Performance Distributed Computing (HPDC-13)*, June 2004.

[48] Chandy, K. and A. Martin, "A Characterization of Product-Form Queuing Networks", *J. ACM*, 30 n.2, p.286-299, April 1983.

[49] Chbat M. W. et al. "Toward wide-scale all-optical transparent networking: The ACTS optical pan-European network (OPEN) project." *IEEE JSAC*, 16(7):1226-1244, September 1998.

[50] Chien A., "Computing Platforms", in Foster and Kesselman (eds.), *The Grid: Blueprint for a New Computing Infrastructure*, Morgan-Kaufman, San Francisco, 1999.

[51] Courtois, P. J. *Decomposability: Queuing and Computer System Applications*, 1977.

[52] Czajkowski K. et al., "Grid Information Services for Distributed Resource Sharing." *Proc. Tenth IEEE Intl. Symp. on High-Performance Distributed Computing (HPDC-10)*, IEEE Press, August 2001.

[53] Czajkowski, A. et al., "Practical Resource Management for Grid-based Visual Exploration." *Proc. Tenth Intl. Symp. on High Performance Distributed Computing (HPDC-10)*, IEEE Press, August 2001.

[54] Czajkowski, K. et al., "A Protocol for negotiating service level agreements and coordinating resource management in distributed systems." *Lecture Notes in Computer Science*, 2537:153-183, 2002.

[55]   Czajkowski, K. et al., "Resource Co-Allocation in Computational Grids." *Proc. Eighth IEEE Intl. Symp. on High Performance Distributed Computing (HPDC-8)*, pp. 219-228, 1999.

[56]   Demaine E. et al., "Generalized Communicators in the Message Passing Interface." *IEEE Trans. Parallel and Distributed Systems*, 12(6):610-616, 2001.

[57]   Deelman, E. et al., "Mapping Abstract Complex Workflows onto Grid Environments." *J. Grid Computing*, 1(1), 25-39, 2003.

[58]   DeRoure D., "The Semantic Grid: A Future e-Science Infrastructure," in Berman F. et al. (eds.), *Grid Computing: Making the Global Infrastructure a Reality*, Wiley, March 2003.

[59]   Deelman, E. et al., "Grid-Based Metadata Services". *16th Intl. Conf. on Scientific and Statistical Database Management (SSDBM04)*, June 2004.

[60]   Deelman, E. et al., "Pegasus Mapping Scientific Workflows onto the Grid." *Across Grids Conf.*, 2004.

[61]   Denning, P., and J. Buzen, "The Operational Analysis of Queuing Network Models", *ACM Computing Surveys*, Vol. 10, No. 3, Sept. 1978, pp. 225-261.

[62]   DeRose L. et al., "An Approach to Immersive Performance Visualization of Parallel and Wide-Area Distributed Applications." *Proc. Intl. Symp. High Performance Distributed Computing (HPDC'99)*, 1999.

[63]   DeFanti T. and R. Stevens, "Tele-immersion", in Foster and Kesselman (eds.), *The Grid: Blueprint for a New Computing Infrastructure*, Morgan-Kaufman, San Francisco, 1999.

[64]   Disney R. and D. Konig, "Queuing Networks: a Survey of their Random Processes", *SIAM Review*, 27:335-403, 1985.

[65]   Dolzer K. et al. "A simulation study on traffic aggregation in multi-service networks." *Proc. IEEE Conf. High Performance Switching and Routing (ATM 2000)*. 2000. 9 pp.

[66]   Duser Michael, et al., "Analysis of a dynamically wavelength-routed optical burst switched network architecture", *IEEE, Journal of Lightwave Technology*, Vol. 20, no. 4, April 2002.

[67]   Dzemydiene H. and P. Pranevicius, "Integration of Aggregate Approach in the Multi-modal Transport Evaluation System", *Proc. Third Intl. Workshop on Databases and Information Systems*, 1998.

[68]   http://www.cordis.lu/ist/grids/towards_a_european_research_area_ for_grids.htm

[69]   http://europa.eu.int/information_society/policy/nextweb/grid/index_ en.htm

[70]   http://www.cordis.lu/ist/rn/grids.htm

[71] http://www.e-irg.org/meetings/2004-NL/EmilBroesterhuizen.html

[72] http://www.ercim.org/publication/Ercim_News/enw59/keynote.html

[73] http://www.celtic-initiative.org/Documents/CELTIC-PurpleBook-V2.pdf

[74] http://www.cordis.lu/cost/src/285_indivpage.htm

[75] Fall, K., "Network Emulation in the Vint/NS Simulator," *Proc. Fourth IEEE Symposium on Computers and Communications*, pp. 244-250, July 1999.

[76] Ferreira L., *Introduction to Grid Computing with Globus*, IBM, Oct. 2003   (http://publib-b.boulder.ibm.com/ Redbooks.nsf/RedbookAbstracts/sg246895.html).

[77] Fitzgerald S. et al., "A Directory Service for Configuring High-Performance Distributed Computations." *Proc. 6th IEEE Symp. on High-Performance Distributed Computing*, pp. 365-375, 1997.

[78] Foster I., A. Roy, V. Sander. "A Quality of Service Architecture that Combines Resource Reservation and Application Adaptation." *8th Intl. Workshop on Quality of Service*, 2000.

[79] Foster, I. et al., "The Anatomy of the Grid: Enabling Scalable Virtual Organizations." *Intl. J. Supercomputer Applications*, 15(3), 2001.

[80] Foster, I. et al., "Grid Services for Distributed System Integration." *Computer*, 35(6), 2002.

[81] Foster, I. et al., "The Physiology of the Grid: An Open Grid Services Architecture for Distributed Systems Integration." *Open Grid Service Infrastructure WG*, Global Grid Forum, June 22, 2002.

[82] Foster I. and J. Nieplocha. "Disk Resident Arrays: An Array-Oriented I/O Library for Out-of-Core Computations." *High-Performance Mass Storage and Parallel I/O*, Wiley, 2002.

[83] Foster, I. et al., "Improving Data Availability through Dynamic Model-Driven Replication in Large Peer-to-Peer Communities." *Global and Peer-to-Peer Computing Workshop*, Berlin, May 2002.

[84] Foster, I. et al., "The Physiology of the Grid: An Open Grid Services Architecture for Distributed Systems Integration." *Open Grid Service Infrastructure WG*, Global Grid Forum, June 22, 2002.

[85] Fox G., "Education and the Enterprise with the Grid," in Berman F. et al. (eds.), *Grid Computing: Making the Global Infrastructure a Reality*, Wiley, March 2003.

[86] Fox G., "Peer-to-Peer Grids," in Berman F. et al. (eds.), *Grid Computing: Making the Global Infrastructure a Reality*, Wiley, March 2003.

[87]   Foster I. et al., "The Grid 2003 Production Grid: Principles and Practice." *13th Proc. Intl. IEEE Symp. on High Performance Distributed Computing*, June 2004.

[88]   Foster, I. et al., "End-to-End Quality of Service for High-end Applications." *Computer Communications*, 27(14):1375-1388, 2004.

[89]   Foster I. et al., "Managing Multiple Communication Methods in High-Performance Networked Computing Systems." *J. Parallel and Distributed Computing*, 40:35-48, 1997.

[90]   Foster, I. et al., "Remote I/O: Fast Access to Distant Storage." *Proc. Workshop on I/O in Parallel and Distributed Systems (IOPADS)*, 1997.

[91]   Foster I. et al., "Wide-Area Implementation of the Message Passing Interface." *Parallel Computing*, 24(12):1735-1749, 1998.

[92]   Foster, I. et al., "Software infrastructure for the I_WAY metacomputing experiment." *Concurrency: Practice and Experience*. Volume 10, Number 7, June 1998.

[93]   Foster I. and N. Karonis. "A Grid-Enabled MPI: Message Passing in Heterogeneous Distributed Computing Systems." *Proc. 1998 SC Conf.*, November, 1998.

[94]   Foster I. and C. Kesselman, "Computational Grids", in Foster and Kesselman (eds.), *The Grid: Blueprint for a New Computing Infrastructure*, Morgan-Kaufman, San Francisco, 1999.

[95]   Foster I. et al., "Distance Visualization: Data Exploration on the Grid." *IEEE Computer Magazine*, 32 (12):36-43, 1999.

[96]   Foster I. "The Beta Grid: A National Infrastructure for Computer Systems Research." *Proc. 1999 NetStore Conf.*, 1999.

[97]   I. Foster et al., "A Distributed Resource Management Architecture that Supports Advance Reservations and Co-Allocation." *Intl Workshop on Quality of Service*, 1999.

[98]   von Laszewski G. and I. Foster. "Grid Infrastructure to Support Science Portals for Large Scale Instruments." *Proc. Workshop on the Distributed Computing Web (DCW)*, Rostock, June 1999.

[99]   Foster I. et al., "Grid Services for Distributed System Integration," *Computer*, June 2002.

[100] Fox G. and D. Walker, "E-Science Gap Analysis," 30 June 2003 (http://www.Grid2002.org/ukescience/ gapresources/GapAnalysis30June03.pdf).

[101] Fox, B. L., and Heine, G. W. "Monte Carlo and Quasi-Monte Carlo Methods in Scientific Computing." In *Lecture Notes in Statistics*, #106. Springer-Verlag, Berlin, 1995.

[102] Frey, J. et al., Condor-G: "A Computation Management Agent for Multi-Institutional Grids." *Proc. Tenth Intl. Symp. on High Performance Distributed Computing (HPDC-10)*, IEEE Press, August 2001.

[103] Frey, J. et al., "A Computation Management Agent for Multi-Institutional Grids." *Cluster Computing*, 5(3):237-246, 2002.

[104] Frey G., "Combinatorial Chemistry and the Grid," in Berman F. et al. (eds.), *Grid Computing: Making the Global Infrastructure a Reality*, Wiley, March 2003.

[105] Fujimoto, R., *Parallel and Distributed Simulation Systems*, Wiley Interscience, New York, 2000.

[106] Fujimoto, R., "Parallel discrete event simulation", *Communic. ACM*, 33(10): 30-53, October 1990.

[107] Gannon D. "Grid Web Services and Application Factories", in Berman F. et al. (eds.), *Grid Computing: Making the Global Infrastructure a Reality*, Wiley, March 2003.

[108] Gelenbe, E., and R. Muntz, "Probabilistic models of computer systems--Part 1 (exact results)", *Acta Informatica* 7, 35 60, 1976.

[109] Gehring J. and A. Reinefeld, "MARS – a framework for minimizing the job execution time ion a metacomputing environment." *Future Generation Computer Systems*, 12(1):87-99, 1996.

[110] Gil, Y. et al., "Artificial Intelligence and Grids Workflow Planning and Beyond." *IEEE Intelligent Systems*, January 2004.

[111] Glynn, P. W. "Stochastic Approximation for Monte Carlo Optimization." *Proc. of the Winter Simulation Conference* (Wilson, J., Henriksen, J. and Roberts, S. (eds), IEEE Press, 1986).

[112] Globus Project, "GridFTP reference," http://www.Globus.org/dataGrid/Gridftp.html .

[113] Simeonidou D. (ed.), "Optical Network Infrastructure for Grid", *Grid High Performance Networking Research Group*, September 2003, http://forge.Gridforum.org/projects/ghpn-rg/ .

[114] http://www.globus.org/research/papers.html

[115] http://grid.uchicago.edu/grappa/bib/

[116] http://tyne.dl.ac.uk/ReDRESS/WebServices/webServices_doc/node66.html

[117] Hao, Q., Tartarelli, S., and Devetsikiotis, M. "Self-Sizing and Optimization of High Speed Multiservice Networks." *Proc. IEEE GLOBECOM (2000)*, vol. 3, pp. 1818–1823.

[118] Hey T., "The Data Deluge: An e-Science Perspective," in Berman F. et al. (eds.), *Grid Computing: Making the Global Infrastructure a Reality*, Wiley, March 2003.

[119] Hingne V. et al., "Towards a Pervasive Grid", *Proc. Intl. Parallel and Distributed Processing Symposium*, April 2003, Nice France.

[120] Hoo, G. et al., "QoS as Middleware: Bandwidth Reservation System Design." *Proc. 8th IEEE Symp. on High Performance Distributed Computing*, pp. 345-346, 1999.

[121] Iamnitchi A. and I. Foster. "A Problem-Specific Fault-Tolerance Mechanism for Asynchronous, Distributed Systems." *Proc. 2000 Intl. Conf. on Parallel Processing*, 2000.

[122] Iamnitchi A. and I. Foster. "On Fully Decentralized Resource Discovery in Grid Environments." *Intl. Workshop on Grid Computing*, Denver, CO, November 2001.

[123] Iamnitchi A., M. Ripeanu and I. Foster. "Small-World File-Sharing Communities." *INFOCOM 2004*, Hong Kong, March, 2004.

[124] *ACM SIGCOMM Internet Measurement Conference 2003*, http://www.icir.org/vern/imc-2003 /program.html.

[125] *Measurement, Modeling and Analysis of the Internet*, http://www.ima.umn.edu/complex/winter /c4.html.

[126] "Grid Computing," *IT Professional Magazine*, Vol. 6 No. 6, March 2004. IEEE Computer Society.

[127] Jacob B. et al., *Enabling Applications for Grid Computing with Globus*, IBM, June 2003 (http:// publib-b.boulder.ibm.com/Redbooks.nsf/RedbookAbstracts/SG246936.html).

[128] Jackson, J., "Networks of Waiting Lines", *Oper. Res.* 5:518-521, 1957.

[129] Jelenkovic, B. and J. Melamed, "Automated TES modeling of compressed video", *Proc IEEE INFOCOM'95*, pp. 746-752, 1995.

[130] Johnston, B. "Implementing Production Grids", in Berman F. et al. (eds.), *Grid Computing: Making the Global Infrastructure a Reality*, Wiley, March 2003.

[131] Juiz C. , R. Puigjaner and H. Perros, "Performance analysis of multi-class data transfer elements in soft real-time systems using semaphore queues", *Performance '99*, August 99, Instabul, Turkey

[132] Karonis N. et al., "High-Resolution Remote Rendering of Large Datasets in a Collaborative Environment." *Future Generation of Computer Systems (FGCS)*, 2003.

[133] Keahey K. et al., "Computational Grids in Action: The National Fusion Collaboratory." *Future Generation Computer Systems*, 18:8, October 2002.

[134] Keahey K. et al., "Dynamic Creation and Management of Runtime Environments in the Grid." *Workshop on Designing and Building Grid Services*, GGF-9, October 2003.

[135] Keahey K. et al., "Grids for Experimental Science The Virtual Control Room," *Proc. Challenges of Large Applications in Distributed Environments (CLADE)*, Honolulu, June 2004.

[136] Kelly, F., *Reversibility and Stochastic Networks*, Wiley, 1979.

[137] Kovalev, M. et al., "Self-Adapting Hybrid Simulation Models of Heterogeneous Telecommunication Networks", *Information Processing*, Vol. 2, No. 2, 2002, http://www.jip.ru/2002/29.pdf .

[138] Kunszt P. and L Guy, "The Open Grid Service Architecture and Data Grids" in Berman F. et al. (eds.), *Grid Computing: Making the Global Infrastructure a Reality*, Wiley, March 2003.

[139] von Laszewski G. et al. "CoG Kits: A Bridge between Commodity Distributed Computing and High-Performance Grids." *ACM Java Grande 2000 Conf.*, San Francisco, June 2000.

[140] von Laszewski G. et al. "Grid-based Asynchronous Migration of Execution Context in Java Virtual Machines." *Proc. EuroPar 2000*, Lecture Notes in Computer Science, Munich, September 2000. Springer.

[141] von Laszewski G. et al., "InfoGram: A Grid Service that Supports Both Information Queries and Job Execution." *Proc. 11th IEEE Intl. Symp. on High-Performance Distributed Computing (HPDC-11)*, 2002.

[142] von Laszewski G., "Commodity Grid Kits - Middleware for Building Grid Computing Environments", in Berman F. et al. (eds.), *Grid Computing: Making the Global Infrastructure a Reality*, Wiley, March 2003.

[143] von Laszewski G. et al., "Real-Time Analysis, Visualization, and Steering of Microtomography Experiments at Photon Sources." *Ninth SIAM Conf. Parallel Processing Scientific Computing*, April 1999.

[144] von Laszewski G. and I. Foster. "Grid Infrastructure to Support Science Portals for Large Scale Instruments." *Proc. Workshop Distributed Computing on the Web (DCW)*, Rostock, June 1999.

[145] Leigh J. et al., "A Review of Tele-Immersive Applications in the CAVE Research Network." *Proc. IEEE Virtual Reality 2000 Intl. Conf. (VR 2000)*.

[146] Lee, J. et al., "Applied Techniques for High Bandwidth Data Transfers Across Wide Area Networks." *Proc. Intl. Conf. on Computing in High Energy and Nuclear Physics*, Beijing, September 2001.

[147] Lee B. and J. Schopf. "Run-time Prediction of Parallel Applications on Shared Environments." *Proc. Cluster 2003*, December 2003.

[148] Lee C., C. Kesselman, S. Schwab. "Near-real-time Satellite Image Processing: Metacomputing in CC++." *Computer Graphics and Applications*, 16(4):79-84, 1996.

[149] Lee, C. et al., "A Network Performance Tool for Grid Computations." *Supercomputing '99*, 1999.

[150] Lucas, M. et al., "(M,P,S)-an efficient background traffic model for wide-area network simulation," *Proc. IEEE Globecom*, Vol. 3, pp. 1572-1576, 1997.

[151] Mann V., "Discovery and Computational Steering," in Berman F. et al. (eds.), *Grid Computing: Making the Global Infrastructure a Reality*, Wiley, March 2003.

[152] Mahinthakumar G. et al., "Multivariate Geographic Clustering in a Metacomputing Environment Using Globus." *Proc. Supercomputing 99*, November 1999.

[153] Morato D. et al. "On linear prediction of Internet traffic for packet and burst switching networks." *Proc. Tenth Intl. Conf. Computer Communications and Networks*. 2001. 6 pp.

[154] Moore R. and C. Baru "Virtualization Services for Data Grids", in Berman F. et al. (eds.), *Grid Computing: Making the Global Infrastructure a Reality*, Wiley, March 2003.

[155] Moab Grid Scheduler, www.clusterresources.com/

[156] Morris, T., and Perros, H.G., "Approximate analysis of a discrete-time tandem network of cut-through queues with blocking and bursty arrivals", *Performance Evaluation* J 17 (1993) 207-223.

[157] Murshed M. and R. Buyya, "Using the GridSim Toolkit for Enabling Grid Computing Education," *Proc. CNDS 2002*, 2002.

[158] Nabrzyski, J. et al. (eds.), *Grid Resource Management*. Kluwer, Fall 2003.

[159] NC Networking Initiative, http://www.ncni.net/ .

[160] NC Research and Education Network, http://www.ncren.net/ .

[161] National LambdaRail, http://www.nationallambdarail.org/ .

[162] *Report of NSF Workshop on Network Research Testbeds*, http://www-net.cs.umass.edu/ testbed_workshop/testbed_workshop_report_final.pdf, November 2002.

[163] Pattnaik P., "Autonomic Computing and GRID," in Berman F. et al. (eds.), *Grid Computing: Making the Global Infrastructure a Reality*, Wiley, March 2003.

[164] *QNA*, http://www.columbia.edu/~ww2040/A1b.html .

[165] Ramesh S. and H. Perros "A multi-layered queuing network model of a client-server system with synchronous and asynchronous messages", *IEEE Trans. Software Engineering.* Vol 26, Nov. 2000.

[166] Ramesh S., G. Rouskas and H. Perros "Computing blocking probabilities in multi-class wavelength routing networks", *ACM Trans. On Modeling and Computer Simulation*, Vol 10, April 2000, 87-103.

[167] Ramesh S. and H. Perros, "A multi-layer client-server queuing network model with synchronous and asynchronous non-hierarchical messages", *Performance Evaluation J*, Vol. 45 (4) (2001) pp. 223-256.

[168] Ranganathan K. and I. Foster. "Design and Evaluation of Dynamic Replication Strategies for High Performance Data Grids." *Proc. Intl. Conf. Computing High Energy Nuclear Physics*, September 2001.

[169] Ranganathan K. and I. Foster. "Decoupling Computation and Data Scheduling in Distributed Data-Intensive Applications." *Proc. 11th IEEE Intl. Symp. HPDC-11*, July 2002.

[170] Ranganathan K. and I. Foster. "Identifying Dynamic Replication Strategies for High Performance Data Grids." *Proc. Intl. Workshop on Grid Computing*, Denver, CO, November 2002.

[171] Ranganathan, K. et al., "Improving Data Availability through Dynamic Model-Driven Replication in Large Peer-to-Peer Communities." *Proc. Global and Peer-to-Peer Computing Workshop*, Berlin, May 2002.

[172] Reed D. et al., "Performance Analysis of Parallel Systems: Approaches and Open Problems." *Proc. Joint Symposium on Parallel Processing (JSPP)*, June 1998.

[173] Reed D. and R. Ribler, "Performance Analysis and Visualization", in Foster and Kesselman (eds.), *The Grid: Blueprint for a New Computing Infrastructure*, Morgan-Kaufman, San Francisco, 1999.

[174] Rhee Y. and H. Perros, "Analysis of an open tandem queuing network with population constraint and constant service times", *European J. of Operations Research.* 92 (1996) 99-111.

[175] Rhee Y. and H. Perros, "An approximate analysis of an open tandem queuing network with population constraint and constant service times", *IIE Trans. 1998* Oct. 30 (10) 973-979.

[176] Ripeanu M., A. Iamnitchi and I. Foster. "Cactus Application: Performance Predictions in Grid Environments." *EuroPar 2001*, Manchester, UK, August 2001.

[177] Ripeanu M., A. Iamnitchi and I. Foster. "Performance Predictions for a Numerical Relativity Package in Grid Environments." *Intl. J. High-Performance Computing Applications*, Vol. 15 (4) 2001.

[178] Ripeanu M. and I. Foster; "A Decentralized, Adaptive, Replica Location Service." *11th IEEE Intl. Symp. on High Performance Distributed Computing (HPDC-11),* Edinburgh, July 2002.

[179] Rizzo, L., "Dummynet: a simple approach to the evaluation of network protocols", *ACM CCR,* January 1997.

[180] Rizzo, L. "Dummynet and Forward Error Correction." *Proc. USENIX Technical Conf.,* June, 1998.

[181] Russel M. et al., "The Astrophysics Simulation Collaboratory: A Science Portal Enabling Community Software Development." *Cluster Computing,* 5(3):297-304, 2002.

[182] Rubinstein, R. Y. "Sensitivity Analysis and Performance Extrapolation for Computer Simulation Models." *Oper. Res.* 37 (1989), 72–81.

[183] Sander, V. et al., "End-to-End Provision of Policy Information for Network QoS." *Proc. Tenth IEEE Symp. on High Performance Distributed Computing (HPDC-10),* IEEE Press, August 2001.

[184] Schopf J. and S. Vazhkudai. "Predicting Sporadic Grid Data Transfers." *11th IEEE Intl. Symp. on High-Performance Distributed Computing (HPDC-11),* IEEE Press, Scotland, July 2002.

[185] Shaffer E. et al., "Real-Time Immersive Performance Visualization and Steering," *ACM SIGGRAPH Computer Graphics Newsletter,* May 2000.

[186] Singh, G. et al., "A Metadata Catalog Service for Data Intensive Applications." *Proc. Supercomputing 2003 (SC2003),* November 2003.

[187] Sirbu M. and D. Marinescu. "A scheduling expert advisor for heterogeneous environments." *Proc. Heterogeneous Computing Workshop,* IEEE Computer Society, 1997.

[188] Smith W., I. Foster, V. Taylor. "Scheduling with Advanced Reservations." *Proc. IPDPS Conf.,* May 2000.

[189] Smith, P. J., Shafi, M., and Gao, H. "Quick simulation: A review of importance sampling techniques in communications systems." *IEEE JSAC* 15, 4 (May 1997), 597–613.

[190] Smith W., I. Foster, V. Taylor. "Predicting Application Run Times Using Historical Information." *Proc. IPPS/SPDP '98 Workshop on Job Scheduling Strategies for Parallel Processing,* 1998.

[191] Smith W., V. Taylor, I Foster. "Using Run-Time Predictions to Estimate Queue Wait Times and Improve Scheduler Performance." *Proc. IPPS/SPDP '99 Workshop Job Scheduling for Parallel Processing,* 1999.

[192] Snelling D. "Unicore and the Open Grid Services Architecture", in Berman F. et al. (eds.), *Grid Computing: Making the Global Infrastructure a Reality*, Wiley, March 2003.

[193] Stockinger, H. et al., "File and Object Replication in Data Grids." *Proc. Tenth Intl. Symp. on High Performance Distributed Computing (HPDC-10)*, IEEE Press, August 2001.

[194] Stockinger, H. et al., "File and Object Replication in Data Grids." *J. Cluster Computing*, 5(3)305-314, 2002.

[195] Stadler, J. S., and Roy, S. Adaptive Importance Sampling." *IEEE J. Select. Areas in Commun.* 11, 3 (Apr. 1993), 309–316.

[196] Stelling P. et al., "A Fault Detection Service for Wide Area Distributed Computations." *Proc. 7th IEEE Symp. on High Performance Distributed Computing*, pp. 268-278, 1998.

[197] Stelling P. et al., "A Fault Detection Service for Wide Area Distributed Computations." *Cluster Computing*, 2:117-128, 1999.

[198] Sulistio A., C. Yeo, and R. Buyya, "Visual Modeler for Grid Modelling and Simulation (GridSim) Toolkit," *Proc. ICCS 2003*, Springer Verlag , June 2003.

[199] Sulistio A. and R. Buyya, "A Grid Simulation Infrastructure Supporting Advance Reservation," *Proc 16th Conf. PDCS 2004*, November 2004.

[200] Sulistio A., C. Yeo, and R. Buyya, "A Taxonomy of Computer-based Simulations and its Mapping to Parallel and Distributed Systems Simulation Tools," *Intl. J. Software*, 34 (7), Wiley, June 2004.

[201] de Supinski B. and N. Karonis. "Accurately Measuring MPI Broadcasts in a Computational Grid." *Proc. 8th IEEE Symp. on High Performance Distributed Computing*, pp. 29-37, August 1999.

[202] Towsley, D., "Queuing Network Models with State-Dependent Routing", *J. ACM*, 27 n.2, p.323-337, April 1980.

[203] Vahdat, A. et al., "Scalability and accuracy in a large-scale network emulator", *Proc. 5th Symp. Operating Systems Design and Implementation*, December 2002.

[204] Vahdat, A. et al., "Modelnet distribution, http://issg.cs.duke.edu/modelnet.html .

[205] Vazhkudai, S. et al., "Predicting the Performance of Wide Area Data Transfers." *Proc. 16th Intl. Parallel and Distributed Processing Symp. (IPDPS 2002)*, April 2002.

[206] Vazhkudai S. and J. Schopf. "Using Disk Throughput Data in Predictions of End-to-End Grid Transfers." *Proc. 3rd Intl. Workshop on Grid Computing (GRID 2002)*, November 2002.

[207] Vazhkudai S. and J. Schopf. "Using Regression Techniques to Predict Large Data Transfers." *The Intl. J. High Performance Computing Applications*, Vol 17, No. 3, August 2003.

[208] Washington A.N. and Harry Perros, "Call blocking probabilities in a traffic-groomed tandem optical network", *Technical Report*, Computer Science Dept., NC State University, 2003.

[209] Watson P., "Databases and the Grid," in Berman F. et al. (eds.), *Grid Computing: Making the Global Infrastructure a Reality*, Wiley, March 2003.

[210] Watson W. "Storage Manager and File Transfer Web Services", in Berman F. et al. (eds.), *Grid Computing: Making the Global Infrastructure a Reality*, Wiley, March 2003.

[211] Washington, A. N., Hsu, C., Perros, H. G., and Devetsikiotis, M. "Approximation techniques for the analysis of large traffic-groomed tandem optical networks." *38th Annual Simulation Symposium.*

[212] Weissman J. and X. Zhao, "Runtime support for scheduling parallel applications in heterogeneous NOWS." *Proc. Sixth IEEE Symp. On High Performance Distributed Computing*, 1997.

[213] White et al., "An integrated experimental environment for distributed systems and networks", *Proc. 5th Symp. Operating Systems Design and Implementation*, 2002.

[214] Williams R., "Grids and the Virtual Observatory," in Berman F. et al. (eds.), *Grid Computing: Making the Global Infrastructure a Reality*, Wiley, March 2003.

[215] Woodside C.M., "Throughput calculation for basic stochastic rendezvous networks", *Performance Evaluation* 9 (1989) 143-160.

[216] Woodside C.M., J.E. Neilson, D.C. Petriu, S. Majumdar, "Stochastic rendezvous network model for performance of synchronous client-server like distributed software", *IEEE Trans. Computers* 44 (1995).

[217] Yang L., I. Foster, J. Schopf. "Homeostatic and Tendency-based CPU Load Predictions." *Proc. IPDPS 2003*, April 2003.

[218] Yang, L. et al., "Conservative Scheduling: Using Predicted Variance to Improve Scheduling Decisions in Dynamic Environments." *Supercomputing 2003*, November 2003.

[219] Yocum, K. et al. "Modelnet: scalability and accuracy in a large-scale network emulator". *Proc ACM SIGCOMM*. August 2002.

[220] Zhou Y., G. Rouskas and H. Perros, "A path decomposition approach for computing blocking probabilities in wavelength routing networks," *IEEE/ACM Trans. on Networking*, 8 (2000) 747-762.

[221] Zhou Y., G. Rouskas and H. Perros, "A comparison of wavelength allocation policies in wavelength routed networks", *Photonic Network Communications Journal*, Vol. 2 (2000), 265-293.

[222] Zhou Y., G. Rouskas and H. Perros, "A comparison of allocation policies in wavelength routing networks", *SPIE Terabit Optical Networking*, Vol 4123, November 6-7, 2000 Boston, pp:64-72.

[223] Zhang X., J. Freschl, and J. Schopf. "A Performance Study of Monitoring and Information Services for Distributed Systems." *Proc. HPDC*, August 2003.

[224] Zhang, H. et al., "Providing Data Transfer with QoS as Agreement-Based Service." *2004 Intl. Conf. on Services Computing (SCC 2004)*, Shanghai, September 2004.

[225] Zhang X. and J. Schopf. "Performance Analysis of the Globus Toolkit Monitoring and Discovery Service," MDS2. *Proc. 23rd Intl. Performance Computing and Communications Workshop (IPCCC)*, April 2004.

[226] Zhou Y., G. Rouskas and H. Perros, "Blocking in wavelength routing networks, Part I: the single path case", *INFOCOM '99*, pp 321-328.

[227] Zhou Y., G. Rouskas and H. Perros, "Blocking in wavelength routing networks, Part 2: mesh topologies", *ITC '99*.

CHAPTER 12

Network Simulator NS2: Shortcomings, Potential Development and Enhancement Strategies

Nino Kubinidze, Ivan Ganchev, Máirtín O'Droma

ECE Department, University of Limerick,
Limerick, Ireland

Abstract. *The basic concept and evolution of the network simulator (NS) is discussed. MPLS Network Simulator (MNS) and Wireless Mobile Networking are described with emphasis on conceptual model and architecture. Current NS compatibility issues and shortcomings are considered and a potential development plan is proposed. NS enhancement strategies in terms of enriching various modules with MPLS and HMIPv6 functionalities are proposed and novel schematic embedding algorithms are shown.*

Keywords: Network Simulator (NS), NS2, MPLS NS (MNS), Base Station (BS), MPLS-aware BS, integrated MPLS & HMIPv6 BS functionality

1. INTRODUCTIONS

Construction of real telecommunication networks for testing purposes with reasonable number of network components may require high cost and long time. Network Emulators and Simulators are inevitable in telecommunications area as they bring the opportunity to create, modify and test simulated networks, justify the cost and prove their ability to survive in real environment. The Network Simulator (NS) is a freely distributed, open-source, discreet event simulator, the main objective of which is to accommodate research in the field of networking.

Series of versions of NS have been issued, starting from the year 1995 with NS1 version v1.0a1, the year 1996, when NS2 version NS-2.0a1 was released, up to now with NS-2.29 as a pending release. According to [1] the major new elements introduced in NS2 are:

- Complex objects have been decomposed into simpler ones, for greater flexibility;
- Configuration interface is object oriented version of Tcl, (OTcl);
- Interface code to the OTcl interpreter is separated from the main simulator.

Renewed architecture of NS2 is capable to support flexibly configurable and programmable simulations of different telecommunication network scenarios, multicast protocols over fixed and mobile wireless local and wide area networks, satellite networks, as well as specific network or protocol components, such as traffic studies, unicast, multicast and hierarchical routing, queue management, packet scheduling, delays etc. New NS patches have been developed by individual researchers supporting IntServ, DiffServ, MPLS, WLAN, GPRS and UMTS network scenarios.

NS2 operates on Windows or Unix (FreeBSD, Linux, SunOS, Solaris) platforms and is written in C++ & Object Tool command language (OTcl) programming languages [1]. Two different languages are used for two different purposes: systems programming language is used for the tasks that require long run-time, namely manipulation of bytes, packet headers, algorithms that run over large data sets. For these tasks iteration time is less important i.e. to run the simulation, find bug, recompile, re-run. C++ is fast to run and slow to change, so it perfectly suits the detailed protocol implementation procedures. However researchers are mostly involved in varying parameters or configurations, creating scenarios etc., where the scripting language is more convenient to use. For these purposes iteration time is more important than the run-time-as configuration usually runs once only at the beginning of the simulation. Since OTcl can be changed very quickly, it is ideal for simulation configuration purposes.

MPLS Network Simulator (MNS) has been implemented by extending the NS. The two versions of MNS support QoS as well as basic MPLS functions, such as label switching, LDP, CR-LDP and various options of label distribution. The MPLS network design is focused on the main components, such as MPLS Classifier, Service Classifier, Admission Control, Resource Manager and Packet Scheduler. MPLS node supports label switching by means of MPLS Classifier and an LDP agent, inserted into the IP node [2, 3].

The main projects currently involved in the development of NS simulator are:

- SAMAN [10] (Simulation Augmented by Measurement and Analysis for Networks), which is mostly concerned with robustness of the protocols and operations to failure. SAMAN elaborates on extending NS simulation tools with:
  - Analytic-pre-processing of simulation scenarios;
  - Application-level traffic models;
  - Tools for measurements of traffic models [10].
- CONSER [11] (Collaborative Simulation for Education and Research), which is applying NS to support research in the NS protocol development and evaluation, and teaching concepts about network protocols.
- ACIRI [12] (i.c.s.i. centre for internet research), which pursues research on the Internet architecture and related networking issues.

2. WIRELESS & MOBILE NETWORKING IN NS

Wireless model of NS was originally ported by CMU's Monarch group's mobility extension to NS [1]. The model creates a basic mobile node, its routing mechanism and the network components for the simulation scenarios. The components include a Channel, Network Interface (NI), Radio propagation model (RPM), MAC, Interface Queue (IQ), Link Layer (LL) and Address Resolution Protocol (ARP). CMU model generates corresponding trace files, including records on node movement. Scenarios may be written to simulate wireless LAN (WLAN) or multi-hop ad-hoc networks. Functionality of a mobile node is split for C++ objects and Otcl. Mobility features like node movement, periodic position updates, topology boundary maintenance are implemented in C++. Classifiers, de-multiplexers (dmux), as well as wireless objects like LL, MAC Channel are done in Otcl.

Multiple WLANs interconnected with wired backbone use a slightly different approach than CMU's mobile node extension to NS. A base station (BS) node is created to play a role of a gateway between wired and wireless domains. BS is a hybrid between hierarchical node and a mobile node, and is created to handle packet delivery from/to wireless domains [1].

Routing scheme used in wired-cum-wireless scenarios is hierarchical. Thus each wireless domain with its base station has a unique address, and packets destined to a particular wireless terminal within a given wireless domain are received by a serving base station and re-transmitted to the destination mobile node. Hierarchical routing is used for saving memory in case of large topology based simulations. When such a topology is broken down into hierarchies, routing tables downsize from n^2 for flat routing to about $log\ n$ for hierarchical routing.

Up to three level of hierarchy can be used in NS, for which the optimum results were found. Routing protocols implemented for WLAN scenarios are DSDV and DSR.

Addressing scheme in wired-cum-wireless scenario is also hierarchical. This means that nodes have hierarchical form of address structure. Comparing to a flat routing where every node is aware of any other node in a topology (n^2 routing table size), in hierarchical scheme nodes are aware of the other nodes, which are in the same hierarchical domain (level). The packets then are forwarded from one hierarchical level to another by means of border routers (sometimes border routers are called anchor points).

Wired-cum-wireless model in NS has a Mobile IP extension, which is based on Sun Microsystem's Mobile IP model [1]. Main architectural components of Mobile IP BS node are: hierarchical classifiers, encapsulator, decapsulator, registering agent (reg_agent) and routing agent. Base station nodes in Mobile IP extension are represented by home agent (HA) and foreign agent (FA), and mobile node is called a mobile host (MH). HA and FA are to be connected to a wired domain on one side and to a wireless domain on the other side. Traffic flows are setup from a MH and a wired node.

Various tools are implemented in NS for obtaining data from the simulated networks. NAM is a Tcl/Tk based animation tool, used to visualise the simulated topology and trace packets in real time. NAM reads data from trace files, produced during the simulation process and uses it as an input to NAM console to visualise the animation of the simulation process. *x*Graph is a plotting program, which reads data from a trace file and represents it in a form of a graph. Perl and AWK[1] programming languages can be used for simulations for various purposes, such as calculating bandwidth, delay and jitter values from trace files.

3. COMPATIBILITY ISSUES AND SHORTCOMINGS OF NS

Together with an opportunity to freely download, install and simulate various types of networks and protocols NS showed a number of incompatibilities and shortcomings, which limited its usage, and/or development of its specific features and protocols for new scenarios in wired, wireless, mobile and especially future heterogeneous networks.

1. One of the major problems in NS is the **lack of documentation** on the new patches contributed by independent researchers (third parties). Thus, when simulating new scenarios, understanding and usage of new APIs, tracing various events and reading data from traced files would constitute a problem. Still, in NS Manual of March 18, 2003 [1], among the sections which are not documented are simulator basics for wireless LANs and CBQ queuing, hierarchical addressing, hierarchical routing and routing in MPLS and ad-hoc networks etc.

2. **MPLS patch is inconsistent.** First version of NS used within this research was NS-2.1b6, for which MNS version 1.0 had to be patched. MNS v1.0 runs basic LDP procedures, and protocols such as CR-LDP, IS-IS, and OSPF-TE were not implemented yet. Due to the lack of a wide range of signaling protocols, it is impossible to accomplish comparison studies of LDP, CR-LDP and RSVP-TE. MNS version 2.0 was installed as an additional patch together with MNS v1.0, in order to accomplish CR-LDP, CR-LSP setup and CoS differentiation features, however APIs of the two different patches of MPLS network simulation did not coincide.

   Version NS-2.1b7 has included in its distribution MNS v1.0, however, MNS v2.0 was not updated for the higher NS release. In NS-2.1b8, MNS v.1.0 has been included and MNS v.2.0 patch was updated. APIs for the following (higher) versions of MNS are not documented and

---

[1] AWK (Aho, Weinberger, Kernighan) is an interpreted programming language, used to extract parts from a large file and format it.

features like ER-LSP and CR-LSP setup and maintenance, re-route models due to link failures (Haskin, Makam and Simple dynamic), TCP/FTP traffic transmission as well as tracing models do not work flawlessly.

3. Various independently contributed **NS patches do not work together**:

   - *MPLS and ATM*: The latter was contributes by the third party for NS version 1 and does not co-operate with MPLS in the higher versions of NS.

   - *DiffServ and IntServ as a hybrid*: Former was contributed by Nortel Networks for NS-2.1b8a distribution and enriched later with multiple schedulers (WFQ, WF2Q+, SCFQ, SFQ, LLQ), new policy and monitoring schemes. The latter was written for NS-2.1b3 and called RSVP implementation, later updated for NS-2.1b5 and NS-2.26. DiffServ / IntServ do not work as a hybrid.

   - *MPLS with DiffServ or IntServ*: MPLS does not run when configured together with DiffServ and/or IntServ.

   - *MPLS with WLAN and/or Mobile IP*: The two latter are contributions from CMU & Sun Microsystems. The attempts to create a hybrid were demonstrated above. WLAN and Mobile IP modules operate with hierarchical addressing and routing structure. MPLS is not compatible with hierarchical routing, because MPLS Classifier is explicitly coupled with unicast Address Classifier, which does not support hierarchical routing [7]. MPLS Classifier together with Address Classifier has an efficient way of defining a route, classifying it as either an updated route, new entry or a no change.

     If being decoupled from Address Classifier a node can only tell if a route has been added or not, through the interface "add-route {}", but it won't be able to classify whether its an update, a new entry or no change case [7].

   - *GPRS and MPLS*: GPRS module was contributed for NS-2.1b7a distribution and contains Mobile Station (MS) and Base Station (BS). GPRS runs with its own specific routing protocols, and agents, which is not recognized by MPLS module.

- *UMTS and MPLS*: UMTS patch has been created within the framework of EU 5th framework project SEACORN for NS-v2.26 release, for which MPLS has not been updated. UMTS model comprises additional nodes such as Radio Network Controller (RNC), BS and User Equipment (UE). Separate UTRAN/UMTS modules are contributed for NS-2.1b9a distribution, containing UE, NodeBs and new routing module. Running routing, addressing and classifier for a joint (hybrid) model will constitute a problem.

- *MPLS and Bluetooth*: The latter was created for NS-2.1b6 release and called BlueHoc. BlueHoc supports features like terminal discovery, connection establishment and QoS negotiation, MAC scheduling schemes etc, but it is not compatible with existing wireless patches and MPLS.

- Large topology simulations (up to 100 nodes) use *large run-time memory* and require additional *work on optimising* newly implemented schemes to reduce memory usage.

- *Graphics is poor*: Current version of xGraph package (v12.1) gives very limited options for graph visualization. Third-party contributions of enhanced xGraph package are Linux-based and have difficulties with running on Windows or UNIX OSs.

4. POTENTIAL NS DEVELOPMENT PLAN

Shortcomings of NS current release and compatibility issues of its supported patches, as well as general trend towards integration of existing wired, wireless and mobile networks and protocols into one common heterogeneous networking environment, outline the need for considerable enhancements of NS simulator in terms of creating new protocols or enriching existing ones with the new features and hooks, which will contribute towards constructive development of the simulator. Newly imported protocols, their specific features and/or networking models should be fully based on their standardized protocol specifications in order to avoid misconceptions and inadequate behaviour of simulated scenarios. The following proposes potential development lines in NS from the perspective of 4G and with the focus on deploying MPLS protocol with

MPLS-specific signaling and traffic engineering (TE) solutions and IP QoS models in wireless and mobile scenarios:

1) Decoupling of MPLS Classifier with its entities (partial forwarding table (PFT), label information base (LIB)) from Address Classifier, as well as compatibility of MPLS-based network to run multiple wired and wireless routing protocols would allow MPLS module to operate as a self-contented and adaptable system.

2) Introducing MPLS & HMIPv6 functionality into a wireless Base Station (as proposed in section 5) will allow the whole system to support end-to-end (E2E) QoS and TE features.

3) Configuring MPLS internals for co-operation with hierarchical addressing, hierarchical routing and multicast routing will increase the range of possible simulation scenarios for testing MPLS features in various networking configurations.

4) Implementation and compatibility solutions for co-operation of various routing protocols, such as OSPF-TE, IS-IS, and MPLS/BGP will give a chance to model MPLS-based Virtual Private Networks (VPNs). Tests could be run for multiple link failures and fast, optimal re-route schemes.

5) Upgrade of MNS v2.0 for the current and higher NS release in terms of specifying and testing APIs for LDP, CR-LDP, LSP etc. will enable researchers to model MPLS-based networks in current NS releases and avail of the advantages of the improved simulator internals.

6) Sustainable integration of MPLS and IP QoS modules into WLAN, GPRS, UMTS specific entities typical to NS, such as Node Bs, RNCs, BSs, Mobile Stations, as well as their supporting routing modules will give a wide range of opportunities to model heterogeneous networking environment and elaborate on issues like mobility and location management, smooth handover, signaling overhead reduction, and E2E delay, jitter, packet loss and throughput issues.

7) Implementing a new wireless MPLS module as per [8, 9] with modified structure of MPLS packet header, wireless label structure

and wireless LDP may pave the way towards establishing a wireless model of MPLS in NS.

8) Implementing wireless MPLS features in ad-hoc network and in Bluetooth network entities of NS (Bluehoc) will create a chance to test MPLS features and MPLS traffic performance in new wireless environments.

9) Implementing an air interface in a user terminal with light MPLS functionality and connecting it to BS with full MPLS[2] will create an opportunity to test the implications of overloading terminal equipment with MPLS features and will contribute towards defining and supporting E2E QoS from one terminal to another within a heterogeneous networking environment.

10) Documenting the comprehensive tutorial for MPLS node internals in NS Manual will be of a great benefit.

5. NS ENHANCEMENT STRATEGIES

Heterogeneous nature of next fourth generation (4G) networks assumes wired, wireless and mobile protocols to work together in an interoperable and efficient manner. Rapid development of various networking tools, such as measuring and testing equipment, simulators and network emulators is essential for supporting and cost-effectively designing 4G networking environment.

The following subsections propose enhancements to NS simulator for the purpose of supporting simulation scenarios of heterogeneous networks. Subsection 5.1 describes a novel network simulator algorithm, suitable for NS2 [4], which represents a hybrid base station node consisting of a wireless node and MPLS node. The model will enable to simulate multiple wireless LANs connected through MPLS enabled wired domains.

Subsection 5.2 proposes more advanced hybrid node, which integrates the functionalities of two technologies (HMIPv6 and MPLS). Schematic embedding of MPLS functionality into a HMIPv6 BS is proposed for

---

[2]   The idea has been proposed during ANWIRE (EU-FP5-IST-2001-38835) summer school at IST in Lisbon, by Prof. R. Rocha.

implementation in NS2 software [5]. The structural parts of the block diagram are explained in detail.

5.1 MPLS-aware Wireless BS Node

In order to provide an interoperability and flexible QoS at the edge between wireless and wired networks, a wireless base station (BS) should be augmented with MPLS functionality and operate as an edge LSR. In this subsection we propose a flowchart of a wireless BS enhanced with MPLS (Figure 1), which can be implemented in NS2 software.

The mobile/wireless part of the diagram is adopted from NS2 [1], where it is described as follows: Link Layer (LL) object, an Address Resolution Protocol (ARP) module, an interface priority queue (IFq), a Medium Access Control (MAC) module and a network interface (NetIF) are all connected to a channel. LL object supports data link protocols and mechanisms such as packet fragmentation and reassembly, queuing, link-level retransmissions, piggybacking etc. LL object for a wireless node has an ARP module connected to it, which finds and resolves the IP address of the next –hop/node into the correct MAC address. The MAC destination address is set into the MAC header of the packet and LL places the packet into the IFq for further processing. IFq gives a priority to routing protocol packets by running a filter over the packets and removing those with a specified destination address. MAC sub-layer provides multiple functionalities such as carrier sense, collision detection and avoidance etc. The NetIF is an interface for a mobile node to access the channel. Each packet leaving the NetIF is stamped with the meta-data in its header with the information of the transmitting interface such as transmission power, wavelength etc. to be used by the propagation model of the receiving NetIF. Radio Propagation model uses Free-space attenuation ($1/r^2$) at near distances and an approximation to Two ray Ground ($1/r^4$) at far distances [1]. Mobile nodes use the unity gain omni-directional antenna.

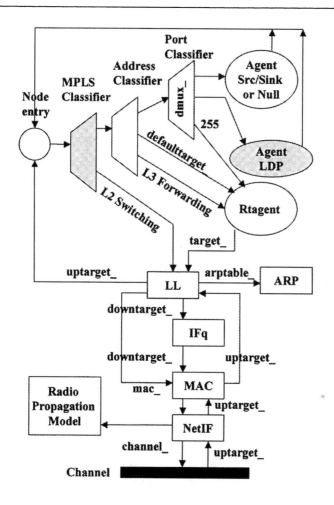

Fig. 1 MPLS-Aware Wireless BS Node

As the packets reach MPLS part of the BS they are treated according to [3], e.g. the node instance variable "node entry_" handles incoming packets and directs them to the MPLS Classifier, which sorts the packets as labeled and unlabeled ones. Labeled packets undergo label swapping and packet switching operations and MPLS Classifier executes L2 switching that sends them directly to the next node. All unlabeled packets are sent to Address Classifier, which is responsible for packet forwarding based on IP destination address and/or type of service they required. If the packets are destined to a mobile node, a "defaulttarget_" may be used in Address Classifier, to direct the packets to the appropriate mobile wireless routing agent. Port Classifier receives packets in cases where the packets destination address is the node itself, so that it could forward them to the corresponding "Agents" that deliver the packets to the receiver. All the packets destined to a mobile node or mobile ad-hoc node pass a port 255 to be attached a Destination Sequence Distance Vector routing agent. A Label Distribution Protocol (LDP) [6] Agent is responsible for receiving and handling LDP messages, as well as selecting a new node, creating a new LDP message and sending it to an Agent attached to a new node.

5.2   MPLS & Hmipv6 Functionality Embedded In BS Node

The first step towards integration and embedding of the two technologies, namely MPLS & HMIPv6, into the wireless BS is shown on Figure 2. Wireless part of the diagram is substituted by dashed lines between Link Layer (LL) object and the Channel and is explicitly described in [4]. Wired part of the diagram has been modified (hierarchical routing has been removed) and enhanced with MPLS capability. As most of the components of an MPLS node are transferred to the wireless hierarchical BS node without modification, a detailed procedure of packet processing through the wired part of the new integrated (hybrid) diagram is shown on Figure2.

Through the node instance variable "node entry_" incoming packets are directed to the MPLS Classifier, which sorts the packets as labeled and unlabeled ones. Labeled packets undergo label swapping and packet switching operations and MPLS Classifier executes L2 switching that sends them directly to the next node. MPLS Classifier of an edge node or a Mobile Anchor Point (MAP) is also responsible of attaching a label to packets that belong to particular LSP, based on the records from Label Information Base (LIB).

All unlabeled packets are sent to Address Classifier, which is responsible for packet forwarding at layer 3, based on IP destination address and/or type of service they required. An encapsulator is attached to an Address Classifier to be used in case when there is no outgoing label for a packet destined to a mobile node, which makes a packet to be processed at IP layer. In case when the CoA of a HA node does not match with the one of a packet, the latter is being encapsulated with IP-in-IP header and sent to a FA domain to be delivered to a mobile terminal (MT). This may occur in case when traffic transits an IP network, which does not support MPLS

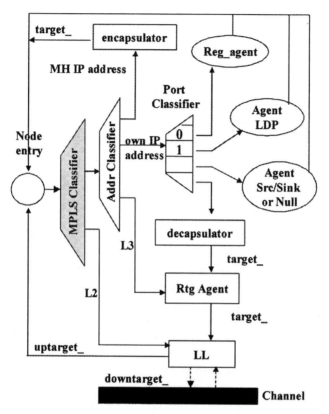

Fig. 2 MPLS & Hmipv6 Integrated Block Diagram

Port Classifier receives packets in cases when the packets destination address is the node itself, i.e. CoA of a packet matches an IP address of a node. Port Classifier is responsible to forward the packets to the corresponding "Agents" that deliver the packets to the receiver. All the packets destined to a fixed node would be forwarded to Source/Sink agent. If the packets are to be delivered to a mobile terminal and the node is a FA, a decapsulator is activated to strip off an IP-in-IP header and packets are forwarded to an appropriate routing agent. This function is used only when packets passed non-MPLS domain. Registering agent (Reg_agent) is responsible to send out a beacon signal to the nodes, set up encapsulator and decapsulator and response to solicitations from MTs. A LDP Agent is responsible for receiving and handling LDP messages, as well as selecting a new node, creating a new LDP message and sending it to an Agent attached to a new node.

6. CONCLUSION

This paper introduced basics of NS network simulator and explained its concept. NS evolution has been discussed and project involved in its extension and development listed. MNS and Wireless Mobile Networking specific internals have been described with emphasis on conceptual model and architecture of both MPLS and Wireless Networking. Current compatibility issues and shortcomings of NS, and potential development plan have been proposed.

Proposals of NS enhancement strategies have been made in terms of enriching various modules of NS simulator with MPLS functionality. Schematic algorithm for embedding of MPLS functionality into a wireless base station and its implementation in NS2 has been shown. This implementation provides a chance to create and analyse different scenarios in future heterogeneous telecommunications networks and define and measure QoS parameters, such as delay, jitter etc.

A novel algorithm for NS network simulator that involves enriching base station nodes with HMIPv6 functionality in combination with MPLS, so that the whole system is capable to support end-to-end QoS and Traffic Engineering features, has been proposed.

REFERENCES

[1] The VINT Project, "The NS Manual",at http://www.isi.edu/nsnam/NS/NS-documentation.html, 2003.
[2] G. Ahn, W. Chun, "Design and implementation of MPLS network simu-lator (MNS) supporting QoS," presented at Information Networking, 2001. Proceedings. 15th International Conference on, 2001.
[3] G. Ahn, W. Chun, "Design and implementation of MPLS network simu-lator supporting LDP and CR-LDP," presented at Networks, 2000. (ICON 2000). Proceedings. IEEE International Conference on, 2000.
[4] Kubinidze N., O'Droma M., Ganchev I., "On Deployment of MPLS Functionality in Next-Generation Wireless Networks," presented at 2nd ANWIRE Workshop on Reconfigurability, Greece, 2003.

[5] Kubinidze N., O'Droma M., Ganchev I., "Integrated algorithm of HMIPv6 and MPLS for NS to support next generation wireless networks," presented at Telecommunications Quality of Service: The busi-ness of success (QoS 2004), The IEE, March 2004.

[6] L. Anderson, P. Dolan, N.Feldman, A.Fredette, B.Thomas, "LDP Speci-fication", RFC 3036, IETF, August 2000.

[7] Haobo   Yu,   "/nfs/jade/vint/CVSROOT/NS-2/tcl/mpls/NS-mpls-node.tcl, v1.4", NS-2 distribution, November 2001.

[8] Chung, Jong-Moon, "Wireless multiprotocol label switching (WMPLS)," presented at Signals, Systems and Computers, 2001. Conference Record of the Thirty-Fifth Asilomar Conference on, 2001.

[9] Jong-Moon Chung;   Kannan Srinivasan, and Subieta Benito, "Wireless Multiprotocol Label Switching (WMPLS)", Internet Draft IETF, August 2002.

[10] SAMAN, at http://www.isi.edu/saman/index.html.

[11] CONSER, at http://www.isi.edu/conser/index.html.

[12] ACIRI, at http://www.icir.org/.

CHAPTER 13

Integrated Simulation of Communication Networks and Logistical Networks

Using Object Oriented Programming Language Features to Enhance Modelling

Markus Becker, Bernd-Ludwig Wenning, Carmelita Görg

Communication Networks, University Bremen,
Otto-Hahn-Allee - NW1, 28359 Bremen, Germany

Abstract. *Integrating simulation of communication networks with a logistical networks simulation is investigated taking advantage of object-oriented languages' features and design patterns. In this paper we present asimulator for logistic processes and the way how it is extended to enable combined simulation of logistical and communication networks. The aim of this integration is to measure the influence that an improved communication has on logistical transport processes.*

1. INTRODUCTION

The dynamic and structural complexity of logistical networks makes it very difficult to provide all information necessary for a central planning including autonomous capabilities for the decentralised coordination of autonomous logistic objects in a heterarchical structure. The autonomy of logistic objects such as cargo, transit equipment and transportation systems needs to be supported by wireless communication networks. These technologies and others permit new control strategies and autonomous decentralised control systems for logistic processes. In this setting, aspects like flexibility, adaptivity and reactivity to dynamically changing external influences while maintaining the global goals are of central interest.

A simulator was developed in the Collaborative Research Centre 637 "Autonomous Cooperating Logistic Processes - A Paradigm Shift and its Limitations" (CRC 637) (see [1]) for simulating logistic networks and their components and to investigate the autonomy of the components. This simulator however had the communication implicitly integrated and it is hidden in the algorithms. As the communication is of high importance to the autonomous logistics it needs to be studied and therefore the implicit communication needs to be made explicit and measureable.

In order to enable the measuring and enforcing the explicit communication, features of object oriented programming languages, such as access control and templates, were chosen as the simulator is implemented in C++ using the Communication Networks Class Library (CNCL, available at [2]). CNCL provides the basis for discrete event simulations. In the following sections we will discuss the advantages and problems of these methods. Furthermore we integrated several design patterns (as suggested by [3]) into the simulator and will point out what design patterns can do for a structured modeling of systems.

2. IMPLEMENTATION

2.1 Access Control

The access to member functions of classes can be restricted. The access descriptors are public, protected and private. Methods with public access descriptor can be called from other classes, methods with protected access descriptor can be called from this class and its derivatives, methods with private descriptor can be called only from this class.

In order to circumvent these restrictions for some classes they may be declared as friend. Additionally it is also possible not to declare whole classes as friend, but just make methods friend. For more information on access control, refer to [4].

2.2 Templates

Templates are an advanced concept of object-oriented programming languages for reuse of code. When the same methods are available for multiple objects, the method is a candidate for a template. In other words, when an algorithm is not only valid for one but multiple data types, it makes sense to use templates. Templates are also called type parameterization. The most well-known application of templates is the C++ standard library, formerly called standard template library (STL), see [5].

2.3 Problems with Templates

Currently there are some problems with the compilers' support for templates at least in the GNU compiler collection ([6]). The G++ versions up to 3.3 can't parse templates with a template return value. This problem is solved in version 3.4 of the G++.

But as of version 3.4 it is still not possible to put the implementation of a template function into the `.cpp` file, it has to be in the `.h` file. This leads to problems with mutual inclusion of header files. Consider the files `a.h`, `a.cpp`, `b.h` and `b.cpp`, where `a.cpp` includes `a.h` and `b.h` and `b.cpp` includes `a.h` and `b.h`. When templates are needed in both classes, as for our setup in the next section, and the template methods in both classes refer to the other class, we have mutual inclusion in the header file, which can't be resolved by current G++ compilers. This might be resolved by refactoring at least one class to two classes which breaks the mutual inclusion of the header files.

2.4 Application of Access Control and Templates

The logistical simulator mainly consisted of classes for logistical objects `Vertex` (e.g. stores, factory), `Edge` (e.g. roads, tracks), `Vehicle` (e.g. lorries, trains) and `Package` (e.g. goods). The communication between them was done by directly calling methods of the other classes, thus hiding the communication that takes place between the components of the logistical network.

The first step to explicit the communication was to integrate `Communication Units` into the logistical components. The

`Communication Unit` might be a UMTS card, WLAN adapter, GSM phone, Bluetooth Dongle or similar devices. As the logistical components could have more than one `CommunicationUnit`, we introduced a `CommunicationUnitManager`, which selects one of the `CommunicationUnits`. Furthermore to connect the `CommunicationUnits` and their Manager, we created a `MetaCommunicationUnitManager`, in which all the functionality to implement the communication is assembled.

The second step was to use the access controls of C++, as described in the earlier section. The access is restricted to methods that are providing information (implicit communication) to the `protected` scheme. This gave rise to compiler warnings, which show the places in the simulator where implicit communication is used.

Introducing template functions as `communicateWithVertexReq()` into the `CommunicationUnits`, `CommunicationUnitManager` and `MetaCommunicationUnitManager` was the third step. This function has the access rights and friends set in such a way that the logistical classes can only access the `CommunicationUnitManager`, this class again can only call the `CommunicationUnit`, which itself in turn calls the `MetaCommunicationUnitManager`. This gives us the possibility to access and restrict the Communication in all three instances. The `MetaCommunicationUnitManager` will finally go back to the `CommunicationUnitManagerRec` (receiving side) and `CommunicationUnit` of the appropriate Vertex by calling the template method `communicateWithVertexInd()`. The reason for the split-up of `CommunicationUnitManagerRec` and `CommunicationUnitManager` is that templates need to be defined in the header in combination with the mutual inclusion of `CommunicationUnitManager` and `MetaCommunication UnitManager`. By the division of the functionality of the sending and the receiving side, we can break the mutual inclusion and work around the compiler's deficiencies, as pointed out in the previous section.

In order to call a method that includes communication, the method's name and its parameters are given as parameters to `communicateWith VertexReq()`. The next section explains the way the

MetaCommunicationUnitManager will map the name of the method to the method itself.

2.5 Communication Function Database

The communication function database provides access to the functions that are implicitly containing communication for the classes that are allowed to communicate. This way we prohibit the usual way to call a function, in order to intercept the function call at a single point in the simulator for gathering statistics regarding the function calls.

The communication function database is a class called CommFunctionDB and the UML diagram is shown in Fig. 2. The class consists of two access methods: registerMethod() to register a function in the database and getMethod() in order to return a function pointer to the function. The functions are stored in a map container with the function name as the key to retrieve the function.

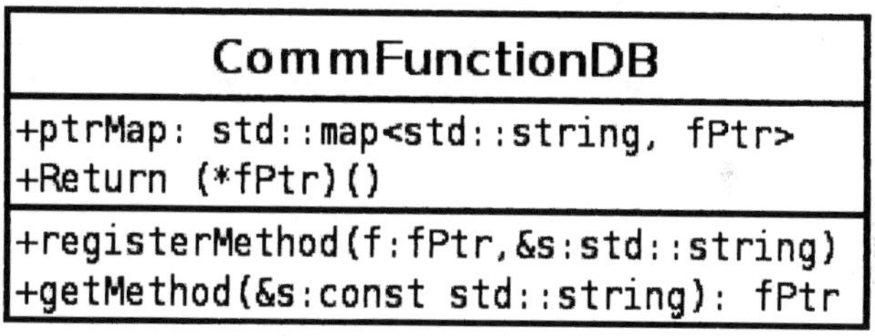

CommFunctionDB
+ptrMap: std::map<std::string, fPtr> +Return (*fPtr)()
+registerMethod(f:fPtr,&s:std::string) +getMethod(&s:const std::string): fPtr

Fig. 1 UML Diagram Of The Commfunctiondb

2.6 Design Patterns

Design Patterns are elements of reusable object oriented software. They are recurring class structures in programmer's work of years and have been ensembled by Gamma et. al. in [3].

One such design pattern is the Singleton. The Singleton ensures that a class has only one instance and provides a global point of access to it. The UML diagram is shown in Fig. 2. It is important for some classes to have only one instance. For example, in the combined simulation of logistics and communication, there should be only one transportation network. The way this is done, is to limit access to the classes constructor, but to add an access method getInstance(). This method returns the object, if the object exists, otherwise the method creates and returns the object. In simulation, this ensures that the class being a singleton cannot be created twice, thus removing a possible source of errors.

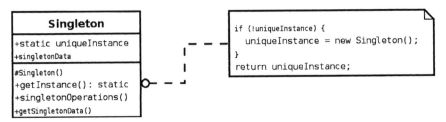

Fig. 2 UML Diagram Of The Design Pattern Singleton

In the combined simulation presented in this report, the singleton pattern is used frequently combined with standard containers, such as `list<>` or `vector<>`. The resulting pattern is called `Manager`. Depending on the data type stored in the containers, e.g. `Vehicle` or `Packet`, the `Manager` is then called `VehicleManager` or `PacketManager`. The managers provide global access to instances of different types. This way the instances are always accessible at the same location.

3. SIMULATION

The simulation, that is based on the aforementioned object-oriented features and design patterns, is carried out using a scenario, that consists of the biggest german cities, which are interconnected by the major freeways in Germany, depicted in Fig. 3. The distances between the cities are according to the real distances using the freeways. The need for transport is created in sources attached to the cities; the amount of this need is in accordance to the size of the city. The offer of transport means (i.e. Vehicles) is as well corresponding to the size of the city. The packets, created in the sources, can have different destinations, but the packets of one source have the same properties, such as e.g. size or the type of communication systems. The destinations are chosen randomly, before the simulation is run. From each city packets are created for 8 different destinations. The way that the packets are discovering their ways through the network, is based on an algorithm derived from the Ad-hoc On-Demand Distance Vector (AODV) routing algorithm [7]. The loading strategy acts in such a way that it loads as many packets as possible for one destination onto the vehicle. The destination of the vehicle is determined by the destination of the majority of packets on the vehicle.

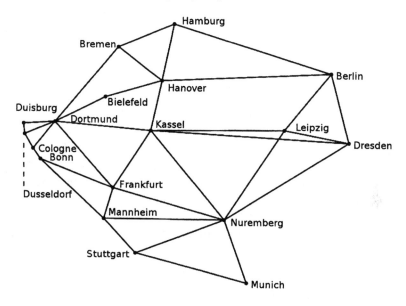

Fig. 3 Simulation Scenario (Transport Network)

Further parameters can be taken from Table 1.

Table 1 Simulation Parameter

Parameter	Value	Unit
Number of nodes	18	
Number of edges	69	
Sources per node	2	
Destinations per source	4	
Transport capacity /Transport means	60	Packets
Transport velocity	120	km / time unit
Transport creation rate / Source	17...1	Packets / time unit
Number of transport means	71	

Resulting from the modifications to the earlier version of the simulator, it is now possible to intercept and record the communication frequency and volume. For the routing algorithm mentioned earlier the following functions are called using the function database:

- getting and setting of the destination of vehicles,
- getting and setting of the destination of packets and
- routing information exchange between the cities.

Table 2 shows some of the results based on the parameters given in Table 1. At a first glance the absolute communication frequency seems rather high, but the simulated time is 30000 time units long and the number of delivered packets is high as well. This results in an average net to communication volume per communication act of about 5 byte. This communication volume does not include addresses, signalling or coding. An average communication frequency per communication unit of about 42 byte results in an average communication volume per communication unit of close to 200 byte.

Table 2 Simulation Results

Parameter	Value	Unit
Absolute communication frequency	213,352,114	
Communication volume	962,518,244	byte
Average communication volume per communication act	4.5114	byte
Total number of transported packets	~5,000,000	
Average communication volume per communication unit	192.5036	byte
Average communication frequency per communication unit	42.6704	

4. CONCLUSION

Object-oriented languages with template mechanisms such as C++ enable a way to a modular design of simulators. Using templates, design patterns and access control it was possible to enhance a simulator for logistical networks with functionality to simulate communication networks in such a way that they are of interest in logistical networks.

REFERENCES

[1] Collaborative Research Centre 637 "Autonomous Cooperating Logistics Processes: A Paradigm Shift and its Limitations", http://www.sfb637.uni-bremen.de/index.php?id=8&L=2

[2] Communication Networks Class Library (CNCL), http://www.comnets.uni-bremen.de/typo3site/index.php?id=31

[3] E. Gamma, R. Helm, R. E. Johnson, J. Vlissides, Design Patterns – Elements of Reusable Object-Oriented Software, Addison Wesley. 1995.

[4] B. Stroustroup, The C++ Programming Language (Special Edition). Addison Wesley. February 2000.

[5] N. Josuttis, The C++ Standard Library: A Tutorial and Reference. Addison-Wesley. 1999.

[6] The GNU compiler collection. http://gcc.gnu.org/

[7] C. E. Perkins, Ad Hoc Networking, Addison-Wesley 2001.

CHAPTER 14

Evaluating Vehicular Networks:
Analysis, Simulation, and Field Experiments

Richard Fujimoto [1], Hao Wu [1],
Randall Guensler [2], Michael Hunter [2]

[1] College of Computing
Georgia Institute of Technology
Atlanta, Georgia 30332-0280

[2] School of Civil and Environmental Engineering
Georgia Institute of Technology
Atlanta, Georgia 30332-0355

Abstract. *As deployment of smart vehicles becomes more widespread, the application of system evaluation methodologies to networked in-vehicle computing systems becomes increasingly more important. Such methods are essential to understand system behaviors as well as to assess alternate approaches toward realizing a rich variety of computing and information services to travelers. We describe our experiences in evaluating vehicular networks using analytic analysis, simulation, and field experiments. Each evaluation method presents its own challenges and offers different strengths and weaknesses. The context for these investigations concerns the exploration of the use of vehicle-to-vehicle and vehicle-to-infrastructure communication to disseminate information in vehicular networks.*

1.   INTRODUCTION

Advances in computing and wireless communication technologies have increased interest in "smart" vehicles – vehicles equipped with significant computing, communication and sensing capabilities to provide services to travelers. Smart vehicles are often associated with the emergence of Intelligent Transportation System (ITS) [1] with the goal of improving safety, congestion, and pollution in surface transportation systems. Smart vehicles create the opportunity to deploy a plethora of new services. Applications using these in-vehicle systems can be generally classified as either safety or non-safety related. Safety applications [18, 35] include collision avoidance, cooperative driving, etc. Non-safety applications include traveler information support [38] [29], toll service, Internet access [11], or entertainment, to mention a few. The USDOT's ITS Vehicle-Infrastructure Integration initiative (VII) is attempting to capitalize on "smart" vehicles by encouraging public-private partnerships where wireless communication devices are installed in the nation's vehicle fleet (private investment) and roadside communication infrastructure is installed along the highways, arterials, and intersections of the transportation system (public investment) [27, 28]. Such a system has the potential to vastly improve safety [18, 35], vehicle mobility, and provide new public and commercial services[11, 29, 38]. These opportunities are near at hand, with potentially 10% of the nation's vehicle fleet instrumented within two years of the commitment to deploy the system [28].

Wireless communication is obviously one of the key enabling technologies for the afore-mentioned applications. Dedicated Short Range Communications (DSRC) has been proposed to support public safety and private operations for vehicle-to-vehicle (v2v) and vehicle-to-roadside (v2r) communications. DSRC together with cellular communications can provide network connectivity to moving vehicles. There are currently multiple DSRC standards programs in progress worldwide. In North America, the Federal Communication Commission (FCC) has allocated 75MHz of spectrum at 5.9GHz for DSRC [2] based on IEEE 802.11a.

A vehicular network is a communication network organizing and connecting "smart" vehicles to each other and with mobile and fixed-location resources. A vehicular network consists of instrumented vehicles (and) network infrastructures. At a minimum an instrumented vehicle is equipped with on-board computing, wireless communication devices, and a GPS device enabling the vehicle to track its spatial and temporal trajectory. Additionally, vehicle instrumentation may include a pre-stored digital map and sensors for reporting crashes, engine operating parameters, etc. One cannot assume that every vehicle will have these capabilities; due to the gradual nature of market penetration, only a fraction of the vehicles on the road will be instrumented, at least for the next several years. The term penetration ratio is defined as the fraction of vehicles on the road that are instrumented. In the remainder of this discussion, if not specified otherwise, the term "vehicle" refers to instrumented vehicles only. Vehicular networks differ from other wireless networks (e.g., sensor networks) due to the fact that users reside within vehicles. Several wireless technologies exist for creating vehicular networks, including Wireless Wide Area Networks (WWAN), Wireless Metro Area Networks (WMAN), Wireless Local Area Networks (WLAN) using roadside base stations, and ad hoc networks using v2v communications. These technologies offer different tradeoffs in cost and performance.

There are several possible network architectures for creating vehicular networks. Three alternatives include a pure wireless v2v ad-hoc network, a wired backbone with wireless last-hop, or a hybrid architecture using v2v communication that does not rely on a fixed infrastructure, but can exploit it for improved performance and functionality when it is available.

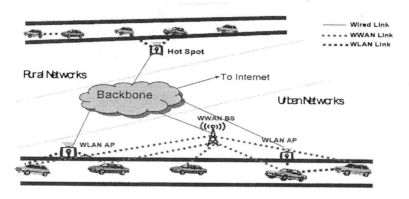

Fig. 1 A Notional Vehicular Network (Note the use of v2v and v2r communications)

Fig. *1* depicts a notional vehicular network one can easily envision in the near future. Rural and urban areas may deploy different network architectures. In urban areas wireless infrastructures such as cellular phones provide nearly ubiquitous connectivity and Wi-Fi deployments continue to become more and more widespread. V2v communications can also be used for direct inter-vehicle information exchange. In rural areas, it might be more economical to rely on v2v communications supplemented by limited infrastructures placed in certain hot spots.

System evaluation methodologies are essential to better understand vehicular networks and to design effective protocols and services. In this paper, we focus on three different methods to evaluate vehicular networks. We first begin with an overview of research in vehicular networks (Section 2). We then introduce vehicular network evaluation methodologies (Section 3). In the next 3 sections, we will describe some of our recent work in evaluating vehicular networks using analytic analysis, simulation, and measurement, respectively. We present conclusions in Section 7.

2. VEHICULAR NETWORK RESEARCH OVERVIEW

Vehicular communications are distinct from other types of wireless communication because of the high mobility of vehicles and the environment in which they operate. Studies concerning vehicular networks have largely focused on ad-hoc v2v communications. The Fleetnet project [9] and its successor Network-On-Wheels (NOW) project, and the California PATH project [4] have investigated radio devices [14], MAC protocols [36] and routing protocols [25] [13] for v2v communications.

Most safety-related services proposed so far use one hop communication. In particular, single-hop broadcast is often proposed for safety applications where the focus is on improving reception reliability [26, 35, 37]. For example, a driver may be notified of a braking vehicle upstream to provide an early warning of possibly hazardous conditions that lie ahead. On the other hand, multi-hop communications can be utilized to extend communication range [30, 33]. For example, the coverage of a roadside service station can be substantially extended through v2v communications. Multi-hop v2v communication may be a viable approach to providing communication services over a long distance, particularly where cellular access is not available, however a sufficient density of vehicles is required. Due to the high mobility inherent in v2v networks, most routing protocols proposed thus far use position-based addressing and forwarding [15], and assume end-to-end connections [25]. An adaptive information propagation scheme is introduced in [29].

A few researchers examine placing WLAN-based Internet gateways along the road [11] [20]. For example, Kutzner et al. [11] discussed how to leverage wired gateways along the road to assist routing in a hybrid structure. However no detailed implementation is given.

There has been some initial work in measuring the performance of v2v and v2r communications. Singh et al. [24] measured the performance of v2v 802.11b communications. They categorize operating environments as suburban, urban and freeway, and report the communication performance in each category. Ott and Kutscher [20] investigated the TCP/UDP performance of a car driving by a roadside WLAN access point. The above two efforts focus on single-hop communications. Möske et al. [19] demonstrated multi-hop v2v communications. However, these studies have not attempted to relate the vehicular environment to communication performance.

3. VEHICULAR NETWORKS EVALUATION METHODOLOGIES

Studying vehicular networks often requires an assessment of various network architectures, proposed protocols, and potential applications. These evaluations must take vehicle traffic conditions, driving behavior, wireless communication characteristics, and protocol/application behavior into considerations. Analytic analysis, simulation, and field experiments provide alternate means for assessing system performance.

While on-road deployments provide the most realistic results, many vehicular network experiments are too difficult and/or dangerous to perform in a "live" setting or entail a prohibitive cost, especially to perform large scale experiments. As a result, live experiments are usually conducted on a small scale. For example, Morsink et al. [18] demonstrated a co-operative collision warning and avoidance system to support longitudinal control of the vehicle using a fleet of three vehicles. Warnings based on acceleration, velocity and inter-vehicle headway data are generated to trigger both driver and automatic vehicle actions. In Section 0, we will present some of our experiments using proof-of-concept smart vehicles.

Understanding the behavior of large-scale systems requires the use of analytic analysis and/or simulation. Analytic analysis [21] [32] offers a way to reach some basic understandings of system behavior and helps to derive meaningful settings for detailed evaluation. Analytic analysis often does not require very detailed data; rather coarse-grained statistical information is often adequate, making it especially attractive where detailed information is not readily available. Rudack et al. [21] derived mathematical models to study: (1) the duration when two vehicles stay within the range of each other; and (2) the duration when one vehicle's neighbor list remains stable. However, analytic analysis normally employs some simplifying assumptions (e.g., a Poisson process) to enable closed-form solution, making it less suitable to analyze complex systems.

Simulation can provide more realistic results and enable the study of detailed behaviors that are difficult or impossible to capture with analytic models. It is necessary to simulate vehicle movement, radio signal propagation, and protocol/application behavior. Often these models already exist, but need to be integrated to model vehicular networks. One approach is to develop a monolithic simulator incorporating relevant aspects of each of these models. This approach requires a significant effort to develop, verify and validate the models. Another approach is to leverage and extend existing diverse simulators, e.g., transportation [8] and wireless communication simulators [23] [16] [5], enabling the re-use of models that have already been developed and verified. The problem then becomes one of integrating these simulators to interoperate and execute in a seamless fashion. The High Level Architecture (HLA) standard (IEEE 1516) [22] developed by the U.S. Department of Defense has defined services for creating federated distributed simulation systems. Following this approach, we have developed a simulation-based test bed by federating two independent commercial simulation packages, QualNet [23] and CORSIM [8]. This will be described later.

In the next several sections we describe recent work in evaluating vehicular networks using analytic analysis, simulation, and measurement techniques, respectively.

4. ANALYTIC ANALYSIS

A set of models to study the spatial propagation of information using v2v communications is described in [32]. We now examine the connectivity problem in a type of vehicular network. The vehicular network consists of instrumented vehicles and WLAN base stations placed along the road.

Assume the WLAN radio range is r. Specifically, we wish to compute the probability at an instant in time that a vehicle can communicate with a roadside base station using at most m hops through other instrumented vehicles. That is, we wish to compute $P_m(x)$: the probability that a vehicle with a distance x from a roadside WLAN access point can reach the access point in fewer than m hops. Intuitively $P_i(x) \geq P_j(x)$ when i>j. This probability requires certain assumptions concerning vehicle traffic conditions.

We model the road as one-dimensional since the width of the road is relatively small compared to the radio range. To simplify the problem, it is assumed that spatial positions of vehicles can be modeled as a one-dimensional Poisson point process where the instrumented vehicle traffic density as λ. A Poisson process captures the randomness of vehicle locations. Usually vehicle locations are not totally random, e.g., a vehicle must keep some distance from the vehicle in the same lane that is in front of it. When considering all the free-flow vehicles in a multi-lane road such as a freeway in a metro area, vehicle locations are usually approximated as Poisson process [21].

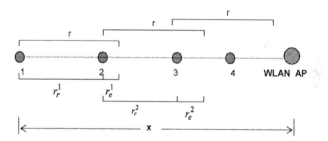

Fig. 2 The One-Dimensional Forwarding Model

The situation is illustrated in Fig. 2. We compute $P_m(x)$ for node 1, which is a distance x from the WLAN access point. We first review the analysis by Cheng and Robertazzi [6] that forms the basis for our model. In the first hop, the rightmost node that node 1 can reach using a single transmission is node 2. Without loss of generality, only the receiving station (node 2) most distant from the transmitter (node 1) will be considered. Node 2 is at a distance r_r^1 (advancement) from node 1. r_e^1 is the empty gap of the current hop, i.e.,

$$r = r_r^1 + r_e^1$$

Conditioned on the event that there is at least one vehicle within the radio range of node 1 towards the access point, the PDF of r_r^1 is

$$f(r_r^1) = \frac{\lambda e^{-\lambda(r-r_r^1)}}{1-e^{-\lambda r}} \quad 0 < r_r^1 < r \tag{1}$$

The PDF of r_e^1 is

$$f(r_e^1) = \frac{\lambda e^{-\lambda r_e^1}}{1-e^{-\lambda r}} \quad 0 < r_e^1 < r \tag{2}$$

where λ is the density of instrumented vehicles.

The next hop can be considered as originating from node 2. In general, suppose r_r^i is the advancement towards the access point in the ith hop and r_e^i is the empty gap in the ith hop, we have

$$r = r_r^i + r_e^i$$

Conditioned on the event that there is at least one receiver towards the access point in the ith hop, the PDF of r_r^i is

$$f(r_r^i) = \frac{\lambda e^{-\lambda(r-r_r^i)}}{1-e^{-\lambda(r-r_e^{i-1})}} \quad r_e^{i-1} < r_r^i < r \tag{3}$$

The PDF of r_e^i is

$$f(r_e^i) = \frac{\lambda e^{-\lambda r_e^i}}{1-e^{-\lambda(r-r_e^{i-1})}} \quad 0 < r_e^i < r - r_e^{i-1} \tag{4}$$

Next we compute $P_m(x)$.

Define $fr(\xi) = \lambda e^{-\lambda(r-\xi)}$

Clearly, for m = 1,

$$P_1(x) = \begin{cases} 1 & 0 < 0 \leq r \\ 0 & \text{else} \end{cases} \tag{5}$$

For m = 2, $P_2(x)$ is the probability that the first i (i<2) hop reaches a node that is located at a distance less than r from the access point. We have

$$P_2(x) = \begin{cases} 1 & 0 < x \leq r \\ (1-e^{-\lambda r})\int_{x-r}^r f(r_r^1)dr_1 = \int_{x-r}^r fr(r_1)dr_1 & r < x \leq 2r \\ 0 & \text{else} \end{cases} \tag{6}$$

where $(1-e^{-\lambda r})$ is the probability that the first hop exists.

For m = 3, $P_3(x)$ is the probability that the first i (i<3) hops reach a node that is located at a distance less than r from the access point. After some manipulations, we have

$$P_3(x) = \begin{cases} 1 & 0 < x \leq r \\ \int_0^r fr(r_1) * \begin{cases} 1 & x - r_1 - r \leq 0 \\ \int_{max(r-r_1, x-r_1-r)}^r fr(r_2) dr_2 & else \end{cases} dr_1 & r < x \leq 3r \\ 0 & else \end{cases} \qquad (7)$$

In general, $P_m(x)$ is the probability that the first i (i<m) hops reach a node that is located at a distance less than r from the access point. After some manipulation, we arrive at

$$P_m(x) = \begin{cases} 1 & 0 < x \leq r \\ \int_0^r fr(r_1) \cdots \begin{cases} \int_{r-r_{m-3}}^r (fr(r_{m-2}) \begin{cases} 1 & x-r-\sum_{n=1}^{m-3} r_n \leq 0 \\ \int_{max(r-r_{m-2}, x-r-\sum_{i=1}^{m-2} r_i)}^r fr(r_{m-1}) dr_{m-1} & else \end{cases}) dr_{m-2} & else \end{cases} \cdots dr_1 & r < x \leq m \cdot r \\ 0 & else \end{cases}$$

(8)

Many quantities of interest can be derived from this model. A set of usage models can be found in [31].

5. SIMULATION

We have developed a simulation-based test bed by federating two independent commercial simulation packages: a wireless network simulator QualNet [23] and a transportation simulator CORSIM [8]. These two simulators were federated using a distributed simulation software package called the Federated Simulations Development Kit (FDK) [17] developed at Georgia Tech that provides services to exchange data and synchronize computations. FDK implements services defined in the Interface Specification of the HLA. In addition to FDK, the test bed includes software called the CORSIM-QualNet Communication Layer (CQCL) that defines the interactions between CORSIM and QualNet and simplifies and streamlines the management of the distributed simulation execution. QualNet conducts a packet level telecommunication network simulation and implements the complete protocol stacks and physical environment for wireless communications. CORSIM is a microscopic traffic simulation model that simulates vehicle interaction, traffic flow, and congestion. The Run-Time Extension (RTE) facility available in CORSIM was utilized to extend the functionality necessary to operate the simulator in a distributed manner. For example, individual vehicle identification is retained when vehicles move between the freeway and arterial simulation modules. These unique vehicle IDs then flow from the traffic simulation to the communications simulation.

Instrumented vehicles in CORSIM are mapped to mobile nodes in QualNet to provide realistic mobility in the wireless network simulations. Common message formats are defined between CORSIM and QualNet for exchanging vehicle status and position information. During initialization, the transportation road network topology is transmitted to QualNet. Once the distributed simulation begins, vehicle position updates are sent to QualNet and are mapped to mobile nodes in the wireless simulation. Due to the large number of update messages, CQCL aggregates messages to reduce communication overhead. At the same time, the information arriving at mobile nodes in QualNet should also be sent to CORSIM as it may affect driving behavior, as in Ohio State University's OKI project [7], however, this has not been implemented at this time.

Extensive geometric and operational data is required to model an area in CORSIM, including traffic flows, geometric layouts, intersection signal control parameters, observed vehicle speeds, travel times, etc.   An extensive set of CORSIM models have been developed at Georgia Tech [12]. Local and state government partners provided the majority of the required operational and geometric data. A calibration effort utilizing field surveys was completed to insure that the CORSIM model provided a reasonable representation of actual operations.

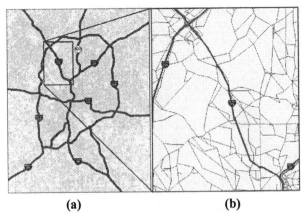

(a)   (b)

Fig. 3 (A) I-75 Corridor Study Area Location in Greater Atlanta Region (B) Corridor Study Area

The test bed has been used extensively to analyze vehicular networks [31] and associated protocols [30]. Here, we describe its use to simulate the propagation of information in a highway crash scenario. The modeled area is the I-75 corridor in the northwest quadrant of Atlanta, Georgia, traversing I-75 from the I-85 interchange to the south to the I-285 interchange to the north (Fig. 3). The modeled area incorporates approximately 7.6 miles of I-75 with a posted speed limit of 55 mph. In addition to the freeway, approximately 100 miles of arterials surface streets are included within the study area.

The scenario involves an accident in the northbound lanes of I-75 near the northern end of the modeled region during the evening rush hour. This accident results in the closure of two lanes (the second and third from the left). A message is generated by one of the vehicles involved in the crash, and the information propagated southward along I-75 using v2v communications. Due to the nature of short-range communications, the networks formed by v2v communications are usually partitioned [32]. Partitioning occurs when, under certain operating conditions, there are groups of instrumented vehicles on the roadway that are separated by more than the communications range. A communication path exists between any two vehicles within the same partition, but information cannot propagate between partitions. When partitions are present, vehicle movement must be relied upon to propagate the information between partitions [32]. This occurs when an instrumented vehicle carrying a message comes within radio range of a vehicle residing in another partition.

The message propagates a distance of 6 miles. Vehicle traffic in either direction can be exploited to relay the message. We employ an idealized data propagation scheme in QualNet. A vehicle is referred to as informed if it carries the message being propagated. When an uninformed vehicle enters the radio range of an informed vehicle, the uninformed vehicle becomes informed. Every instrumented vehicle is assumed to have the same radio transmission range (250 meters in this experiment). The message transfer time from informed to uninformed vehicles is a function of the communications system. Here, t_r is set at 4 ms. We assume a vehicle requires a specific amount of time (t_r) to receive and process a message before it is available for further retransmission. The major metrics collected are the End-to-End (E2E) delay and the number of partitions traversed by the message. The results below are the aggregation of 50 samples with different random seeds.

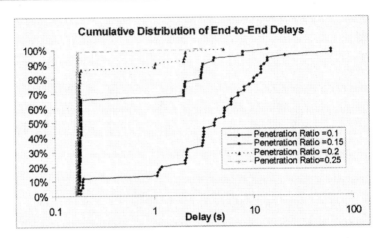

Fig. 4 Cumulative Distribution of End-to-End Delays

The average E2E propagation delays for penetration ratios of 0.1, 0.15, 0.2, 0.25 are 7.62624, 1.49832, 0.36216, and 0.1708 seconds, respectively. The shortest E2E delay for the 6 mile propagation distance was observed to be 0.164 seconds given the parameters used here where an E2E path exists. When the penetration ratio reaches 0.25, the average E2E delay is very close to this minimum and the message can quickly propagate through the designated area. Further penetration will not result in much faster propagation. When the penetration ratio is low (<0.2), the majority of the propagation time is spent moving information from one partition to another since intra-partition vehicle forwarding is relatively fast. Fig. *4* shows the cumulative distribution of E2E delays. When the penetration ratio is 0.1 or 0.15, a large variance in delay was observed. This is the result of a high probability of network partitioning. Once the penetration ratio reaches 0.2, an E2E path usually exists and thus a rapid propagation occurs most of the time. When the penetration ratio is 0.25, it is very likely that an E2E path exists.

The above results show the feasibility of using v2v communications to propagate crash information along the I-75 freeway in the Atlanta metro areas, as well as other roadway systems with similar traffic characteristics as Atlanta during peak or high traffic density periods, even with a relatively low penetration ratio (0.2). The propagation performance depends largely on the density of instrumented vehicles along the E2E path, which is a function of the traffic density and fleet penetration ratio. With a sufficient fleet penetration ratio and traffic flow rate, information can quickly propagate through the system. The propagation delay variance is often relatively large unless the penetration ratio is relatively high (>= 0.2).

6.   MEASUREMENT

We have conducted several field experiments using instrumented vehicles. Each vehicle is equipped with a laptop computer running Red Hat Linux 9, an ORINOCO 802.11b gold card with a 2.5dB omni-directional external antenna placed on the roof of the vehicle, and a Garmin 72 GPS receiver (with WAAS correction). Additional equipment will be added to the system based on specific requirements. For example, a vehicle tracking and monitoring application was demonstrated at the Digital Government Conference in 2005 by adding a digital camera to the system. These systems have served as our experimental platform to test wireless communication performance as well as proposed protocols and applications [34]. We now present measurements of wireless communication performance between a moving vehicle and a fixed roadside station in a highway environment.

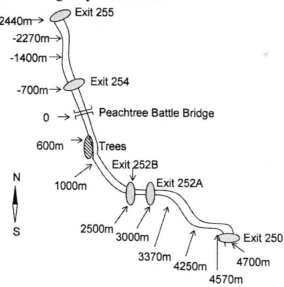

Fig. 5 Road Illustration

The experiments were conducted in the northwest sector of Atlanta, GA along I-75 between Exit 250 and Exit 255 as shown in Fig. 5. The distance shown is the direct distance to Peachtree Battle Bridge, an overpass between Exit 252B and Exit 254 of I-75. Negative distances represent the roadway to the north of Peachtree Battle Bridge, while positive distances correspond to the south. All the experiments were conducted between 2pm and 5pm under non-congested and non-inclement weather conditions. We used IPerf [3] in conjunction with GPS readings to document network performance measurements. All communications are conducted using broadcast at 2Mbps that allows us to explore basic communication performance because there is no RTS/CTS/ACK and retransmissions.

The fixed roadside station was also equipped with the same equipment as the in-vehicle system. The roadside station was placed above the median on Peachtree Battle Bridge. An instrumented vehicle traveled on the road segment shown in Fig. 5. The moving vehicle constantly broadcasted packets of 1470 bytes. The data packets were sent as rapidly as possible at an average rate of about 150 packets/s. The vehicle was driven in the rightmost lane whenever possible. We have experimented with different configurations of the roadside station: it was situated in separate experiments in either the northern (facing Exit 254) or southern (facing Exit 252B) side of the bridge, and the external antenna was either placed on a table (0.7m) or on a tripod (1.8m). Here we examine the impact of different configurations of the roadside station on communication performance. The results presented in this section are the aggregation of 5 laps. The distance is plotted the same as in Fig. 5, with the origin located at Peachtree Battle Bridge. The success ratio is defined as the fraction of packets transmitted by the vehicle that are successfully received by the roadside station. Each data point represents a distance scale of 30m.

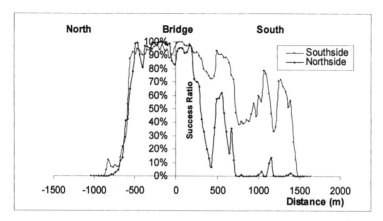

Fig. 6 The impact of the placement of the roadside station when the vehicle was driving south and the antenna was placed on a table

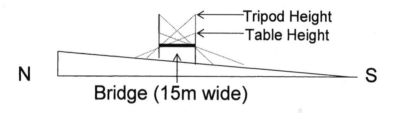

Fig. 7 Side View of the Road

We first placed the roadside station in different sides of Peachtree Battle Bridge while the antenna was placed on the table. This is to examine the effect of the bridge body (about 15m) on communication performance. As illustrated in Fig. 6, the body of the bridge has little impact on the communication performance to the north of the bridge. However, it significantly affects communication performance to the south of the bridge. This is due to the elevation of the roadway. In the area surrounding Peachtree Battle Bridge, the elevation of the road gradually declines from the north to the south as shown in Fig. 7. Therefore, when the vehicle is traveling in the north of the bridge (in either direction), the roadside station can almost always maintain line-of-sight (LOS) with the vehicle even when the roadside station is placed on the southern side of the bridge. However, when the vehicle is traveling in the south of the bridge, the bridge significantly blocks the signal when the roadside station is placed on the northern side of the bridge.

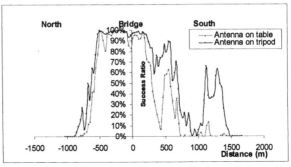

(a) roadside station in the northern side of the bridge

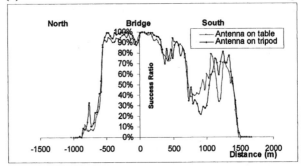

(b) roadside station in the southern side of the bridge

Fig. 8 The impact of the height of the antenna for the roadside station when the vehicle was moving south

We next experimented by raising the height of the roadside station antenna by placing it on a tripod (1.8m). As shown in Fig. 8 (a), when the roadside station is in the northern side of the bridge, raising the antenna height significantly improves the reception to the south of the bridge due to the increased area with LOS as shown in Fig. 7. However, as demonstrated in Fig. 8 (b), when the roadside station is in the southern side of the bridge, raising the antenna only has minor influence on the communication because it does not significantly improve LOS.

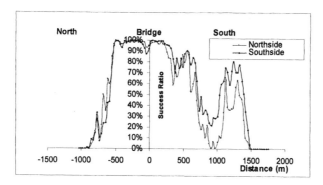

Fig. 9 The impact of the placement of the roadside station when the vehicle moved south and the antenna was on a tripod

Fig. *9* shows that the placement of the roadside station only has minor impact on communication performance once the antenna is on a tripod. Comparing Fig. *9* and Fig. *6*, it is obvious that the bridge body is no longer a significant factor once the antenna is elevated. This is easily explained by examining Fig. *7*. These results emphasize the importance of properly placing the roadside station to maximize communication performance, and specifically, the importance of maintaining a good LOS between the base station and the vehicles.

7. CONCLUSION

In this paper, we describe our practices in evaluating vehicular networks using analytic analysis, simulation, and field experiments. Each evaluation method offers certain strengths and weaknesses, and is useful for different purposes. We have illustrated the usage of each to address different problems under different assumptions. Effective system evaluation typically requires a combination of several methods. For example, analytic models can be developed to provide basic understandings and derive effective settings for detailed simulation study, as illustrated in [31].

As more and more efforts are devoted to studying the benefits brought about by smart vehicles, system evaluation methodology becomes increasingly important to help understand the system and assess proposed services. In the area of analytic analysis, we expect vehicle traffic flow theory [10] will play a more important role in the study. In simulation studies, information needs to flow from the communication simulator to the transportation simulator to study the impact of information services on driving behavior and vehicle traffic conditions. While non-safety related services can be modeled using the simulation test bed presented in Section 0, much more detailed simulations, e.g., time scales as low as milliseconds, are required for safety-related applications. More complex smart vehicles as well as roadside infrastructures need to be developed to support advanced applications. For example, it is expected that future smart vehicles will be equipped with multiple wireless interfaces (e.g., cellular, 802.11x, DSRC, etc). Mechanisms are required to effectively exploit these interfaces to improve overall communication performance while minimizing the cost.

REFERENCES

[1]   Developing an Information Technology-Oriented Basic Research Program for Surface Transportation Systems, National Science Fundation and US Department of Transportation, Chicago, IL, 2000.

[2]   Standard Specification for Telecommunications and Information Exchange Between Roadside and Vehicle Systems - 5GHz Band Dedicated Short Range Communications (DSRC) Medium Access Control (MAC) and Physical Layer (PHY) Specifications, 2003.

[3]   ALANR. Iperf, 2004.

[4]   California PATH project. http://www.path.berkeley.edu/.

[5]   Chang, X., Network simulations with OPNET. in Winter Simulation Conference, (1999).

[6]   Cheng, Y.-c. and Robertazzi, T.G. Critical Connectivity Phenomena in Multihop Radio Models. IEEE Trans. on Communications, 37. 770-777.

[7]   Dogan, A., Korkmaz, G., Liu, Y., Ozguner, F., Ozguner, U., Redmill, K., Takeshita, O. and Tokuda, K., Evaluation on Intersection Collision Warning System Using Inter-vehicle Communication Simulator. in 7th International IEEE Conference on Intelligent Transportation systems, (2004).

[8]   Federal Highway Administration. Traffic Software Integrated System User's Guide, 2003.

[9]   FleetNet. www.et2.tu-harburg.de/fleetnet.

[10]   Gerlough, D.L. and Huber, M.J. Traffic flow theory: a monograph. Washington: Transportation Research Board, National Research Council, 1975.

[11] Kutzner, K., Tchouto, J.-J., Bechler, M., Wolf, L., Bochow, B. and Luckenbach, T., Connecting Vehicle Scatternets by Internet-Connected Gateways. in MMC'2003, (2003).

[12] Lee, J., Hunter, M., Ko, J., Guensler, R. and Kim, H.K., Large-Scale Microscopic Simulation Model Development Utilizing Macroscopic Travel Demand Model Data. in 6th Transportation Speciality Conference, (Toronto, Ontario, Canada, 2005).

[13] Lochert, C., Hartenstein, H., Tian, J., Fubler, H., Hermann, B. and Mauve, M., A Routing Strategy for Vehicular Ad Hoc Networks in City Environment. in IEEE IV'2003, (2003).

[14] Lott, M., Performance of a Medium Access Scheme for Inter-vehicle. in SPECTS'02, (2002).

[15] Mauve, M., Widmer, J. and Hartenstein, H. A Survey on Position-Based Routing in Mobile Ad Hoc Networks. IEEE Network, 15 (6).

[16] McCanne, S. and Floyd, S. The {LBNL} Network Simulator, 1997.

[17] McLean, T., Fujimoto, R. and Fitzgibbons, B. Next Generation Real-Time RTI Software Workshop on Distributed Simulations and Real-Time Applications, 2001, 4-11.

[18] Morsink, P., Cseh, C., Gietelink, O. and Miglietta, M., Design of an application for communication-based longitudinal control in the CarTALK project. in IT Solutions for Safety and Security in Intelligent Transport (e-Safety), (2002).

[19] Moske, M., Fubler, H., Hartenstein, H. and Franz, W., Performance Measurements of a Vehicular Ad Hoc Network. in VTC/Spring, (2004).

[20] Ott, J. and Kutscher, D., Drive-thru Internet: IEEE 802.11b for "Automobile" Users. in INFOCOM'04, (2004).

[21] Rudack, M., Meincke, M. and Lott, M., On the Dynamics of Ad Hoc Networks for Inter Vehicle Communications (IVC). in ICWN'02, (2002).

[22] Russo, K.L., Shuette, L.C., Smith, J.E. and McGuire, M.E. Effectiveness of Various New Bandwidth Reduction Techniques in ModSAF. in Proceedings of the 13th Workshop on Standards for the Interoperability of Distributed Simulations, 1995, 587-591.

[23] Scalable Network Technologies. QualNet, 2004.

[24] Singh, J.P., Bambos, N., Srinivasan, B. and Clawin, D., Wireless LAN Performance under varied stress conditions in vehicular traffic scenarios. in IEEE VTC Fall 2002, (2002).

[25] Tian, J., Stepanov, I. and Rothermel, K. Spatial Aware Geographic Forwarding for Mobile Ad Hoc Networks, University of Stuttgart, 2002.

[26] Torrent-Moreno, M., Jiang, D. and Hartenstein, H., Broadcase Reception Rates and Effects of Priority Access in 802.11-Based Vehicular Ad-Hoc Networks. in VANET 2004, (2004).

[27] Werner, J. More Details Emerge about the VII Efforts, Newsletter of the ITS Cooperative Deployment Network, http://www.ntoctalks.com/icdn/vii_details_itsa04.html, 2004.

[28] Werner, J. USDOT Outlines the New VII Initiative at the 2004 TRB annual meeting, Newsletter of the ITS Cooperative Deployment Network, http://www.nawigts.com/icdn/vii_trb04.html, 2004.

[29] Wischhof, L., Ebner, A., Rohling, H., Lott, M. and Hafmann, R., Adaptive Broadcast for Travel and Traffic Information Distribution Based on Inter-Vehicle Communication. in IEEE IV'2003, (2003).

[30] Wu, H., Fujimoto, R., Guensler, R. and Hunter, M., MDDV: A Mobility-Centric Data Dissemination Algorithm for Vehicular Networks. in 1st ACM workshop on vehicular ad hoc networks, (2004).

[31] Wu, H., Fujimoto, R., Hunter, M. and Guensler, R., An Architecture Study of Infrastructure-base Vehicular Networks. To appear in MSWiM'05, (2005).

[32] Wu, H., Fujimoto, R. and Riley, G., Analytical Models for Information Propagation in Vehicle-to-Vehicle Networks. in IEEE VTC 2004-Fall, (2004).

[33] Wu, H., Lee, J., Hunter, M., Fujimoto, R., Guensler, R. and Ko, J., Simulated Vehicle-to-Vehicle Message Propagation Efficiency on Atlanta's I-75 Corridor. in Transportation Research Board 2005 annual meeting, (2005).

[34] Wu, H., Palekar, M., Fujimoto, R., Guensler, R., Hunter, M., Lee, J. and Ko, J., An Empirical Study of Short Range Communications for Vehicles. To appear in 2nd ACM workshop on vehicular ad hoc networks, (2005).

[35] Xu, Q., Mak, T., J.Ko and Sengupta, R., Vehicle-to-Vehicle Safety Messaging in DSRC. in ACM VANET 2004, (2004).

[36] Xu, Q., Sengupta, R. and Jiang, D., Design and Analysis of Highway Safety Communication protocol in 5.9 GHz Dedicated Short Range Communication Spectrum. in IEEE VTC'03, (2003).

[37] Yin, J., ElBatt, T., Yeung, G. and Ryu, B., Performance Evaluation of Safety Applications over DSRC Vehicular Ad Hoc Networks. in VANET 2004, (2004).

[38] Ziliaskopoulos, A.K. and Zhang, J., A Zero Public Infrastructure Vehicle Based Traffic Information System. in TRB 2003 Annual Meeting, (2003).

CHAPTER 15

A Monte Carlo Type Simulation Approach For Performance Evaluation In Optical Burst Switched Networks

Selin Parlar, Ercan Topuz

Istanbul Technical University

Abstract. *Burst switching appears to be the most promising technique for meeting transperancy and bandwidth-on-demand requirements in wavelength division multiplexed, ultra high capacity, backbone optical networks. Modeling and simulation of such networks present challenges which are rather difficult, and in extreme cases impossible to meet using analytical or discrete event approaches. This is due to the absence of sufficiently accurate models for the former, and due to the required exessively large computational resources for the latter approach. In this paper a different modeling approach is presented, which under the assumptions of stationarity of distributions allows for implementation of Monte Carlo sampling for calculating overall burst loss probability performance of JIT protocol in optical burst switched networks. Simulation results are presented for NSFNet and Pan-European Network, which demonstrate the applicability of the proposed approach.*

1. INTRODUCTION

The growth of existing services and the introduction of new ones are creating a constant demand for increasing the traffic flow in telecommunication networks. This increase is relatively slow for some services, but rather quick for others like data and video traffic via Internet. The range of services is very diverse in terms of channel capacity and channel occupancy requirements. The use of optical fiber channel has had a great impact on terrestrial telecommunication networks by helping to remove the transmission bandwidth bottleneck, and developments in hardware and software have resulted in the introduction of intelligence within networks.

Over the last two decades the telecommunication infrastructure has been evolving toward increasingly heterogeneous but interconnected ultra high speed backbone optical networks. The speeds of state-of-the-art electronic multiplexers have reached 40 Gbps levels which are still three orders of magnitude less than the bandwidth capability provided by the fiber. Optical Networks (ON) takes advantage of Wavelength Division Multiplexing (WDM) in order to better exploit the transmission bandwidth capacity of the fiber channel. Current interest is focused on All Optical Networks (AON) wherin a lightpath is established between ingress and egress nodes to transmit data in the optical domain without undergoing in Optical-Electrical-Optical (O-E-O) conversions at the routers and cross-connects on intermediate nodes. Deployed versions of all optical WDM networks utilise Optical Circuit Switching (OCS), or λ-switching technique for establishing dedicated, longlived lightpaths between source and destination nodes. The main drawback of OCS is that the provided bandwidth is usually much greater than that required to accommodate the average traffic loads between the source and destination nodes, so that the capacity provisioning in the network may be rather inefficient. This weakness of λ-switching may be overcome by utilising packet level switching techniques in the optical domain. However, related technologies are still at their infancy and the field level implementation of Optical Packet Switching (OPS) is not expected to take place in the near future. Optical Burst Switching (OBS) has been proposed as an intermediate solution [1-4] for making better use of network resources when compared to OCS.

2.   OPTICAL BURST SWITCHED NETWORKS

The architecture of an OBS network consist of three main segments: access layer, assignment layer, and optical layer. Access layer is an intermediate medium where IP packets from different users are transmitted to ingress nodes. In assignment layer, packets arriving to the ingress nodes are collected and assembled into bursts. Timer-based, burstlength-based, and mixed timer/threshold-based burst assembly algorithms have been proposed for determining both a maximum burst length and the end of the aggregation process.

OBS provides an AON for the data plane. This is achieved by reserving required network resources via the burst header, referred to as the control packet (CP), that is usually sent over a dedicated λ-channel of each fiber, undergoes O-E-O conversion and processed electronically at each node on route. The CP of each burst carries information such as the offset value, priority, and the path determined by the routing algorithm. The CP precedes the burst by some offset time which is determined so as to compensate for the processing delays plus the switch setup times at each intermediate node on the path. Offset time allows for switch configurations and routing decisions to be made prior to the arrival of the data burst so that an attempt can be made to temporarily establish a lightpath between ingress and egress points for the transmission of the burst. In OBS, the bandwidth for a data burst is reserved in a one way process without waiting for confirmation of an end-to-end reservation. It may therefore happen that a reservation request for some output port may arrive at some intermediate nodal switch whilst this port is still engaged due to an earlier reservation. At such instances, depending on the hardware/software capabilities built in the core nodes, one or more additional attempts may be made using contention resolution techniques such as burst segmentation, alternate routing, wavelength converting, and buffering, and the burst is dropped in case these fail to satisfy the reservation request. If the burst arrives successfully to the egress node, it is decomposed into packets, and the packets are delivered to users.

For terabit burst switching, the following three main protocols have been proposed: Just-Enough-Time (JET) [1], Horizon [2] and Just-In-Time (JIT) [3]. In this paper, we will consider the simulation of an OBS network utilising a variation of JIT protocol developed by the JumpStart Group in the U.S.A. [3]. This is so far the only protocol for which a successful implementation has been demonstrated, albeit on a testbed experiment scale, utilising off-the-shelf hardware [5]. In JIT protocol, immediate reservation (IR) scheme is used wherein the wavelength reservation starts immediately after the processing of the CP finishes. IR scheme can be implemented with low complexity since it does not require a global time synchronisation in the network. This feature of the JIT protocol is essential for the applicability of the Monte Carlo technique for modeling and simulation of OBS networks.

A salient feature of OBS networks is that a clear distinction can be made between network's data and control planes. Data plane is an ultra high speed AON which provides fully transparent data channels between source and destination node pairs. Whereas, control plane which, apart from some network management functions, is responsible for provision of transperancy in the data plane, generally involves one conventional OCS channel per fiber operating at much lower speeds. The intelligence built into the network resides in the control plane. A snapshot of the control plane would reveal all relevant information about the current state of the network's data plane, including those concerning routing and wavelength assignment,burst collision, contention resolution, burst dropping processes. It should therefore, be possible to consider the control plane for modeling burst-level end-to-end performance of an ultra high capacity OBS network, without having to address the formidable numerical task of having to simulate the traffic flow in the data plane. Guided by this observation, we introduced [6,7] a Monte-Carlo sampling approach for modeling the control plane to estimate steady-state Burst Drop Probability (BDP) in OBS networks under JIT protocol.

3.   MODEL DESCRIPTION

We consider a bidirectional mesh network comprised of N nodes connected with L links providing up to W concurrent WDM λ-channels in the data plane in either direction. All nodes are provided with add-drop and full wavelength conversion capabilities. We consider Jumpstart version of JIT protocol which does not require a global time syncronisation within the network and assume that the network has reached steady state conditions and that all processes have stationary distributions. These assumptions prepare the setting for the Monte Carlo sampling approach proposed in this paper wherein the time is treated as a parameter which can be "set" locally at each node in the manner to be described. On the other hand, the additional assumptions listed below which do not restrict the applicability of our approach are introduced to the model only for purposes of convenience and can easily be removed in a more or less straightforward manner:

- Least cost, fixed routing and random wavelength assignment schemes are used.
- Nodes are provided with sufficiently large electronic buffering capabilities. Length-based burst aggregation algorithm is utilised and single priority class, single length bursts are considered in Monte Carlo batch runs.

- Burst arrivals are exponentially distributed, traffic matrix is symmetrical and uniform.
- Nodes are not equipped with optical buffers.

Our model may be divided into two subparts: burst injection, and contention resolution / burst dropping. Burst injection is governed by the current traffic offered to the network as described by the burst arrival process and the internodal weights, as well as by the λ assignment scheme and the protocol. Burst dropping algorithm, on the other hand, involves identification of potential burst collision events on the intermediate nodes and determination whether or not they can be avoided utilizing provided contention resolution techniques.

In modeling the burst injection process for the on-the-fly unicast signaling and bandwidth reservation policy of Jumpstart JIT protocol [3], we consider instantaneous observations made at the inputs of the links connected to the edge nodes of the network at sampling time, T_s. Let us denote the mutually exclusive events of the presence and absence of an injected burst at the input of the first link on the w-th λ-channel of the route connecting a source-destination pair by $b_{i,j,w}(T_s)$ where i denotes the source node and j a node directly connected to i. When we assign values "1" and "0" to these events, respectively, then the total number of accepted bursts, $N_A(T_s)$, that the network is engaged in transmitting at the sampling time T_s is obtained simply by summing b over all indices. A λ-channel and two "time" windows are attached to each injected burst: T_{idle} and T_{busy} which are, respectively, determined via the burst arrival statistics and by summing up burst length with the route dependent offset time of its CP. Channel occupation ratio U, will be defined as,

$$U = 1 / (1 + T_{idle}/T_{busy}) \qquad (1)$$

Modeling of burst dropping process involves not only the observation instants but certain, locally synchronized time window around them, and the probabilities of burst collision events need to be supplied externally. To achieve this each injected burst is associated with a state vector containing data on its status (transmitted, dropped etc.), source/destination, Route-Wavelength Assignment (RWA) info (list of links and the assigned λ-channels on each link) and on the time windows: T_{busy}, T_{idle}.. T_{idle} is updated at each node on the route to account for the shortening of the offset time. Burst Vectors (BVs) are examined for identification of possible burst collision events. A burst offered to a route uses the same λ-channel in each link until it exits the network or until the first nodal switch where on the same link the same λ-channel is assigned to two (or more) bursts and hence, a potential burst collision event is identified. In the latter case, the respective bursts are shifted to a common local time reference (T_{ref}) generated according to uniform distribution. In case there is no overlapping between their busy periods, the bursts will not collide. In this case the idle period of the "preceeding" burst is updated to account for the increased traffic load on the channel. We will refer to these events as "burst interposition". If, on the other hand, there is an overlap between their busy periods, then it is decided that the "following" burst is being blocked, and an unused λ-channel is assigned to this burst if available, and its state vector is updated accordingly; otherwise, it is dropped.

Monte Carlo Simulations (MCSs) attempt to estimate a performance parameter of a system model via running a series of random experiments [8]. In the present case, the random experiment is defined as an attempt to transmit the set of injected bursts $N_A(T_s)$ over the OBS network. The outcomes of this experiment can be either "the burst is transmitted" or "the burst is dropped". A MC run of size N_s corresponds to a repetition of the experiment N_s times. Then counting the total number of bursts that are accepted by the network (N_A) and the ones that are dropped on route (N_D), an estimator for BDP can then be obtained, ($\hat{BDP}$), as,

$$\hat{BDP} = \frac{N_D}{N_A} \tag{2}$$

By repeating MCS of the same size (N_s) one can calculate the standard deviation (σ) of the estimator which provides a measure for the level of confidence that can be attached to the calculated result.

4.   NETWORK LEVEL BDP PERFORMANCE

There are very few reports in the literature on performance calculations for OBS at the network level [9]. Considering BDP performance calculations of ultra high capacity backbone OBS networks, one is faced with two major difficulties. Firstly, the aggregate traffic volumes on such networks are several orders of magnitude larger than those that can be processed with Discrete Event (DE) network simulation tools [10]. As a result of this, only simplified scenarios have been considered in DE OBS network simulation examples reported in the literature [11,12]. In this work we demonstrate that utilizing MC techniques the computational burden can be considerably reduced under certain constraints applicable to some realistic OBS networks. The second difficulty is related to the need for determining a measure for the overall "burst traffic load" of a network. In the literature netwok load is usually specified in Erlangs per wavelength [9]. In this paper we will choose to follow a different approach in order to better represent the quasi-quantized nature of traffic flow in OBS networks. For given topology, burst traffic, and OBS protocol one would expect BDP to be displayed against "load", which is expressed in such a way that similar load conditions would yield similar values for BDP in different networks. This is a rather complicated problem which, to the best of our knowledge, has not been addressed in the literature. However, in the case of OCS networks it is possible to estimate the Maximum Number of Connections (MNC) that can be provided by the network for given topology and traffic matrix [13,14],

$$MNC \cong L \ W \ / <nl>_{min} \qquad (3)$$

where it is assumed that all links (L) support W λ-channels in both directions and $<nl>_{min}$,which needs to be determined via independent MCS, denotes the minimum of the average hops in establishing connections in an empty network using a fixed, minimum-hop routing scheme. Although MNC may serve as a convenient measure for assessing connection blocking probability performance in circuit switched networks, attempts for extending this approach to OBS networks are bound to fail, since following this course it is not possible to account for the confounding effects of burst collision events and of the contention resolution techniques used in the network. Under the above outlined simplifying assumptions used in this paper we have shown [14] that the average value of channel occupation ratio U defined in (1) can be used as a first-order measure for representing the effects of the above factors. Based on the above observations, we have chosen to express the overall BDP in OBS networks in parametric form as functions of both the average number of accepted bursts (N_A) and the average value of the channel occupation ratio (U).

5.   SIMULATION RESULTS

To demonstrate the applicability of the proposed MCS approach for simulating overall BDP in realistic networks we have considered two scenarios: NSF Net (NSFN) and Pan-European Net (PEN). The topologies of NSFN [9] and PEN [15] used in the simulations are shown in Fig. 1 and 2. These networks have similar characteristics: NSFN: $N = 16$, $L = 25$, $W =$

64; PEN: $N = 15$, $L = 24$, $W = 64$, where N, L and W refer to the number of nodes, links and λ-channels, respectively.

In the simulation experiments the average value of the number of accepted bursts (N_A) is varied between 350 and 700, and the values of 0.05, 0.1, 0.32 and 0.95 were considered for the channel occupation ratio U. As noted above, we assume that bursts injected from a node are equally likely to be destined to any of the other nodes, and that a fixed, shortest path routing scheme is used. The present version of our code is implemented in Matlab for quick prototyping purposes. The calculations were performed on standard PC platforms (P4, 2.6 GHz). The required computer time ranged from about 12 hours for $N_A \cong 500$ to about 24 hour for $N_A \cong 700$. Due to excessive computational times required, the simulations were limited to a total number of 50 000 bursts, which corresponds to an MC run of size $N_s = 100$ for $N_A \cong 500$. It should be noted that due to the above mentioned limitation our simulations have not reached convergence, as can be inferred from the typical example depicted in Table 1. Therefore, we could not give confidence limits for computed BDP depicted in Fig.3 and 4.

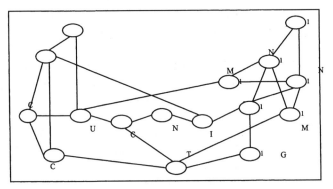

Fig. 1 16 Node NSF Net (NSFN) Topology

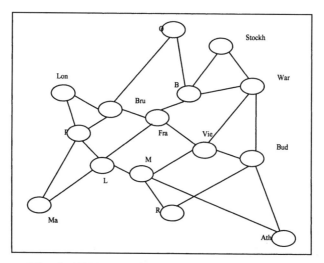

Fig. 2 15 Node Pan-European Network (PEN)

The results we obtained for the BDP performance of NSFN [7] and PEN are given in Fig.3 and Fig.4. From these figures it can be seen that BDP increases with increasing traffic volume parametrized with N_A and U.

Increasing N_A for constant U results in more λ-channel reservations requests on the links; hence, in greater numbers of potential burst collision events. On the other hand, increasing U for constant N_A leads to increase in burst collision probability, and hence to more λ-conversion attempts.

Table 1. Convergence Test for MCS

NSFNet, $N_A \cong 500$			
N_s	Total Bursts	$< B\hat{D}P >$	σ
20	10 000	0.34	0.09
40	20 000	0.25	0.09
60	30 000	0.25	0.08
80	40 000	0.29	0.06
100	50 000	0.25	0.05

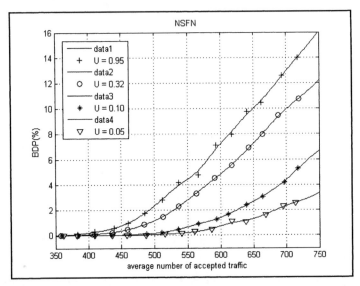

Fig. 3 $_{<B\hat{D}P>}$ **vs. <N_A> for Various Values of U for NSFN.**

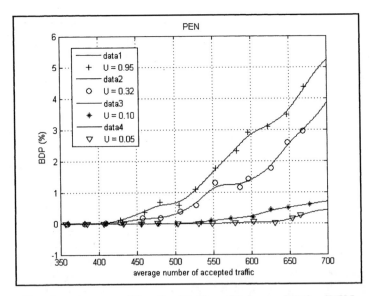

Fig. 4 $_{<B\hat{D}P>}$ **vs. <N_A> for Various Values of U for PEN.**

Table 2. Average Number of Directly Transmitted,Interposed and λ-Converted Bursts in NSFN and PEN.

$N_A \cong 500$	$U = 0.95$		$U = 0.30$		$U = 0.10$	
	NSFN	*PEN*	*NSFN*	*PEN*	*NSFN*	*PEN*
Directly transmitted	263	276	264	278	265	272
interposed only	0	0	21	26	57	71
λ-converted only	237	225	158	149	87	74
interposed and λ-converted	0	0	63	52	100	84
$< B\hat{D}P >$ (%)	2.40	0.75	1.20	0.37	0.20	0.01

This is better demonstrated by the sample output given in Table 2 which, for $N_A \cong 500$ displays the average number of directly transmitted, only interposed, interposed and λ-converted and only λ-converted bursts in both topologies, for various values of U. Directly transmitted bursts are the ones which do not interfere with other bursts in transit and are delivered to their destinations using the wavelength channel assigned to them at their source nodes. The class of bursts designated as "only interposed" is very similar to the class of directly transmitted bursts, in that they are also transmitted along their routes utilizing the λ-channel assigned to them at the outset. The only differences between the two classes is that the bursts belonging to the latter class successfully pass from at least one burst collision identification test, and hence result in a modification of burst collision probability on this λ-channel over some links. The remaining two classes involve bursts which have encountered one or more burst collision events along their routes. Considering the corresponding BDP given in Table 2 for each U, it becomes clear that, as noted in the literature [9], provision of full λ-conversion capability, albeit difficult and costly to implement at the present state, is a very effective means for contention resolution in OBS networks. We further note, that Table 2 confirms that, for constant N_A, the increase in burst collision probability with increasing the channel occupation ratio U results in deterioration of network's overall BDP performance.

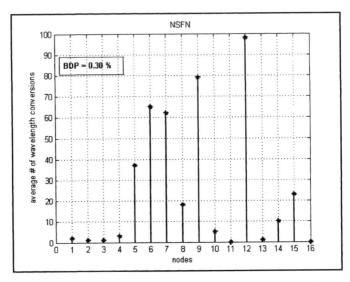

Fig. 5 Average Number of Total λ-Conversions in NSFN for $N_A \cong 510$ and $U \cong 0.1$.

On the other hand, a comparison of the simulation results given in Fig.3 and Fig.4 reveals that, under the assumed symmetrical traffic matrix and fixed, shortest path routing scheme, the overall BDP performance of PEN is significantly better than that of NSFN. Since the number of nodes (N) and links (L) are almost the same, and the number of λ-channels (W) is identical for both networks, the variation of performance should necessarily be due to the differences between the topologies of NSFN and PEN, given, respectively, in Fig.1 and Fig.2. This result is due to the fact that the NSFN topology is more sparsely connected than PEN, and thus, less suited for accomodating uniform traffic, as can be seen from comparison of Fig.5 with Fig.6, which display the average number of total λ-conversions made at the nodes of both networks.

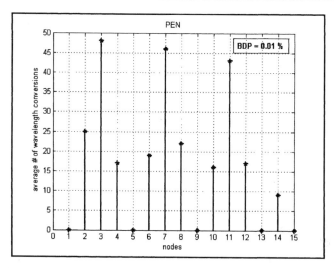

Fig. 6 Average Number of Total λ-Conversions in PEN for $N_A \cong 510$ and $U \cong 0.1$.

From Fig.5 and Fig.6 we infer that the λ-conversion attempts made in the NSFN amount to 407, which is considerably more than the 260 attempts made in PEN. Consulting Table 2 one can calculate the average number of λ-conversion attempts that these bursts undergo on their routes as about 2.2 and 1.6 for NSFN and PEN, respectively. Clearly, as λ-channel requests increase, it may not be possible to accommodate all requests on some links which get overloaded and, as a result of this, BDP will also increase. The links which get overloaded first under the assumed traffic distribution and routing scheme are the bottlenecks of the network, which severely limit BDP performance at high load levels. In the present case, the bottlenecks in both topologies can be easily identified with the aid of Fig.5 and Fig.6. From Fig.5 and Fig.1 we can conclude that the interconnected links between nodes 12, 9, 7, and 6 are the potential bottlenecks in NSFN, whereas Fig. 4 and Fig.2 indicate that the interconnected links between nodes 3, 7, and 11 are the potential bottlenecks in PEN. The total number of λ-channel requests on these links is nearly twice as much in NSFN as compared to PEN, and hence the latter topology is expected to provide a better BDP performance under the assumptions used in our calculations. It should be noted, apart from providing explanation for the differences in BDP performance of NSFN and PEN, given in Fig.3 and Fig.4, respectively, above considerations may also be used as guidelines in cost-effective design modifications for improving the BDP performance via suppliying additional λ-channels on selected links, and also for cutting cost without appreciably deteriorating BDP performance via providing λ-conversion capability not at all nodes but rather only at a selected subset of them [7].

6. CONCLUSIONS

We have shown that BDP performance of an OBS network can be calculated via MC experiments under the assumptions that (i) all events are described with stationary distributions and, (ii) Wavelength Assignment (WA) protocol does not utilize globally synchronized time data. In the proposed approach the temporal behavior of burst traffic is captured via introducing load dependent burst collision probabilities and representing load as a function of the number of bursts injected (N_A) and of the channel occupation ratio (U). The presented approach is sufficiently general and may easily be adapted to address scenarios involving different burst aggregation processes, traffic distributions, routing, and contention resolution schemes. Moreover, it may provide advantages over the DE simulation techniques in that,

(1) Simulation time is bit rate independent.

(2) MC size can be increased by simply running the same scenario on any mix of multiple platforms.

On the other hand, it should also be pointed out that our approach suffers from poor convergence characteristics of MCS. The present version of our code is implemented in Matlab for concept validation and quick prototyping purposes, and as noted in connection with Table 1, its run time requirements are unreasonably high for performing calculations using sufficiently large MC sizes to achieve convergence within, say, 2σ confidence bounds. Ongoing work, which will soon be reported in a follow-up paper, is focused on investigations with different bursts aggregation schemes and traffic distributions as well as on reducing the run time requirements of our code via transforming it into C and implementing acceleration techniques.

REFERENCES

[1] M. You and C. Qiao, "Just-Enough-Time (JET): A High Speed Protocol for Bursty Traffic in Optical Networks", *Digest of the IEEE/LEOS, pp.26-27, August 1997.*

[2] J.S. Turner, "Terabit Burst Switching", *J.High Speed Networks, vol.8, no.1, pp.3-16, Jan.1999.*

[3] I.Baldine, G.N. Rouskas, H.G. Perros, and D. Stevenson, "JumpStart: a Just-In-Time (JIT) Signaling Architecture for WDM Burst-Switched Networks", *IEEE Communications Magazine, vol. 40, no. 2, pp. 82-89. February 2002.*

[4] Y. Chen, C. Qiao, and X. Yu, "Optical Burst Switching: a New Area in Optical Networking Research", *IEEE Network Magazine, vol. 18, pp. 16-23, May-June 2004.*

[5] I. Baldine, M. Cassada, A. Bragg, G.K. Edwards, and D. Stevenson, "Just-in-time Optical Burst Switching Implementation in the ATDnet All-Optical Networking Testbed", *Proc. IEEE Globecom 2003, pp. 2777-2781, December 2003.*

[6] S. Parlar and E.Topuz, "Determination of Burst Drop Probability in Optical Burst Switched Networks", *in Proc. IEEE ICTON 2005, Barcelona, Spain, pp. 429-432, July 2005.*

[7] S. Parlar and E.Topuz, "Performance Evaluation in Optical Burst Switched Networks", *Complex Computing Networks, Springer in Physics Series, Vol. 104, pp. 177-184, Jan 2006.*

[8] W.H. Tranter, K.S. Shanmugan, T.S. Rappaport, and K.L. Kosbar, *Principles of Communication Systems Simulation with Wireless Applications.* New Jersey: Prentice Hall, 2004.

[9] J. Teng and G.N. Rouskas, "Wavelength Selection in OBS Networks Using Traffic Engineering and Priority-Based Concepts", *IEEE J. On Select. Areas in Comm., vol.21, no.8, August 2005.*

[10] A. Bragg and H. Perros, "Modeling and Analysis of Ultra-High Capacity Optical Networks", *Advanced Simulation Technologies Conference (ASTC), Virginia, April 2004.*

[11] S. Ahmad and S. Malik, "Implementation of OBS Framework in Ptolemy Simulator", *E- Tech., pp. 47-52, 2004.*

[12] A. Louridas, K. Panagiotidou, and N. J. Gomes, "Simulation of Optical Burst Switching Protocol and Physical Layers", *London Comms. Symposium, September 2002.*

[13] S. Kumar, "Interest-based Routing in Intelligent Optical Transport Networks", *MSc Thesis, Ohio State University, USA, 2002.*

[14] S. Parlar, "Determination of Blocking Probabilities in Optical Burst Switched Networks with Monte Carlo Simulation", *MSc Thesis*, Istanbul Technical University, Turkey, 2005.

[15] M. Casoni and M.L. Merani, "Resource Management in OBS Networks: Performance Evaluation of an European Network", *Proceedings WOBS-1, October 2003.*

CHAPTER 16

Simulation of Radio Channel and Modulation Schemes Using Markov Chains

Leandro de-Haro-Ariet [1], Ignacio Álvarez Salcidos [1], Manuel García-Sánchez [2]

[1] Dpt. Señales, Sistemas y Radiocomunicaciones. Universidad Politécnica de Madrid

[2] Dpt. Teoría de la Señal y Comunicaciones. Universidad de Vigo.

Abstract *A Markov chain based modeling and simulation approach is presented for estimating performances of radio channels and modulation schemes.The model parameters, i.e. the number of Markov states and transition matrix elements can be derived in a statistically consistent way, from measured data.Once the model parameters are set time series generated via simulation can be used for performance prediction. Example outputs are given which demonstrate that the predicted results for BER in a particular radio channel are in good agreement with the measured average value.*

1. INTRODUCTION

Digital channel concept [1] can be used to speed up the simulation of physical layer when it is necessary. Figure 1 shows a simplified block diagram of a simulator. One of the most accurate approaches is the simulation of the radio channel, as well as the modulation scheme using Markov chains defined with the aid of the measured data, as proposed in this paper. It should be noted that, this solution implies the need for measurement data and its statistical analysis for the purpose of defining Markov chains.

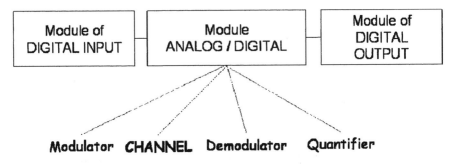

Fig. 1 Digital Channel Concept.

Digital channel characterization using Markov chains is based on the following steps:
1. Channel measurements
2. Markov matrix definition (number of states and transition matrix definition)
3. Time series calculation.

2. CHANNEL MEASUREMENTS

The channel is characterised by a statistical distribution e.g. computed by the measurement of channel variation. Figures 2 and 3 show a channel measurement system based on the use of a VNA and some results obtained for different separations between the transmitter and the receiver.

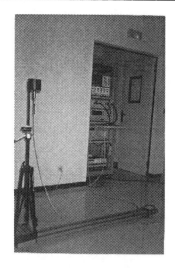

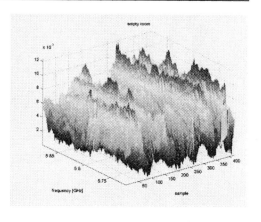

**Figure 3. Spatial Variation of the
Channel Frequency Response**

**Figure 2. Indoor VNA-Based
Channel Sounder**

3. MODEL DESCRIPTION

3.1 Markov Matrix Definition

Figure 4 defines the mathematical elements of a digital channel: input values, output values, output error probability and transition matrix.

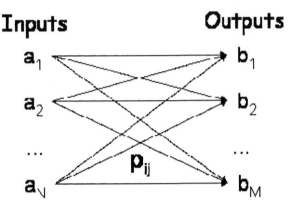

Fig. 4 Transition Matrix Definition

Output, state probability, error probability or transition probability may help to classify the digital channels according to a number of criteria.
When considering the output the digital channel can be classified in:
a) Binary; the two valid outputs are just "0" and "1"
b) non binary

In case the state probability is taken into account, the digital channel can be classified as:
a) stationary, when the state probability does not depend on time
b) non stationary

When viewing the properties of output error probability, the digital channel can be classified in:
a) symmetric; the error probability depends only of the state
b) non -symmetric.

Finally, according to the properties of the transition matrix, the digital channels can be classified as:
a) channel without memory
    Following figure 4 nomenclature, the transition matrix between input and output is thus, $p_{ij}(t) = P\{b(t) = b_j / a(t) = a_i\}$
b) channel with memory
    The transition matrix between input and output is thus, $p_{ij}(t, S) = P\{b(t) = b_j / a(t) = a_i, S(t)\}$

3.2   Definition of Markov States

The number of states of Markov chain gives an idea of the different performance measures for the received signal. Each state also provides a bit error probability.

The number of states depends on the channel environment and the modulation scheme. The statistical consistency of the model requires that it shall be selected to match the probability density function of the measured response. Expressing the error probability of signal as a probability density function, the states "quantas" can be obtained minimizing the quantification noise according to PCM criteria [2].

Once the number of states is decided, a set with several "quantas" (states) is defined

$$Q = [q_0, q_1,..., q_m]$$

where the quantas are expressed in terms of the quantification noise of probability density function as,

$$q_a = \frac{\int_{Q_a} x\, fdp(x)dx}{\int_{Q_a} fdp(x)dx}$$

The output error interval assigned to each state is adjusted to match the model to the results obtained from the measurement. Therefore, the matrix of probability of error for each state gives the following value :

$$P = [p_0, p_1,..., p_m] = [q_0, q_1,..., q_m]$$

3.3   Determination of the Transition Matrix

The transition matrix between states will give the dynamic performance describing the rate of change between the states.

The transition matrix is obtained directly from measurement in terms of the output error interval assigned to each state, and the frequencies of the state changes that occur during the measurement.

Once the state ranges and the error probability are computed

$$V = [v_0, v_1,..., v_{m+1}]$$
$$P_{eb} = [p_1, p_2, p_3,..., p_n]$$

the transition matrix between the states is obtained as,

$$T = \begin{bmatrix} p_{11} & p_{12} & K & p_{1N} \\ p_{21} & p_{22} & K & p_{2N} \\ M & & O & M \\ p_{N1} & p_{N2} & K & p_{NN} \end{bmatrix} \qquad p_{ij} = \frac{N_{ij}}{N_j}$$

4. MODEL IMPLEMENTATION

When the Markov chain parameters are determined as outlined above, the simulator can generate a time series for prediction of the channel performance.

Figure 5 and figure 6 show the flux patterns used to describe the algorithm of Markov chain. It is necessary to decide on the output and to obtain the next state, which depend on the output error probability and on the transition matrix, respectively.

In the present model, measurement processing and simulation has been implemented in MATLAB® [3].

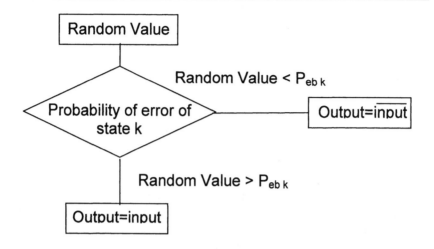

**Fig. 5 Flux Pattern to Define Output Value in the
Markov Simulator**

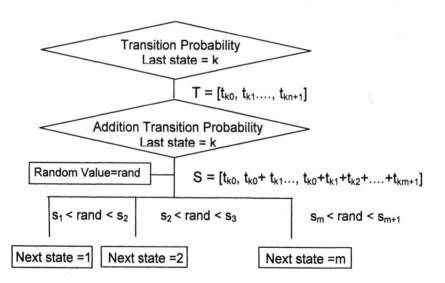

**Fig. 6 Flux Pattern to Define State Transition in the
Markov Simulator**

5. SIMULATION RESULTS

Channel measurements have been done registering time series of the amplitude and phase of the signal at 2.4 GHz within a computer laboratory while people were present and moving. Hence, the measurements include the effects of moving people. The measurement shall be long enough to include all the state transitions in order to provide sufficient accuracy in the computation of state transition probability. An easy rule is to determine the measurement duration in such a way so as to guarantee the occurrence of at least 100 transitions between the states. The measurements reported here include 3000 samples with a noise power level of -60 dBi. Figure 7 illustrates the full process of generating a Markov chain model via channel measurements.

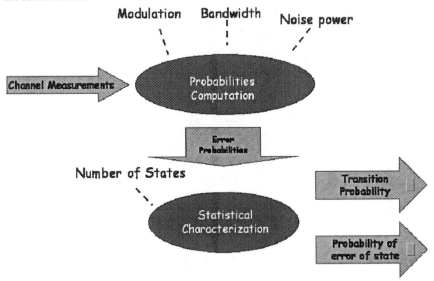

Fig. 7 From Measurements to Markov Chain Computation.

Table 1 and 2 give typical examples of the results obtained via simulation together with the results of the measurements. Table 1 contains the results obtained for the elements of the transition matrix, both from measurements and from simulated time series. The differences between element values of the measured and calculated results provide a means for deciding on the number of digits which can be considered to be significant for a particular application.

Table 2 contains average and extreme values of BER obtained from time series simulation for three different choices of the number of states. It can be seen that the simulated results approach the measured value as the state number increases.

Table 1. Comparison of Transition Matrices Obtained from Measurements or from Simulations.

Transition matrix from original measurements			Transition Matrix from simulations		
0.9896	0.0104		0.9897	0.0103	
0.7573	0.2426		0.7669	0.2330	
0.9666	0.0295	0.0039	0.9673	0.0290	0.0037
0.6582	0.2953	0.0465	0.6557	0.2971	0.0472
0.4520	0.3151	0.2329	0.4384	0.3151	0.2465

Table 2. Channel Error Estimation from Simulation. Measured Value is $44 \cdot 10^{-4}$.

	BER - Average	BER – Max.	BER – Min.
2 states	$47 \cdot 10^{-4}$	$53 \cdot 10^{-4}$ (+6)	$38 \cdot 10^{-4}$(-9)
3 states	$45.7 \cdot 10^{-4}$	$50 \cdot 10^{-4}$(+4.3)	$40 \cdot 10^{-4}$(-5.7)
7 states	$43.9 \cdot 10^{-4}$	$46 \cdot 10^{-4}$ (+2.1)	$42 \cdot 10^{-4}$(-1.9)

6. CONCLUSIONS

A simulation approach is presented for predicting the performance of radio channels and modulation schemes. In the proposed model a Markov chain is used whose parameters are derived from measured data. Our preliminary results demonstrate that the simulated results are in good agreement with the measured values.

Present work is concentrated on extending the presented approach to allow for inclusion of several Markov chains, and also on assessing statistical consistency of the calculated results via running simulation and measurement campaigns in several environments.

REFERENCE

[1] M.C. Jeruchin, P. Balaban and K.S. Shanmugan. Simulation of communication systems. Plenum Press 1992 New York.
[2] Stuart P. Lloyd "Least squares quantization in PCM" IEEE Transactions Information Theory, vol 28, no.2 March 1982
[3] MATLAB Release 13. The Mathworks Inc.

CHAPTER 17

A Component Approach to Optical Transmission Network Design

Marko Lackovic[1] and Cristian Bungarzeanu[2]

[1] Ericsson Research & Development Center, Krapinska 45, HR-10000 Zagreb, Croatia

[2] EPFL STI-TCOM, Station 11, CH-1015 Lausanne, Switzerland

Abstract. *The article describes the philosophy of component approach to system design, illustrated on the optical transmission network example. The approach is based on the object oriented paradigms of Cosmos and Nyx tools. Cosmos serves as the support for system and component development, behavior description, and simulation and analytic calculation development. Nyx in turn enables description of optimization procedures using general heuristic search techniques. The combination of these two development environments and their component approach enables reusability and shortening of the new application development time.*

1. INTRODUCTION

Inherent similarity between the telecommunication models implies that component approach by adding reusable features or functional parts is more appropriate than implementing the whole solution in one large monolithic model at once. Component based model is easier to develop and verify, because the feature and functionality description can be implemented and tested stepwise. In addition, reusability of the model code shortens similar model development process.

Cosmos (*COmplex System Modeling, Optimization and Simulation*) [1] is developed as the all-purpose tool that provides modeling, optimization and simulation capabilities for the wide range of components and systems. It fully utilizes component approach to system modeling. The key parts of the Cosmos tool are structure description, simulation model, analytic procedures and simulation mechanism, as depicted in Figure 1.

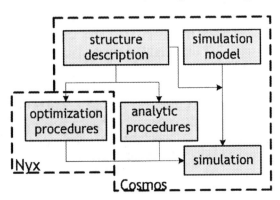

Fig. 1 Interaction Between Analytic Procedures and Simulation

Design of telecommunication networks (especially optical transmission network) often includes large number of design variables. The choice of values for each of the design variables often leads to NP complexity of the design problem. NP problems are usually viewed as complex in both time and space (computer memory) terms. In addition to the NP complexity, solution space in the design of telecommunication transport networks is not continuous. Network design process thus often requires an optimization procedure. These two facts together lead to several possible options in regard to the network design:

- *Enumeration of all solutions*, a procedure limited to small number of design variables,
- *Integer linear programming and derivatives*, with the efficiency close to that of enumeration techniques,
- *Tailored heuristic search techniques* requiring significant knowledge about the problem space, or
- General heuristic search technique (GHST).

We have developed an optimization framework Nyx, that serves as the optimization addition to the Cosmos tool (Figure 1).

The article focuses on describing key architectural features of Cosmos and Nyx, with the emphasis on the component based approach to transmission network modeling, dimensioning and upgrade based on performance constraints, such as communication availability.

2. OBJECT-ORIENTED APPROACH TO SYSTEM MODELING

Component design approach can be easily carried out using object-oriented paradigm (OOP). Cosmos is entirely coded in C++, and uses object-oriented approach in modeling and simulation.

The systems to be modeled can be described by their features (e.g. length of the optical fiber) and functionality (e.g. transmission capability of the optical fiber). This property is implemented by the OOP encapsulation feature. OOP classes (general templates) and objects (classes' instances) consist of the attributes and functionality (methods), thus giving appropriate basis for model development (Figure 2).

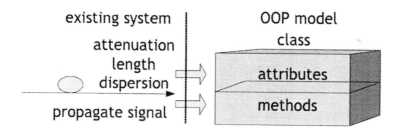

Fig. 2 Encapsulation Mechanism

Adding functionality in OOP terms requires developing a class hierarchy, starting from more abstract classes followed by deriving more specialized ones that represent some system in more details. This mechanism is denoted as inheritance. Each derived (inherited) class contains all functionalities and features of the basic class (Figure 3).

Inheritance is the basis of the OOP code reusability as more abstract classes are expected to be used in different model development processes sharing the same basic model (class). Inheritance is accompanied by polymorphism which enables application of the functions developed for base classes (=developed earlier) to all classes inherited from the base class (=developed later), as shown in Figure 3.

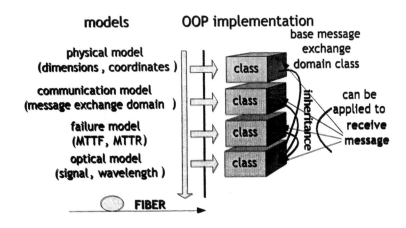

Fig. 3 Inheritance and Polymorphism Mechanism

3. COSMOS STRUCTURE

Cosmos kernel contains basic features and functionality common to all models. These features can be divided into structure (topology) description, and simulation mechanisms.

3.1 Structure Description

Structure description rules formalize the way of system presentation by using data structures from the simulation kernel. *Cosmos* is an object-oriented kernel, and structure description can be made using three classes being modules, layers and systems. These classed make the logical hierarchy starting from the module as the lowest and ending with the system as the highest logical term.

The complete system structure comprises set of modules, and set of module connections which together make the graph structure. The fourth kernel class port is necessary to complete graph presentation. It enables the description of connection between modules.

Figure 4 depicts module-layer-system relationship. Simulation application contains one system object (i.e. one system is simulated at a time). All Cosmos classes have the same base class, which contains common attributes and methods.

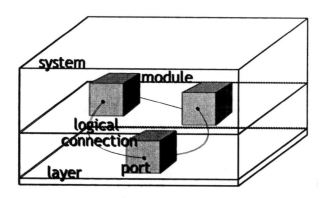

Fig. 4 Module–Layer–System Relationship

Layers merely perform logical aggregation of the modules, and they exist only on the system level. Modules can be directly added to the system, or to another module, what allows subgraph creation. Ports can be added only to the modules. The only port characterization is their direction. Direction of a logical connection (connecting ports) is determined by the direction of edge ports. The logical connection does not implement any physical characteristics (i.e. transmission medium). Every impact of transmission to the simulation has to be modeled by an interconnecting module.

3.1.1 Subgraphs

The module itself can contain a graph, simplifying the system presentation on higher levels of abstraction. The example of the combined subgraphs and layers is a network presentation employing different TCP/IP layers, depicted in Figure 5.

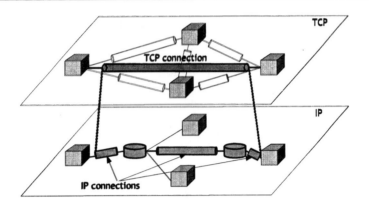

Fig. 5 Subgraphs

Each TCP entity can be connected directly to all other entities on the TCP layer. This connection is described using a module representing TCP logical connection. Each TCP connection module can be further decomposed, disclosing IP connections used to create end-to-end communication. Structure of the TCP connection will be dynamically changed as IP connections (hops) change during data exchange. This decomposition is shown on the IP layer, which contains the modules that are not the part in the TCP layer (e.g. IP routers). This approach allows several views of the same network topology.

3.2 Behavior Description

System functionality, or behavior of the modeled system under different external conditions and input data depends on functionality of its components (modules), and interaction between them.

Each module has its own functionality, which can be divided into two parts relevant to the execution time, being initialization, and run time code. The initialization code of each module in the system is run only once during simulation initialization. System initialization initializes all modules directly contained in the system. Each module initializes all contained sub modules. The run time code is executed during simulation, and its execution time is determined by the simulation mechanism associated with the simulation domain used in simulation (e.g. event expiration in an event driven simulation).

Subgraphs are important for the module behavior implementation using of its submodules' behaviors. Functional decomposition of the complex module behavior can be achieved by presenting this module as a (functional) aggregation of its submodules. In that way user can avoid extensive run time code and simplifies the implementation, testing and maintenance process.

3.3 Simulation Domains and Simulation Mechanisms

Simulation domain describes the way the modules communicate. There are three basic types of communication between the modules implemented in *Cosmos* including direct method calls, message exchange (ME), and events.

Direct method calls are the simplest way of communication where one module interacts with another by directly calling its methods. Message exchange domain is implemented in the same way as direct method calls. It is limited to two generic communication methods (*sendMessage* and *getMessage*), what makes it more suitable for network communication. The user provides methods' code for each module. The events represent the artificial way of communication via simulation mechanism, and are usually used for scheduled external interactions with the system.

Message exchange domain is the only one that has to be directly implemented in a module (get and send message capability). These features are contained in ME domain class, and they can be added to any inherited module class (Figure 6).

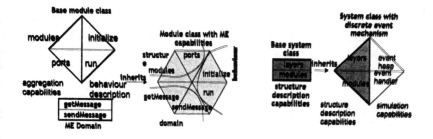

Fig. 6 Message Exchange Domain

Fig. 7 Discrete Event
Simulation Mechanism

Simulation mechanisms serve as a simulation engine. They are implemented in a system, and have to be carefully chosen according to the nature of simulation. Telecommunication network simulation mostly uses a discrete event simulation, implemented in the classed derived form the base system class, as shown in Figure 7.

Event handler and event heap implement the discrete event simulation mechanism. Modules produce events during the initialization phase. Events are put on the event heap. Scheduling mechanism takes an event with the closest expiration time and makes an appropriate action on the module that created the event. This can be done directly, by calling the run method, or by using the event handler, which calls the method associated with simulation domain. In the first case the run method calls another method or performs some action. In the message exchange domain in the latter case, the *sendMessage* method would be called (Figure 8).

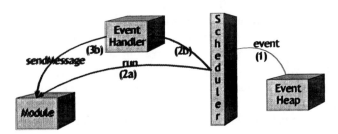

Fig. 8 Discrete Event Simulation Mechanism

3.3.1   *Analytical Algorithms*

Analytical algorithms are implemented in the system, with the modules directly contained in the system as the input data. They are added to the inherited system classed, derived from the base event system class. The analytic algorithm addition contains some basic network algorithms, such as the shortest path algorithms, or all-paths algorithms. These algorithms can be used for new algorithm development.

4.   OPTIMIZATION KERNEL NYX

Nyx optimization kernel represents generic object oriented framework for implementing optimization procedures based on general heuristic search techniques. It allows fast implementation of optimization procedures. In addition, it allows creation of adaptive optimization procedures.

The architecture of Nyx allows application of different optimization procedures to the same optimization problem, what enables choice of the best possible optimization procedure for the given optimization.

4.1 General Heuristic Search Techniques

Majority of proposed heuristic methods have been tailored to a particular problem, but three among them have become particularly popular recently: genetic algorithm (GA) [2], simulated annealing (SA) [3] and tabu search (TS) [4]. Their popularity lies mainly in the applicability to the wide variety of problems.

Three modern heuristic search techniques differ from the traditional methods in the following aspects:

- Work with encoded solutions, not the solutions themselves, and
- Fitness function usage, not other auxiliary knowledge, such as derivatives.

These two facts make listed techniques applicable to problems for which there is a limited knowledge, i.e. only fitness function. This is the reason why we term such techniques general heuristic search techniques (GHST).

4.1.1   Genetic Algorithms (GA)

Genetic algorithms are based on analogy with the processes of the natural genetics. The structure of a simple genetic algorithm is shown in Figure 6. GAs work with population of possible solutions, denoted in GA terminology as chromosomes.

```
Genetic Algorithm {
  k = 0;
  Initialize_Population( );
  foreach Chromosome in Population {
    Evaluate_Fitness( Chromosome );
  }
  do {
    k = k + 1;
    Selection( );
    Crossover( );
    Mutation( );
    foreach Chromosome in Population {
      Evaluate_Fitness( Chromosome );
    }
  } while( stop_criterion_not_reached );
}
```

Fig. 9 Genetic Algorithm

In each iteration of the algorithm, population experiences utilization of three main GA operators. During *selection*, best solutions (chromosomes) are selected to be members of a subsequent generation. Depending on a selection type, this is achieved with more or less influence of stochastic. *Crossover* implements the exchange of genetic material between two randomly chosen chromosomes. In addition, parts of the solutions to be exchanged are also chosen randomly. Finally, *mutation* operator is related to a random alteration of gene value, which can be optimization variable, or its part. In this context, the role of mutation is secondary - to recover lost genetic material. It is often said that GA uses stochastic transition rules meaning that genetic operators, namely crossover and mutation, are a pplied with defined probabilities. Generation gap is the fraction of a population, in the interval (0,1), which takes part in the reproduction procedure creating a new generation.

In each GA application user should define how to generate the initial population and how to stop the algorithm. The initial population could be generated at random, or as genetic material from a previous procedure. The termination of GA running could be done simply by counting if a prescribed number of steps are reached, or by testing if a termination criterion is fulfilled.

4.2 Nyx Structure

GHSTs are unique in their ability to perform optimization with limited (rather low) amount of knowledge about the problem. As previously stated, they only require fitness (objective/optimization) function. This is the basis for the Nyx optimization kernel. Following this fact, two basic components of the Nyx kernel are *optimization module* (OM), and *problem module* (*PM*).

The interaction between PM and OM is illustrated in Figure 10. Problem module is not aware of the optimization procedure. Its only role is to receive encoded solution and asses its quality in terms of optimization goal. On the other side, problem module does not need to know anything about the problem being optimized, except its quality, which will be determined by the fitness function contained within the problem module.

Clearly, the interaction between OM and PM is defined mostly by the type of solution encoding, as both OM and PM should agree on the manner the encoded solution is sent to the PM for quality assessment.

Basic goal of the optimization kernel Nyx is fast development of optimization applications. The user is responsible for defining the problem by using predefined set of rules. Defined problem is then attached to Nyx. Due to the fact that OM offers set of optimization procedures (such as enumeration, GHST), the Hyx user will have at disposal all of the methods defined within the OM.

4.2.1   Nyx Modules

Nyx is composed of problem module (PM), optimization module (OM), and user Interface Module (UIM). Connectivity and communication between these modules is shown on the.

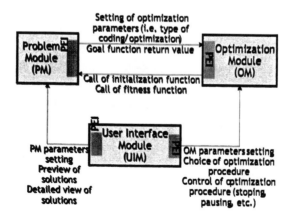

Fig. 10 Communication and Basic Function of Nyx Modules

Behavior of both PM and OM is modified by their respective parameters. The parameters that can be static or dynamic are encapsulated within the module. In order to allow access to the parameters each module (OM or PM) must implement *Property Exchange Interface* (PEI). The concept of communication between modules by means of PEI is shown in Figure 10.

PEI allows access to the list of all parameters within the module. It also allows getting and setting the value of the parameter with the given name. In addition to its value, each parameter has the access mode that defines permissions, i.e. read only or full access to the property value.

4.2.2   User Interface Module (UIM)

UIM implements the user interface, through which end user interacts with the OM or PM. UIM has following main functions:

- Setting and reading of OM and PM parameters,
- Control and management of the optimization (communication with OM). This implies dynamic change of optimization procedure, setting the terminating criterion, inspecting state of the optimization etc.,
- Displaying details and decoding of particular solution (communication with PM).

These functions are actually implemented in either OM or PM, and UIM presents only a proxy that calls required methods in the corresponding modules.

4.2.3   Problem Module (PM)

Basically, problem module is a user defined module containing knowledge on problem which is being optimized. The problem knowledge is contained within the fitness function, which is the main part of any PM. Main functions of PM are:

- Implementation of fitness function for specific problem, and
- Setting of OM parameters.

Fitness function takes the encoded solution as the single input parameter. In general, encoded solution is an array of objects, meaning of which is known only to the PM (fitness function defined within the PM) responsible for quality assessment. The main consequence is that problem module must select encoding scheme out of the available set.

A note should be made regarding the possibility of setting the OM parameters within the problem module, as its definition is of users concern. This is mainly conceived in order to allow the choice of solution encoding by the problem module, but it also allows dynamic behavior of the optimization procedure. For example, a user can within its problem module, in addition to solution encoding, select optimization procedure and, depending on the quality of the solution, modify desired optimization parameters. Even more, user can change optimization procedure. A user thus, has the possibility of creating adaptive optimization procedure.

4.2.4   Optimization Module (OM)

Main functions of OM are:

- Implementation of general optimization procedures, that can be selected from the UIM, and PM,
- Optimization function (procedure) will call fitness function of selected PM, and will expect return value,
- Behavior of optimization procedure is modified by different parameters. Parameters depend on the specific procedure. Optimization procedure parameters can be set either from UIM, or PM. In the latter case, PM should be aware of optimization parameters.

One of the important issues in the design of problem module is the selection of appropriate encoding. Optimization procedure is defined by the given encoding type. Operators of the optimization procedure will be defined differently for different encoding scheme. For example, neighborhood of a solution will be defined differently for binary and permutational encoding. In the same manner, genetic operators have to be defined differently for the different encodings.

Selection of encoding defines the set of available optimization procedures. Similarly to optimization problems, optimization procedures are also registered, but within the catalog of methods (optimization procedures), with encoding type being the main key.

4.2.4.1 Encoding Schemes

The encoding scheme provides a solution representation for a given optimization problem in a way suitable for an optimization procedure. The problem module implementer chooses encoding scheme which describes solution of the stated problem in the best possible manner.

All encoding schemes within Nyx, are derived from the base coding class that contains basic operators. Of particular importance are the operators that allow exploration of the current solution neighborhood. This inherently supports application of GHST based on neighborhood search.

Nyx includes following encoding schemes:

- *Binary* - solutions are coded as arrays of binary values,
- *Integer* - solutions are coded as arrays of unsigned (non-negative) integer values. Each member of an array is additionally specified by the minimal (default 1) and maximal value.
- *Permutational integer* - solutions are coded as arrays of unsigned (non-negative) integer values so that there are no two equal values. Minimal value is 1, and maximal is equal to the code word length.

4.2.5   Creation of Problem Module

Generic procedure for a problem module creation comprises following steps:

1. Choice of encoding, or implementation of new encoding, and
2. Choice of optimization procedure, or implementation of new procedure.

The first step in the creation of problem module is choice of encoding scheme. There are two scenarios; use of existing encoding scheme, or definition of a new encoding scheme. In the first case, required actions are implementation of problem module, and its registration within the problem catalog.

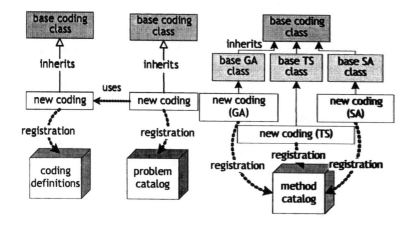

Fig. 11 Creating Problem Module with New Coding

In the second case, following actions are required (see Figure 11):

1. Creation of new encoding scheme, based on the base coding class,
2. Registration of encoding within the encoding catalog,
3. Creation of problem module for the encoding scheme,
4. Registration of the problem module within the problem catalog,
5. Creation of optimization methods using newly defined encoding scheme, and finally
6. Registration of optimization procedure within the method catalog.

Creation of new encoding scheme includes implementation of basic operators as defined in base coding class. In addition, methods for accessing particular members of a code word should be defined.

New optimization procedures and corresponding operators must be implemented for a new encoding scheme. This principle is illustrated in Figure 11. It is pragmatic to define abstract class for an optimization procedure which will contain minimal set of operators required by the implementation of the optimization procedure for an encoding scheme. For example, since each GA requires at least crossover, selection and mutation operators, basic GA class will contain these methods.

5.   DESIGN PROCEDURES

The goal of the network design procedure is to model and dimension the transmission network wit certain performance constraints, such as communication availability and/or packet loss ratio. In order to perform performance calculation the network has to be fully defined in terms of resources (equipment) and interconnections between these resources.

The network is a classical optical circuit switched network employing wavelength division multiplexing.

5.1   General Description of the Design Process

The aim of the network modeling procedure is to create network structure, including links and nodes, and to form transport entities needed for performance calculation. Figure 12 depicts network modeling procedure. The procedure includes preprocessing of input data, routing and wavelength assignment, node structuring, and transport entity creation.

Input data contains topology specification (Cost 266 topologies [5]), and traffic data needed to form the traffic matrix. Topology information contains the list of nodes and links and allows creation of connectivity matrix, and link length matrix. At this point of the design process, nodes in the topology represent some geographical points. Geographical distances between the nodes are calculated using the Haversine formula [6]. Geographical distances are corrected (made longer) in order to represent link lengths [5].

Traffic model uses link lengths and traffic data to generate traffic matrix containing the traffic for each node pair. The traffic volume is divided by the channel capacity to obtain required number of channels to support each communication. The traffic model serves in this step to generate capacity requirements.

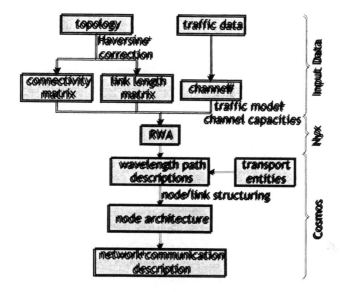

Fig. 12 Network Modeling Procedure

The modeling procedure also includes selection of the node architecture, including passive or automatically reconfigurable architecture (the foundation of ASON [7]).

The second step of the design procedure is to determine physical paths used by each communication and to assign wavelengths. Physical path is determined by the sequence of nodes, fibers and wavelengths. The process is in the literature known as routing and wavelength assignment (RWA). The choice of protection and restoration (P&R) scheme is included in this step.

Physical paths, fiber information and wavelength vectors are sufficient for node structuring assuming that the node architecture is determined. Structured nodes and links serve as the input data for creation of transport entities and logical hierarchy.

In the case of 1+1 or 1:1 protection, for each working lightpath, a protection (or backup) lightpath is determined. The protection lightpath is routed over path that is link or node independent to the path over which working lightpath is routed. The shortest path algorithm is used in this case too; first in order to determine shortest path, and then to determine shortest link and/or node independent to the shortest path.

We can utilize the optimized routing for 1+1 or 1:1 P&R scheme, with the objective of the shortest sum of primary and spare path lengths. k shortest independent path pairs are enumerated by using algorithm presented in [8]. A pair of independent paths is used to route working and protection light paths. A pair of independent paths is selected for each logical channel in the optimization procedure, depending on the optimization goal. Sets of k shortest paths or k shortest independent path pairs are optimization variables in possible optimization procedures.

5.2 Simulation Availability Model

Many models fail to include the communication disruption due to equipment failures. We will incorporate the availability model (on-off model, as depicted in Figure 13) in each network component to simulate the component failures. Communication disruption requires traffic re-routing, and potential congestion.

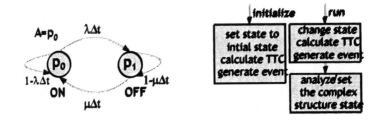

Fig. 13 On-Off Availability Model

Fig. 14 Steps of the Monte Carlo Availability Simulation

In this modeling case availability (A) is equal to probability that the module will be in the *on* state (marked as p_0). Transitions between the states are described by three parameters including transition probability, and probability density function of the time to transition variable.

A state and transition diagram is described by the states class, which uses arbitrary number of states N (in availability model $N = 2$). It contains $N \times N$ matrix with change probability, mean time to transition value and probability density function (PDF) for each transition.

The states class contains a random number generators, generating the time to change to the next state, taking into account all parameters and PDF type. This causes creation of a new event (availability model implies discrete event simulation mechanism) with expiration time set to current time + time to change. Upon event expiration, targeted module changes its state and a new transition time is calculated. Change of the module state potentially changes the state of some complex structure (like the module presented as logical aggregation), or the whole system. The state of the complex structure, like the transport entities describing communication, is defined by the states of the module it contains. In this way we can calculate the communication availability by calculating the mean value of a large number of computer-generated experiments what is a common Monte Carlo simulation approach. The steps of the Monte Carlo simulation are shown in Figure14.

5.3   Network Analysis and Upgrade

We can perfrom performance analysis of the network solution obtained in the initial network design/dimensioning process depicted in Figure 12. This performance analysis requires a preformance model, such as availability or packet loss ratio model. These models can be based on analytical calculation and/or simulation, like the avalability calculation based on the on-off availability model presented in the previous chapter.

Figure 15 depicts the iterative design process with the availability and/or packet loss ratio constraint. The network solution is attributed with the required model(s), and communication performance regarding availability of packet loss ratio can be calculated. These calculations are described in more details in [9, 10].

If the communication performance regarding some performance parameter is not satisfying (i.e. the unavailability is larger than the preset worst unavailability), the initial design/dimensioning is modified for the communication to meet the constraing. An example of such iterative design process based on analytical cell loss ratio calculation is given in [10].

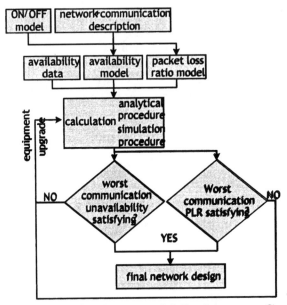

Fig. 15 Iterative Design Process with Performance Constraint

6. CONCLUSION

A well-structured model often comprises reusable parts for other models, starting from the simulation mechanism itself (e.g. event driven simulation) which is common to the components sharing the same simulation domain. Many simulations are performed by an ad-hoc tools tailored for the single purpose.

The described object oriented approach to modeling employed in the Cosmos tool and its optimization supplement Nyx offer more functionality than the used ones in this specific modeling process, as they (can) also serve in other modeling procedures. This reduces the development time of new applications, and allows creation of a base of user modules that can be reused using the same environment, being the simulation mechanism/domain, structure of analytical and optimization procedures.

We have illustrated the Cosmos/Nyx potential by a small example of optical transmission network design and optimization.

Acknowledgement

This work has been partially conducted within the Cost 285 Action and FP6 e-Photon/One project and with the partial support of the Swiss State Secretariat for Education and Research.

REFERENCES

[1] M. Lacković, R. Inkret, "Network Design, Optimization and Simulation Tool Cosmos," Proc. Workshop on All-Optical Networks, WAON 2001, June 13-14, 2001, Zagreb, Croatia, pp. 37-44.

[2] D. E. Goldberg, "Genetic Algorithms in Search, Optimization, and Machine Learning," Reading, Massachusetts, Addison-Wesley, 1989.

[3] E. Aarts, J. Korst, "Simulated Annealing and Boltzman Machines," John Wiley & Sons, 1990.

[4] F. Glover, M. Laguna, "Tabu Search," Kluwer Academic Publishers, USA, 1997.

[5] R. Inkret, A. Kuchar, B. Mikac (eds.), "Advanced Infrastructure for Photonic Networks: Extended Final Report of COST Action 266," FER, University of Zagreb, Zagreb, 2003.

[6] R. W. Sinnott, "Virtues of the Haversine," Sky and Telescope, Vol. 68, No. 2, 1984, pp. 159.

[7] ITU-T Recommendation G.8080/Y.1304, Architecture for the Automatic Switched Optical Network (ASON).

[8] D. Eppstein, "Finding the k Shortest Paths," Proc. of 35th IEEE Symp. Foundations of Comp. Sci., Santa Fe, 1994, pp. 154-165.

[9] M. Lacković, B. Mikac, "Analytical vs. Simulation Approach to Availability Calculation of Circuit Switched Optical Transmission Network," Proc. of ConTel, June 11-13, Zagreb, Croatia, 2003, pp. 743-750.

[10] M. Lacković, C. Bungarzeanu, "Planning Procedure and Performance Analysis of Packet Switched All-optical Network," Proc. of ONDM 2003, February 3-5, 2003, Budapest, Hungary, pp. 253-271.

CHAPTER 18

Fast Dimensioning of Packet-Switched All-Optical Networks

Cristian Bungarzeanu[1], Marko Lackovic[2]

[1] EPFL, STI-TCOM, Station 11, CH-1015 Lausanne, Switzerland

[2] Ericsson Research & Development Center, Krapinska 45, HR-10000 Zagreb, Croatia

Abstract. *This paper addresses the basic concepts of the packet switching in the optical domain and describes an analytical approach to evaluate the end-to-end performance of networks employing slotted (fixed length) optical packets. Thus, for a given topology and traffic matrix, the end-to-end cell loss ratio is analytically computed assuming an uncorrelated traffic. A network dimensioning method relying on this approach is also presented.*

1. INTRODUCTION

In recent years, communication networks faced the increasing pressure of bandwidth-intensive applications, which generate a high amount of traffic over long distances. Today only photonic technologies can provide the capacity required by the ever-increasing traffic demands. High-capacity point-to-point transmission links are today available. However, switching has evolved for a very long time only in the electronic domain. The deployment of new switching techniques, such as fast optical packet switching, enables the building of a network environment where the bandwidth can be used in a flexible way, so as to better match the service features [1]. Wavelength division multiplexing (WDM) has emerged as a major physical layer technology. New all-optical switch architectures based on WDM delay-lines and tuneable optical wavelength converters (TOWC) are presently under study [2].

In this context, there is a growing need for planning tools to deal with the main issues of optical networking in an automated manner. This paper addresses a planning methodology for large-scale WDM transport networks relying on all-optical packet switching. This tool has been implemented within the network planing environment CANPC (Computer-Aided Network Planning Cockpit) [3]. CANPC offers an intuitive user interface that allows the network planner to graphically specify the topology of a network and the traffic relations between network nodes.

The effectiveness of our dimensioning approach is illustrated on a 20-nodes pan-European transport network.

2. NETWORK MODELLING

To some extent, the optical packet switching in WDM networks can be compared to ATM switching in multi-channel transmission groups [4]. Therefore, by analogy with ATM we call the fixed size optical packets cells. The node architecture is illustrated in Fig. 1 [4]. It includes the following blocks:
- a cell encoder where a WDM demultiplexer (DMUX) selects the cells arriving at n fixed wavelengths, $\lambda_1...\lambda_n$, and TOWCs which address free space in the fibre delay-line output buffers
- a nonblocking space switch to access the desired outlet as well as appropriate delay-line in the output buffer
- cell buffers realised with fibre delay-lines

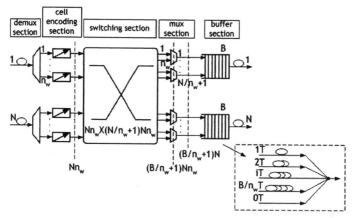

Fig. 1 Switch Architecture

The size of the space switch is $nN \times N(B/n+1)$ where B is the number of cell positions in the buffer, N the number of in- and out-lets and B/n is the number of delay-lines (B is a multiple of n so that B/n is an integer).

It is assumed that the cells arrive synchronously, so that in each time slot (cell duration) a number of cells situated between 0 and nN have to be distributed to the N outlets. The probability a_k^j that $k = r_1 + r_2 + ... + r_N$ cells are directed to a given outlet j is given by

$$a_k^j = \sum_{\text{all distributions } r_i} \alpha_k^j(r_i)$$

$$\alpha_k^j = \prod_{i=1}^{N} \binom{n}{r_i} (u_i \rho_{ij})^{r_i} (1 - u_i \rho_{ij})^{n - r_i}$$

(1)

where u_i is the input load per wavelength i.e. the probability that the time slot at inlet i carries a cell and ρ_{ij} is the forwarding probability from the inlet i to the outlet j [5].

The model based on a Markov chain described in [4] allows the computation of the cell loss ratio (CLR) for any outlet. An end-to-end CLR associated to a demand m is computed as

$$CLR_m = 1 - \prod_{l \in p_m} (1 - CLR_l \cdot \eta_{ml})$$ where p_m is the route followed by the

demand m and η_{ml} is the weight of the demand m in the link l ($\sum_m \eta_{ml} = 1$).

The first stage of the dimensioning process consists in routing all demands. This allows the computation of forwarding probabilities and the loads of each link. Routing can be performed manually or automatically. A shortest path routing is available by running Bellman-Ford algorithm [6]. Two options are available:
- minimise the connection length
- minimise the number of intermediate nodes.

The initial number of fibres for a link l is computed by dividing the total traffic that shares the link to n and round the result up.

The goal of the planning process is to determine the minimum number of fibres for each link, which determine for all demands CLRs inferior to a ceiling value, which in our case is 10^{-4}. The network topology and the matrix of traffic demands are known. The dimensioning procedure is as follows:
1. Input network data (topology, traffic demands)
2. Perform routing
3. Allocate initial capacity (fibres)
4. Compute the forwarding probabilities
5. Perform the computation of link loads and the end-to-end CLRs
6. *If* a CLR is higher than the ceiling value *then* add 1 fibre to the link with the highest CLR and *go to* 4.
   else
7. End of dimensioning.

3. PAN-EUROPEAN NETWORK

The planning tool described previously has been used to dimension the 20-nodes pan-European network [7] whose topology is shown in Fig. 2. All nodes are labelled with the names of major European cities followed by a number in brackets ranging from 5 to 10. The traffic demand between two nodes is the product of these two associated numbers and a weight coefficient, which in our case is 0.01. Thus, the traffic demand between Paris and London is 2.72 Erlangs.

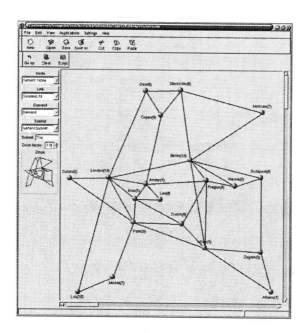

Fig.2 Pan-European All-Optical Transport Network

The results described thereafter correspond to 8 wavelengths per fibre
and 2 delay-lines per outlet.

Table-1 Link Zurich-Milan

	Iteration 1	Iteration 2
Total traffic	7.83 E	7.83 E
Fibres	1	2
Traffic/wavelength	0.978 E	0.489 E
CLR$_{\text{Zurich-Milan}}$	8.12×10^{-4}	10^{-11}

Table 1 shows the evolution of the load of the link "Zurich-Milan" over
two iterations. At the first iteration with only one fibre, the load is so high
that the CLR of the demand "Zurich-Milan" is 8.12×10^{-4}, greater than the
ceiling value of 10^{-4}. After the addition of a new fibre, the load is halved
and the CLR of the demand "Zurich-Milan" become less than 10^{-11}.

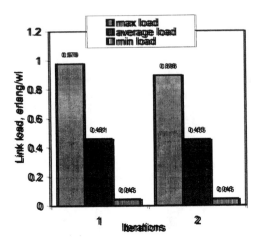

Fig. 3 Load Statistics

Fig. 3 shows the evolution of the load statistics after doubling the capacity of the link "Zurich-Milan". Although the network mean load is decreased to a little extent, the average CLR over all the demands drops by more than an order of magnitude as it results from Fig. 4.

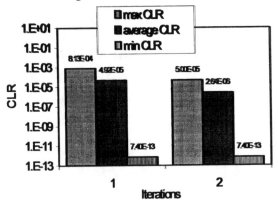

Fig. 4 End-To-End CLR Statistics

4.   CONCLUSION

The software tool described above performs a fast dimensioning of an existing network or a new architecture under various load conditions. This planing approach allows the test of various switch configurations (buffer capacity) for different routing strategies in order to find an optimum that suits the user particular needs.

Acknowledgement

This work has been performed within the COST Action 285 "Modelling and simulation tools for emerging telecommunications networks" with the support of Swiss State Secretariat for Education and Research.

REFERENCES

[1] A.A. Pattavina, M. Martinelli, G. Maier and P. Bo, "Techniques and technologies towards all-optical switching," *Optical Networks*, Vol. 16, No. 2, April 2000.

[2] M. Sotom, A. Jourdan, P.A. perrier and L. Berthelon, "Photonic network and node architectures for flexible multigigabit transport network," *Proc. of ICC'93Workshop on Optical Networks and Nodes*, Geneva, Switzerland, May 1993.

[3] tcomwww.epfl.ch/research/canpc/canpc.htm

[4] S.L. Danielsen, B. Mikkelsen, C. Jorgensen, T. Durhuus and E. Stubkjaer, "WDM packet switch architectures and analysis of the influence of tuneable wavelength converters on the performance," *J. of Lightwave Tecnology*, Vol. 15, No. 2, February 1997.

[5] G. Castanon, "Design model for transparent packet switched irregular networks," *Proc. of 26th European Conference on Optical Communications (ECOC 2000)*, Munich, September 3-7, 2000.

[6] A. Kershenbaum, *Telecommunications network design algorithms*, McGraw Hill 1993.

[7] T. Cinkler, "Heuristic algorithm for wavelength routing in static multi-hop DWDM networks, " Combined COST 266 and COST 267 Workshop " Optical signal processing in photonic networks", Berlin, April 5-7, 2000.

CHAPTER 19

Quality Assessment of Modeling and Simulation of Network-Centric Military Systems

Osman Balci [1], William F. Ormsby [2]

[1] Department of Computer Science
660 McBryde Hall, MC 0106
Virginia Tech
Blacksburg, Virginia 24061, USA

[2] Naval Surface Warfare Center
Dahlgren Division, Code T12
17320 Dahlgren Road
Dahlgren, Virginia 22448, USA

Abstract. *Modeling and simulation (M&S) of network-centric military systems poses significant technical challenges. A network-centric military system (also known as network-centric operations, network-centric warfare or FORCEnet) is a system of systems aligning and integrating other systems such as battlefields, computers, databases, mobile devices, people (users), processes, satellites, sensors, warriors, shooters, and weapons into a globally networked distributed complex system. Characteristics of a network-centric military system are described using a layered architecture. Challenges for M&S of network-centric military systems are presented. The paper focuses on the quality assessment challenge and advocates the use of a quality model with four perspectives: product, process, project, and people. A hierarchy of quality indicators is presented for network-centric military system M&S. An approach is described for conducting collaborative assessment of M&S quality using the quality indicators.*

Key Words and Phrases: modeling, network-centric military system, quality assessment, simulation, validation, verification.

1. INTRODUCTION

Quality is a critically important issue in almost every discipline and is sometimes referred to as "Quality is Job 1." Whether we manufacture a product, employ processes or provide services, quality often becomes a major goal. Achieving that goal is the challenge. Many associations have been established worldwide for quality, e.g., American Society for Quality (http://www.asq.org/), Australian Organization for Quality (http://www.aoq.asn.au/), European Organization for Quality (http://www.eoq.org/), and Society for Software Quality (http://www.ssq.org/). Manufacturing companies have quality control departments, business and government organizations have Total Quality Management (TQM) programs, and software development companies have Software Quality Assurance (SQA) departments to be able to meet the quality challenge [Balci 2004].

The U.S. Department of Defense (DoD) is the largest sponsor and user of Modeling and Simulation (M&S) applications in the world. DoD uses many types of M&S applications (such as continuous, discrete-event, distributed, hardware-in-the-loop, software-in-the-loop, human-in-the-loop, Monte Carlo, parallel, and synthetic environments bringing together simulations and real-world systems) for the purpose of acquisition, analysis or training. DoD M&S applications typically are large-scale and complex and cost millions of dollars to develop over many years, e.g. [MDA 2005]. One category of such DoD large-scale complex M&S applications exists in the problem domain of network-centric military systems (NCMSs) described in Section 0. Assuring the quality of diverse types of large-scale and complex DoD M&S applications poses significant technical and managerial challenges.

M&S applications are mostly made up of software or are software based. Software is inherently complex and very difficult to engineer. Under the current technology, we are incapable of developing a reasonably large and complex software product and guaranteeing its 100% accuracy. Accuracy is considered just one of many quality characteristics of an M&S application and is judged by conducting Verification and Validation (V&V). As advocated by Balci et al. [2002b], we can increase our confidence in the accuracy of large-scale and complex M&S applications by employing a quality-centered assessment approach.

The purpose of this paper is to present such a quality-centered assessment approach for large-scale and complex NCMS M&S applications. Section 2 describes the characteristics of a NCMS using a layered architecture. Section 3 lists a number of technical challenges for NCMS M&S. Section 4 presents a NCMS M&S quality assessment approach based on four perspectives: product, process, project, and people. Concluding remarks are given in Section 5.

2. NETWORK-CENTRIC MILITARY SYSTEM CHARACTERISTICS

A network-centric military system (NCMS) (also known as network-centric operations, network-centric warfare or FORCEnet) is a system that aligns and integrates other systems such as battlefields, communities of interest (COIs), computers, databases, mobile devices, organizational entities, people (users), processes, satellites, sensors, software, warriors, warfighters, shooters, and weapons into a globally networked, distributed complex system, scalable across the spectrum of conflict from seabed to space and from sea to land [FORCEnet 2005].

A layered architectural view of a NCMS is depicted in Fig. 1. The universe of discourse consists of many systems such as the ones listed in Fig. 1. Example COIs include Command and Control COI, Finance COI, Intelligence COI [e.g., Central Intelligence Agency (CIA), Defense Intelligence Agency (DIA), National Imagery and Mapping Agency (NIMA), National Reconnaissance Office (NRO), National Security Agency (NSA)], Logistics COI, Personnel COI, and other Institutional and Expedient COIs. Computers include cluster computers, desktops, handhelds, laptops, servers, and supercomputers.

The processes include the following: (a) DoD Enterprise Management Processes [e.g., requirements generation process; planning, programming, budgeting, and execution process; acquisition process; readiness process; policy management process; military department management processes (organize, train, equip); reserve management processes; agency management processes; inspector general processes; operational test and evaluation processes], (b) DoD Business Processes (e.g., installations and environment, human resources, strategic planning and budget, accounting and finance, logistics, acquisition, technical infrastructure), and (c) DoD Warfighter Processes [e.g., combatant CDR processes (plan and conduct operations, experiment and assess), force application, protection, focused logistics, battlespace awareness, joint command and control, stability operations, homeland security, strategic deterrence].

The weapons include the following: missiles, radars, ships, submarines, tanks, unmanned aerial vehicles (UAVs), and warplanes.

The transport layer includes telecommunications networks such as communications, mobile ad hoc networking, multicast networking, wireless networks, wireless sensor networks, and wireline networks (Fig. 1). One emerging communications technology is Voice over Internet Protocol (VoIP) for transmitting voice, fax, telephone calls, and other information over a data network like the Internet. VoIP enables carrying both voice and data over one line. Worldwide Interoperability for Microwave Access (WiMAX) is another emerging broadband wireless access technology.

The protocols layer consists of many technologies including Common Open Policy Service (COPS), Directory Enabled Networking (DEN), Heterogeneity Aware Peer-to-Peer (P2P), HyperText Transfer Protocol (HTTP), Internet Protocol Security Policy (IPSP), Internet Protocol Version 6 (IPv6), Network Data Management Protocol (NDMP), Simple Object Access Protocol (SOAP), Transmission Control Protocol / Internet Protocol (TCP/IP), and Universal Description, Discovery, and Integration (UDDI).

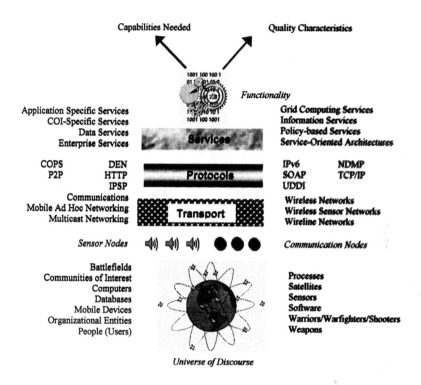

Capabilities Needed   Quality Characteristics

Functionality

Application Specific Services	Grid Computing Services
COI-Specific Services	Information Services
Data Services	Policy-based Services
Enterprise Services	Service-Oriented Architectures

Services

COPS	DEN	IPv6	NDMP
P2P	HTTP	SOAP	TCP/IP
	IPSP	UDDI	

Protocols

Communications	Wireless Networks
Mobile Ad Hoc Networking	Wireless Sensor Networks
Multicast Networking	Wireline Networks

Transport

Sensor Nodes   *Communication Nodes*

Battlefields	
Communities of Interest	Processes
Computers	Satellites
Databases	Sensors
Mobile Devices	Software
Organizational Entities	Warriors/Warfighters/Shooters
People (Users)	Weapons

Universe of Discourse

Fig. 1 Layered Architectural View of a Network-Centric Military System

The services layer includes the services listed in Fig.1 Application specific services include tracking, identification, data mining, sensor management, and threat analysis. Enterprise services include messaging (e.g., chatting, instant messaging, e-mail), mediation, registry, discovery, storage, and security. Information services include information assurance, information exchange, information metadata, information modeling, information processing, information security, information transfer, database services, models, pedigrees, and metrics. Policy-based services include Class of Service (CoS), Common Information Model (CIM), Common Open Policy Service (COPS), Directory Enabled Networking (DEN), Policy Core Information Model (PCIM), Quality of Service (QoS), and Service Level Agreement (SLA). Service-Oriented Architectures (SOAs) include Common Object Request Broker Architecture (CORBA), Distributed Component Object Model (DCOM), Utility Computing / On-Demand Computing, and web services.

The capabilities needed from a NCMS include the following:

- Common, Consistent Knowledge
- Distributed, Collaborative Planning and Execution
- Dynamically Managed, Interoperable, High-Capacity Connectivity
- Enterprise-Wide Integrated Information
- Time Sensitive Decision Making
- Communications and Networking
    - High Altitude Airborne Communications Relay and Router
    - Multi-band, Multi-Beam Antennas
    - Maritime Optical Communications
    - Dynamic, seamless, mobile inter/intra networking
    - Cognitive Networks
    - Integrated Autonomous Network Management
    - End-to-End QoS/CoS Enabled Networks

- Intelligence, Surveillance, and Reconaissance (ISR)
    - Automated and Autonomous Sensor Networking and Management
    - Horizon Extension Surveillance System (HESS)
    - Surveillance Warfighter Array of Reconfigurable Modules (SWARM)
    - Automated Management and Control of Intelligence and Cryptologic Assets.

- Common Operational and Tactical Picture (COTP)
    - Compose, Manage, and Distribute the COTP
    - Decision Support for Dynamic Target Engagement (using Fusion Engines and Intelligent Agents)

- Information Assurance
  - o Assured Authenticity of Data and Information
  - o Collaboration among Multiple Security Domains
  - o Development and Sharing COTP across Multiple Security Domains
  - o Trusted Authentication Mechanisms

The capabilities provided are subject to quality assessment by using indicators such as dependability, interoperability, maintainability, scalability, and survivability. The dependability indicator is assessed by availability, reliability, safety, and security.

3. NETWORK-CENTRIC   MILITARY   SYSTEM   M&S CHALLENGES

M&S can be used for a NCMS for many objectives including the following:

1. Guide the military personnel, commanders, subordinate commands, stakeholders, developers, and others in the acquisition of a NCMS.

2. Evaluate NCMS designs to carry out a specified mission or function against specific criteria.

3. Assess how well a set of capabilities can be provided by a particular design of a NCMS.

4. Evaluate several proposed operating policies or procedures in architecting a NCMS.

5. Predict the effects of emerging new technologies (e.g., VoIP, WiMAX) on the NCMS performance.

6. Assess the interoperability of the NCMS elements.

7. Simulate the input that goes into a NCMS under operational test and evaluation.

8. Facilitate the elicitation of stakeholder needs, identification of new threats, determination of new capabilities needed, and generation of new requirements for a NCMS.

9. Provide significant economical benefits through the reuse of models and simulations already developed.

10. Help reduce the NCMS engineering and integration risks.

Conducting an M&S project under any one of the objectives stated above poses an onerous task with many technical challenges, some of which are described below.

NCMS architecture is typically represented under different views including the following [FORCEnet 2005]:

- *Operational View* represents the tasks and activities, operational elements, and information flows required to accomplish or support a military operation.

- *System View* is a description, including graphics, of systems and interconnections providing for or supporting warfighting functions.

- *Technical View* represents the minimal set of rules governing the arrangement, interaction, and interdependence of system parts or elements, whose purpose is to ensure that a conformant system satisfies a specified set of requirements.

- *All Views* show the scope, purpose, intended users, environment, and analytical findings.

Each view provides a different perspective for analysis. Engineering a simulation model for each view as well as for an integrative view of all views is an extremely complex task. The computational power required for such a simulation model dictates the need to use distributed simulation on a supercomputer with thousands of processors.

Incorporating different levels of granularity in model representation poses another serious technical challenge. It is often desirable to simulate system architecture at different levels of granularity, e.g., soldier level, tank level, battalion level or combat level. It is extremely challenging to build a model at a finer level of granularity and to enable model execution at a higher level of granularity.

A NCMS is a system of systems. Each system has its own architecture and exists within its own unique problem domain such as combat systems, computer network systems, sensor systems, weapon systems, and wireless communication systems. Engineering a simulation model of such a system of systems requires technical domain expertise crossing the boundaries of many disciplines.

Another technical challenge is substantiating that the simulation model has sufficient quality so that it can be certified for a set of intended uses. The quality assessment challenge is addressed in the next section.

4.   NETWORK-CENTRIC   MILITARY   SYSTEM   M&S QUALITY ASSESSMENT

Assessing the quality of a NCMS M&S application poses serious technical and managerial challenges for engineers, analysts, and managers. *M&S application quality* is the degree to which the M&S application possesses a desired set of characteristics. Quality assessment is situation dependent and the desired set of characteristics changes from one M&S application to another.

M&S application quality is not assessed to conclude with a binary decision, where 1 implies "perfect quality" and 0 implies "totally imperfect quality". M&S application quality must be judged as a degree on a scale from 0 to 100. Figure 2 shows notional relationships among M&S application quality, utility, and cost during the development life cycle. While the M&S application is being developed, its quality and utility improve and its development cost rises. The M&S application utility continues to increase as its quality continues to improve, but levels off after a point. The M&S application development cost continues to rise to provide better quality and utility. However, after a point, further development cost does not significantly improve the utility. The intersection point of the utility and cost curves in Figure 2 and their shapes are notional and are expected to change from one M&S application development to another.

M&S application quality should be assessed using a nominal score that corresponds to a numerical interval score. An example nominal score set is given in Table 1.

An M&S application can be certified for a particular quality characteristic (indicator) under a given set of intended uses. Example quality indicators include accuracy (which is assessed by conducting verification and validation), interoperability, fidelity, dependability, performance, supportability, and usability. The intended uses must be well defined [Balci and Ormsby 2000].

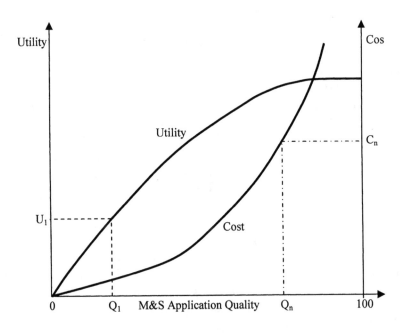

Fig. 2 Notional Relationships among M&S Application Quality, Utility, and Cost

Table 1 Example Nominal Score Set for M&S Application Quality

Nominal Score	Numerical Score	Description
Excellent	[90 .. 100]	M&S application exceeds the requirements for a particular quality indicator (e.g., accuracy) under a given set of intended uses.
Very Good	[80 .. 89.99]	M&S application meets all of the requirements for a particular quality indicator under a given set of intended uses.
Satisfactory	[70 .. 79.99]	M&S application satisfies most of the requirements for a particular quality indicator under a given set of intended uses.
Ordinary	[60 .. 69.99]	M&S application meets some of the requirements for a particular quality indicator under a given set of intended uses.
Marginal	[50 .. 59.99]	M&S application fails to meet most of the requirements for a particular quality indicator under a given set of intended uses.
Deficient	[40 .. 49.99]	M&S application is deficient in meeting the requirements for a particular quality indicator under a given set of intended uses.
Unsatisfactory	[25 .. 39.99]	M&S application is unacceptable with respect to a particular quality indicator under a given set of intended uses.
Superficial	[0 .. 24.99]	M&S application is useless with respect to a particular quality indicator under a given set of intended uses.

The certification decision cannot be made with 100% certainty due to many factors including the complexity of the problem domain, lack of data, reliance on human judgment, and lack of qualified subject matter experts. Therefore, the certification decision should be made with a confidence level similar to the one in statistics. In statistics, a $100(1 - \alpha)\%$ confidence interval is constructed to estimate the unknown population mean of a random variable as $[a, b]$, where α is the significance level, $1 - \alpha$ is the confidence level, a is the lower limit, and b is the upper limit of the interval. For $\alpha = 0.05$, the confidence interval is interpreted as "we are 95% confident that the true value of the population mean is contained within the interval $[a, b]$." Similarly, a certification decision should be stated as "we are highly confident that the M&S application possesses sufficient accuracy for the prescribed set of intended uses." The confidence level should be stated as a nominal value such as "extremely", "highly", or "slightly".

M&S application quality assessment is considered to be a confidence building activity. The more comprehensive and detailed the assessment is, the more confidently the certification decision can be reached. Four major perspectives or four Ps influence the quality as depicted in Fig.3 The quality assessment can be approached from any one of the four Ps, but a combination of all four will provide the best balance and result in a much higher level of confidence in making a certification decision.

Fig. 3 Four Perspectives of M&S Application Quality

M&S application quality should be assessed hand in hand with the model development by way of assessing a particular development life cycle stage's

  a.   output work **product** (or artifact),
  b.   **process** used in creating the output work product,
  c.   quality of the **people** employed, and
  d.   **project** characteristics (e.g., capability maturity, documentation, planning, risk management).

Fig. 4 presents a hierarchy of indicators for assessing a NCMS M&S application quality from the "product" perspective. This is a sample hierarchy showing the higher-level indicators, which should be decomposed further depending on the characteristics of a particular NCMS M&S application [Balci 2004].

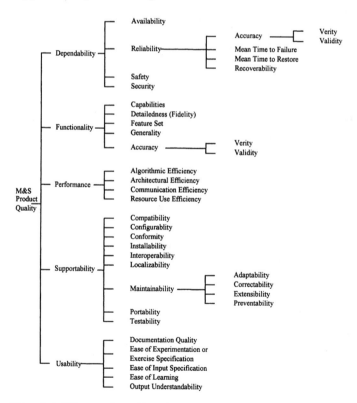

Fig. 4 A Hierarchy of Indicators for Assessing M&S Product Quality

Fig. 5 presents a hierarchy of indicators for assessing a NCMS M&S application quality from the "process" perspective. The indicators chosen represent the process-related areas of the Capability Maturity Model Integration (CMMI) for Systems Engineering and Software Engineering (CMMI-SE/SW) [SEI 2001b]. These indicators need to be further decomposed to provide a desired level of measurement. Other process quality indicators can be added to this hierarchy depending on the characteristics of a particular NCMS M&S application.

Fig. 5 A Hierarchy of Indicators for Assessing M&S Process Quality

Fig. 6 presents a hierarchy of indicators for assessing a NCMS M&S application quality from the "project" perspective. The indicators chosen represent the project-related areas of the CMMI-SE/SW [SEI 2001b]. These indicators need to be further decomposed to provide a desired level of measurement. Other project quality indicators can be added to this hierarchy depending on the characteristics of a particular NCMS M&S application.

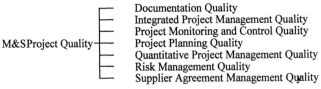

Fig. 6 A Hierarchy of Indicators for Assessing M&S Project Quality

Fig. 7 presents a hierarchy of indicators for assessing a NCMS M&S application quality from the "people" perspective. The indicators chosen represent the process areas of the People Capability Maturity Model (P-CMM) [SEI 2001a]. These indicators need to be further decomposed to provide a desired level of measurement. Other people quality indicators can be added to this hierarchy depending on the characteristics of a particular NCMS M&S application.

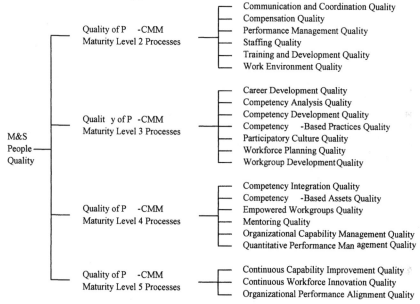

Fig. 7 A Hierarchy of Indicators for Assessing M&S People Quality

Collectively, M&S application quality assessment based on the four Ps generates a large number of quantitative and qualitative indicators under a wide and deep hierarchy. Assessment of the indicators mandates subject matter expert (SME) evaluation, and requires the integration of disparate evaluations. Planning and managing such measurements and evaluations require a unifying methodology with computer-aided assistance and should not be performed in an *ad hoc* manner. Such a methodology has been developed by Balci [2001] and computer-aided assistance has been provided by the Evaluation Environment (EE) software system [Orca 2005].

EE is a Web-based client/server distributed software system [https://www.orcacomputer.com/ee/] structured based on the Java 2 Platform, Enterprise Edition (J2EE) industry-standard architecture. It enables collaborative evaluations by engineers, analysts, and SMEs who are geographically dispersed. EE uses 128-bit encrypted Secure Sockets Layer (SSL) technology to provide secure communication between the EE server and the EE user. EE runs under the IBM WebSphere Application Server, IBM DB2 Universal Database, and IBM HTTP Server. It uses open technology standards such as XML, XSLT, SVG, DHTML, and PDF. EE has its own XML markup language called EEML for project data import/export, archive/restore, and automatic report generation in PDF.

Based on areas of needed expertise, SMEs are employed for constructing a hierarchy of indicators, relative criticality weighting of the indicators, and assigning scores for the leaf indicators. Considering the functional and non-functional requirements specified for the NCMS M&S application, a hierarchy of indicators is created for assessing its quality. Figure 8 shows an example EE project with a hierarchy of indicators presented above. The indicator hierarchy created forms an acyclic graph since an indicator may have more than one parent. The hierarchy should be examined to determine if it possesses sufficient comprehensiveness and depth.

Relative criticality weighting of indicators and SMEs are performed by using the Analytic Hierarchy Process (AHP) [Balci et al. 2002a]. AHP is commonly used in multicriteria decision-making field and consists of the following three steps: (1) perform pairwise comparisons, (2) assess consistency of pairwise judgments, and (3) compute the relative weights.

SMEs are assigned to assess certain leaf indicators in the hierarchy. Since EE is a web-based software system, SMEs can be geographically dispersed. An SME assigns a nominal score (e.g., good) corresponding to an interval score (e.g., [80, 89]) for each leaf indicator. The leaf indicator scores are automatically aggregated throughout the entire hierarchy. EE automatically generates a report in PDF format. Further details of how to conduct an EE project are provided by Balci et al. [2002a].

5. CONCLUDING REMARKS

The DoD Directive Number 8100.1 [DoDD 2002] establishes the Global Information Grid (GIG) Overarching Policy, identifies the GIG Architecture as the DoD Information Technology (IT) architecture required by the IT Reform Act, and directs the DoD Components to ensure that architectures they develop are compliant with the GIG Architecture. The U.S. military systems of the future are required to be network-centric and be engineered in compliance with the GIG architecture. M&S can be used for assessing how well a complex military system design complies with the GIG architecture for acquisition purposes as well as for many other purposes. However, development and quality assessment of M&S applications for such military system architectures and designs poses significant technical and managerial challenges. This paper outlines a strategy for addressing the quality assessment issue.

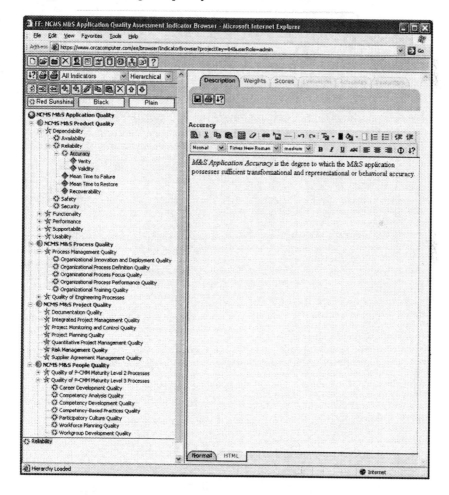

Fig. 8 An EE Project for Collaborative Assessment of NCMS M&S Application Quality

REFERENCES

[1]   Balci, O. (2001), "A Methodology for Certification of Modeling and Simulation Applications," *ACM Transactions on Modeling and Computer Simulation 11*, 4 (Oct.), 352-377.

[2]   Balci, O. (2004), "Quality Assessment, Verification, and Validation of Modeling and Simulation Applications," In *Proceedings of the 2004 Winter Simulation Conference* (Washington, DC, Dec. 5-8). IEEE, Piscataway, NJ, pp. 122-129.

[3]   Balci, O., R. J. Adams, D. S. Myers, and R. E. Nance (2002a), "A Collaborative Evaluation Environment for Credibility Assessment of Modeling and Simulation Applications," In *Proceedings of the 2002 Winter Simulation Conference* (San Diego, CA, Dec. 8-11). IEEE, Piscataway, NJ, pp. 214-220.

[4]   Balci, O., R. E. Nance, J. D. Arthur, and W. F. Ormsby (2002b), "Expanding Our Horizons in VV&A Research and Practice," In *Proceedings of the 2002 Winter Simulation Conference* (San Diego, CA, Dec. 8-11). IEEE, Piscataway, NJ, pp. 653-663.

[5]   Balci, O. and W. F. Ormsby (2000), "Well-Defined Intended Uses: An Explicit Requirement for Accreditation of Modeling and Simulation Applications," In *Proceedings of the 2000 Winter Simulation Conference* (Orlando, FL, Dec. 10-13). IEEE, Piscataway, NJ, pp. 849-854.

[6]   DoDD (2002), "Global Information Grid (GIG) Overarching Policy," Department of Defense Directive Number 8100.1, Sept. 19.

[7]   FORCEnet (2005), "Naval Network Warfare Command FORCEnet Homepage," http://forcenet.navy.mil/

[8]   MDA (2005), "Missile Defense Agency Website," http://www.mda.mil/

[9]   Orca (2005), "Evaluation Environment," Orca Computer, Inc., Blacksburg, VA, https://www.orcacomputer.com/ee/

[10]   SEI (2001a), "People Capability Maturity Model (P-CMM) Version 2.0," Software Engineering Institute, Carnegie Mellon University, Pittsburgh, PA, July, http://www.sei.cmu.edu/cmm-p/

[11]   SEI (2001b), "Capability Maturity Model Integration (CMMI) for Systems Engineering and Software Engineering (CMMI-SE/SW) Version 1.1," Software Engineering Institute, Carnegie Mellon University, Pittsburgh, PA, Dec., http://www.sei.cmu.edu/cmmi/

CHAPTER 20

Using Fuzzy Value Tree Analysis to Support
The Verification, Validation, and Accreditation of
Models and Simulations

Siegfried Pohl

Institute for Technology of Intelligent Systems (ITIS e.V.)
Werner-Heisenberg-Weg 39
D-85577 Neubiberg, Germany

Abstract. *The process of accrediting a model can roughly be divided into first gathering and evaluating results of conducted verification and validation (V&V) activities, and then aggregating these (raw) results into a total value: the accreditation decision. Especially the second part is not trivial, because results of V&V activities are quite manifold and at first sight not comparable with each other. Consider for example the evaluation of subject matter expert statements in contrast to real numbers as an outcome of statistical tests.*

Classical decision analysis is one possibility to support the accreditation decision by using a structured, scientifically justified approach. For that purpose, mainly the concept of value tree analysis is used. In previous papers, the current state-of-the-art concerning decision analytic methodologies supporting VV&A was reviewed. Also a critical examination, describing gaps and deficiencies was conducted. A concept was drafted, how fuzzy value tree analysis, a branch of decision analysis incorporating fuzzy set theory, can be used to overcome some of these deficiencies, as there are: First, modeling subject matter experts statements, second, quantifying not only the value, but also the knowledge resp. ignorance about the model under study, and finally, being able to distinguish between compensational and non-compensational attributes in the value tree.

Fuzzy value tree analysis consists of the steps
– establishing the value tree,
– assigning values to attributes,
– aggregating the attribute values, and
– interpreting the result.

A raw draft of the concept, as well as a definition of attribute values in a fuzzy value tree analysis was dealt with in previous publications. The paper at hand takes these publications as a starting point and presents a concept how attribute values in a fuzzy value tree analysis can be aggregated, i. e. the aggregator is defined.

Starting from establishing requirements and constraints, these postulations are transformed into mathematical properties the aggregator must adhere to. After scanning the literature for possible fuzzy set theoretic operators, it is proven that the defined aggregator fulfills the mathematical properties established and therefore all requirements and constraints. The paper closes with illustrative examples.

1. INTRODUCTION

Verification, Validation and Accreditation (VV&A) deals with the credibility assessment of models and simulations. Idealizing, the process of accrediting a model can be regarded as a decision (Will the model – as it is – be accepted for the intended purpose – as it is – or not?). Proposals exist to use concepts from decision analysis [1, 2], or other aspects of VV&A [3] to support the accreditation process. Especially interesting from the VV&A point of view is the field of value tree analysis, which provides techniques and methods to assess and measure abstract concepts a priori not measurable. However, this technique is not flawless. To overcome some of these flaws and to further extend the set of available methods for VV&A decision support, the application of fuzzy value tree analysis was proposed [4], which is an extension of classical value tree analysis utilizing fuzzy set theoretic methods.

In [4] the basic idea was illuminated together with the identification of a set of tasks that need to be accomplished in order to use fuzzy value tree analysis to support VV&A. In [5] – as a first step – a definition for attribute values in a fuzzy value tree analysis was drafted. The paper at hand focuses on the definition of an aggregating function in a fuzzy value tree analysis. I. e. the function defined will be used to aggregate multiple attribute values into a total one.

The paper is sectioned as follows: First, the contents of [4] are shortly reviewed, in which the proposal to use fuzzy value tree analysis was first stated. Section 3 reviews [5], i.e. reviews the definition of attribute values.

The main section 4 defines the aggregator, followed by illustrative examples in section 5. The paper closes with a summarization and an outlook on further research.

2. PROBLEM AND EXISTING METHODS OF RESOLUTION

(This section briefly reviews [4].)
Value tree analysis is a concept from the decision analysis domain, which itself is a part of business administration. According to [6, 7], value tree analysis consists of:
- – establishing the value tree,
- – assigning values to the attributes,
- – aggregating the attribute values into a total one, and
- – interpreting the results.

In [1, 2, 8, 9] the application of value tree analysis to VV&A is proposed, and furthermore, a concept to support the accreditation decision and the V&V process is established.

However, applying classical value tree analysis is not flawless, as a critical review in [4] shows. The following deficits were identified:
- – In a value tree analysis it is only possible to handle the knowledge about the model under consideration. If, e. g. due to budgetary constraints, certain V&V activities cannot be carried out, the corresponding leaves in the value tree remain empty. Only the question: "What is known about the model?" is answered, while the question "What do we not know about the model?" cannot be dealt with.

- Classical value tree analysis uses simple additive weighting [6, 10, 11] as the aggregating function, which implies that all attribute value are mutually compensational. That in turn implies that bad attribute values can always be compensated by better ones. But in reality, every model will exhibit certain feature that need to be fulfilled in order for the model to be accredited. The distinction between compensational resp. non compensational attributes cannot be modeled with the help of attribute weights and simple additive weighting. For a counterexample see [4].
- The final deficit identified is that by utilizing intervals as attribute values, results of V&V activities cannot be represented satisfactorily. Especially, representing the semantics of SME statements by a single interval is not feasible. The next step in [4] was the proposition of fuzzy value tree analysis as a possibility to overcome the identified deficiencies.

The basic idea of fuzzy value tree analysis is the enhancement of classical value tree analysis by:
- using fuzzy sets instead of real numbers or intervals as attribute values,
- using fuzzy set theoretic operators as the aggregating function.

Fuzzy value tree analysis has advantages in comparison to classical value tree analysis:
- Fuzzy set are predestined to model colloquial statements. Applied to VV&A: Fuzzy sets are predestined to model SME statements as a result of V&V activities.
- There exist fuzzy set operators modeling compensational as well as non compensational aggregation behavior. These operators can furthermore be arbitrarily combined with each other. Also, the degree to which the compensational or non compensational part is in effect, can be specified exactly.

Finally, [4] proposed a roadmap, what needs to be done in order to apply fuzzy value tree analysis to VV&A:

(1) The first issue is the definition of attribute values in a fuzzy value tree analysis. Decisions need to be made, how membership functions, representing results of V&V activities are designed, and how the property of being compensational or non compensational is modeled.

(2) Next, a structured approach to transfer results of V&V activities to attribute values needs to be established. In fuzzy set theoretical literature this is referred to as "elicitating the membership function" [12].

(3) An aggregating function needs to be defined, considering not only the membership functions, but also attribute weights and the degree to which an attribute value is compensational.

(4) The final step in every decision support tool using fuzzy techniques is the interpretation of the results. In the literature [13, 14] several proposals exist, which need to be reviewed.

Implementing these concepts must be conducted with the focus in mind that users without any decision theoretical or mathematical background shall be able to use them. How these concepts work internally needs to be hidden from the user.

3. DEFINITION OF ATTRIBUTE VALUES

(This section shortly reviews [5].)
The gist of [5] was the definition of attribute values to be used in a fuzzy value tree analysis:

Definition 1 Let $X := [0, 1]$. *Let F_X be the set of all continuous, differentiable fuzzy sets in the universe X with image X. Then, an attribute value is a quintuple:*

$$(\mu_v, \mu_k, a_v, a_k, \omega) \in \{F_X, \emptyset\} \times F_X \times \{[0,1], \emptyset\} \times [0,1] \times N.$$

The sets μ_v and μ_k represent the value and knowledge of the model with respectto the attribute under consideration.

The real numbers a_v and a_k represent a compensation factor for value and knowledge with the interpretation:

  $a = 0 \leftrightarrow$ attribute value cannot be compensated at all,
  $a = 1 \leftrightarrow$ attribute value can be compensated fully.

The ω represents the attribute's weight, depicted as a natural number.

The universe $X = [0, 1]$ is divided into three subinterval with the following interpretation: Abscissa values out of

– [0,1 - α] represent little

– [1 - α, α] represent moderate

– [α, 1] represent high

value of the model with respect to the attribute under consideration resp. Knowledge about the model with respect to the attribute under consideration. The value α ∈ [0.5, 0.75] is called crossover point. For details see [15].

Examples of value functions are depicted in figure 1 (μ_1 ≅ high, μ_2 ≅ moderate, μ_3 ≅ low value), examples for attribute values can be found in [5].

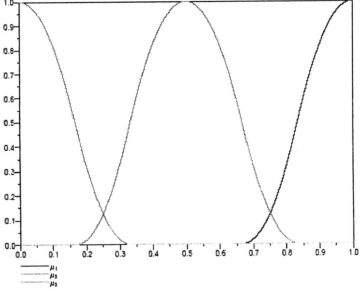

Fig.1 Fuzzy Sets as Value Functions

4. THE AGGREGATING FUNCTION

This main section deals with the definition of an aggregating function used in a fuzzy value tree analysis to support VV&A. First, the problem to be solved is reviewed:

Let all mathematical designators be as in Definition 1.

Let

$$(\mu_{vi}, \mu_{ki}, a_{vi}, a_{ki}, \omega_i), i \in N_n$$

be attribute values. Divide the problem and define a function

$$f : (\{F_X, \emptyset\} \times \{X, \emptyset\} \times N)^n \rightarrow \{F_X, \emptyset\} \times \{X, \emptyset\}$$
$$\{(\mu_1, a_1, \omega_1), ..., (\mu_n, a_n, \omega_n)\} \rightarrow (\mu, a),$$

to be applied twice: First to the value functions

$$f((\mu_{v1}, a_{v1}, \omega_1), ..., (\mu_{vn}, a_{vn}, \omega_n)) =: (\mu_v, a_v),$$

second to the knowledge functions

$$f((\mu_{k1}, a_{k1}, \omega_1), ..., (\mu_{kn}, a_{kn}, \omega_n)) =: (\mu_k, a_k).$$

The results are merged into the aggregated attribute value:

$$(\mu_v, \mu_k, a_v, a_k).$$

An aggregated attribute weight does not need to be computed, because attribute weights are assigned to value tree knots before the evaluation process. The procedure to define the aggregator is as follows:

- First, requirements and constraints the aggregator must adhere to are stated colloquially.
- At the same time, mathematical axioms are used to express these requirements and constraints in a mathematical notation. The reason is that in the literature fuzzy operators are categorized by mathematical axioms and not colloquial descriptions.
- The literature is searched for fuzzy set operators, which fulfill all axioms established. If there is more than one operator in question, the simplest one is picked and used as the aggregator.
- Finally, one proves mathematically that the defined aggregator fulfils all necessary mathematical axioms, therefore all colloquially formulated requirements and constraints stated in step 1.

The technical strategy is to divide the problem:

- A compensational part of the aggregator is defined. This function shall explicitly allow bad attribute values to be compensated by better ones.
- A non compensational part is defined. This function shall not allow that bad attribute values can be offset by better ones.
- In a third step the compensational and non compensational parts are connected by a parameterization, using the compensation factor as the parameter.

4.1 Requirements and Constraints

The idea of this section is to formulate the requirements and constraints the aggregator must adhere to in a colloquial way, and to transform them into mathematical terms. As already stated, the reason is that in the literature fuzzy set operators are defined by mathematical properties and not colloquial descriptions.

The following list of requirements and constraints – and their mathematical counterparts – seems plausible:

(1) The aggregator shall be able to handle an arbitrary amount of input parameters. *f must be defined for all $n \in N$.*

(2) The results shall be independent from the order of the attribute values. *f must be commutative.*

(3) The aggregator must be able to handle the situation that attribute values are the empty set. This is the case, if there is no information at all available about attribute under consideration.
    The elements of the pre-image of f must be enhanced by the empty set:

$$\left(F_X \times X \times N\right) \to \left(\{F_X, \emptyset\} \times \{X, \emptyset\} \times N\right)$$

(4) The aggregator shall respond to a small change in an input value only with a small change in the result. It must be omitted that small changes in an input value result in large modifications of the result. *f shall be continuous in every parameter.*

(5) If all input values are identical, this value is to be computed as the result. *f shall be idempotent.*

(6) If in one input parameter the value (or knowledge) is increased, the value (knowledge) of the result must not be decreased. If in one input parameter the value (or knowledge) is decreased, the value

(knowledge) of the result must not be increased. *f shall be monotone.*

(7) The aggregated compensation factor shall be computed as the minimum of all compensation factors being processed. Hereby it is omitted that a non compensational value looses this property during aggregation.
It shall hold:

$$\left(\_,a\right):=f\left(\left(\mu_1,a_1,\omega_1\right),...,\left(\mu_n,a_n,\omega_n\right)\right)$$

With

$$a:=\begin{cases}\emptyset & a_i=\emptyset\forall i\in N_n,\\ \min\{a_i|i\in N_n\wedge a_i\neq\emptyset\} & otherwise\end{cases}$$

This postulation implies that f must neither be injective nor strict monotone in the second input parameter.

(8) The aggregated value function must not represent a higher (lower) value than the highest (lowest) processed value function. Analogue the knowledge function. *f shall be stable.*

(9) The aggregator must reflect the compensation factor in that sense, that it is not a binary value, but a real number out of the interval [0, 1]. In addition:

(9a) Being fully compensational ($a = 1$) implies that the attribute value under consideration is processed by an averaging process.

(9b) Being fully non compensational ($a = 0$) implies that the aggregated value function must not represent a higher value than this attribute.

These requirements must also hold for the aggregated knowledge function. *Requirement 9b implies that f must neither be injective nor strict monotone in the first input parameter.*

As can be seen, apart from requirement 9, all colloquially stated requirements can be expressed by a mathematical counterpart. This leads to the following – still incomplete – definition:

Definition 2 Let $X := [0, 1]$. Let

$$\left(\mu_i,a_i,\omega_i\right),i\in N_n$$

be triples, consisting of a fuzzy set $\mu_i \in \{F_X, \emptyset\}$, a compensation factor $a_i \in \{X, \emptyset\}$, and an attribute weight $\omega_i \in N$. A function f shall be established, *whichis commutative, stable, idempotent, continuous, monotone, not injective and not strictly monotonic with the property:*

$$f : (\{F_X, \emptyset\} \times \{X, \emptyset\} \times N)^n \to \{F_X, \emptyset\} \times \{X, \emptyset\}$$
$$\{(\mu_1, a_1, \omega_1), \dots, (\mu_n, a_n, \omega_n)\} \to (\mu, a)$$

Furthermore, requirement 9 and the properties

$$\mu_1 = \dots = \mu_n = \emptyset \Rightarrow \mu = \emptyset,$$

and

$$a = \begin{cases} \emptyset, & a_i = \emptyset \forall i \in N_n \\ \min\{a_i | i \in N_n \wedge a_i \neq \emptyset\} & \text{otherwise} \end{cases}$$

must hold.

4.2 Compensatory Part

According to [16], fuzzy set theoretic operators can be divided into t-norms, t-conorms and averaging operators, whereas the last mentioned are interpreted as compensational in the decision analytical literature. Among averaging operators, the additive, multiplicative and multi-linear models are discussed [17]. Nevertheless, the additive one is the only having been applied satisfactorily to fuzzy decision analysis. The method is discussed in [18], examples can be found in [19].

It must be remarked that in the situation at hand only attribute values and not attribute weights are represented by fuzzy sets.

Theorem 1 Compensatory Part of the Aggregator
Let pairs

$$(\mu_i, \omega_i), \quad i \in N_n$$

be given, consisting of a fuzzy set $\mu_i \in \{F_X, \emptyset\}$ and an attribute weight $\omega_i \in N$. Then the mapping

$$f_1 : (\{F_X, \emptyset\} \times N)^n \to \{F_X, \emptyset\}$$
$$\{(\mu_1, \omega_1), \dots, (\mu_n, \omega_n)\} \to \mu$$

with

$$\mu := \begin{cases} \emptyset, \forall i \in N_n : \mu_i = \emptyset \\ \left(\sum_{i \in N_n \wedge \mu_i \neq \emptyset} \omega_i \mu_i \right) / \left(\sum_{i \in N_n} \omega_i \right) \end{cases} \quad \text{otherwise}$$

is commutative, stable, idempotent, continuous, strictly monotonic, injective and fulfils requirement 9a.

Proof. Commutativity, stability, idempotency, strict monotony and injectivity follow from the same properties for point wise addition and multiplication.

By definition only continuous fuzzy sets are used as attribute values, therefore these properties hold for f_1.

Remark: The function f_1 is injective and strictly monotonic, which must not hold for the final aggregator. It must be beard in mind that the non compensational part of the aggregator does not exhibit these properties.

4.3 Non-Compensatory Part

As already stated, fuzzy set theoretic operators can be divided into t-norms, t-conorms and averaging operators. For the non compensational part of the aggregating function t-norms are in question, because these are interpreted as non compensational in the decision analytic literature [18]. The only t-norm fulfilling the requirement of stability is the minimum operator. Because the minimum is neither injective nor strictly monotonic, it is chosen.

Nevertheless, the minimum operator cannot be applied without modification: In [5], the abscissa of the fuzzy sets in use is furnished with subintervals incorporating different interpretations:
 – $[0, 1 - \alpha]$ represents low value (low knowledge),
 – $[1 - \alpha, \alpha]$ represents moderate value (moderate knowledge)
 – $[\alpha, 1]$ represents high value (high knowledge).

To correctly convert requirement 9b, this interpretation must be regarded:
 – Using the minimum operator in subinterval $[0.5, 1]$ is possible without modification.

- In subinterval [0, 0.5] the situation is contrary: High ordinate values represent low value of the model resp. low knowledge. One has to reverse the effect of the minimum operator and use the maximum in [0, 0.5].

The maximum and minimum operators are combined with a linear combination, using the abscissa value as the parameter. In principal, also a case differentiation could be used. For fuzzy sets $\mu_1, \ldots, \mu_n$ set

$$\tilde{\mu}(x) = \begin{cases} \max\{\mu_1(x),\ldots,\mu_n(x)\} & \forall x \in [0,0.5] \\ \min\{\mu_1(x),\ldots,\mu_n(x)\} & \forall x \in [0.5,1] \end{cases}$$

But, as depicted in figure 2, the resulting function $\tilde{\mu}$ is not necessarily continuous at $x = 0.5$, contradicting requirement 4.

The following definition is derived:

Theorem 2 Non Compensatory Part of the Aggregator

Let $\mu_1, \ldots, \mu_n \in \{F_X, \varnothing\}$ be given. The function:

$$f_2 : \{F_X, \emptyset\}^n \to \{F_X, \emptyset\}$$
$$(\mu_1,\ldots,\mu_2) \mapsto \{(x, \mu(x)) | x \in [0,1]\},$$

with

$$\mu(x) := x \cdot \min\{\mu_i(x) | i \in N_n \wedge \mu_i \neq \emptyset\}$$
$$+ (1-x) \cdot \max\{\mu_i(x) | i \in N_n \wedge \mu_i \neq \emptyset\} \quad \forall x \in [0,1]$$

and

$$f(\mu_1,\ldots,\mu_n) := \emptyset, if \quad \mu_i = \emptyset \ \forall i \in N_n$$

is commutative, stable, idempotent, continuous, monotone, not injective and not strictly monotonic. Furthermore requirement 9b holds.

Proof. Analogue to theorem 1. The continuity of f_2 follows from the continuity of linear combinations.

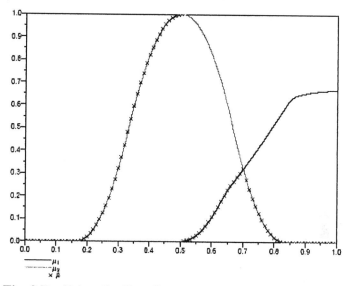

Fig. 2 Realizing the Non Compensational Part by a Case Differentiation

4.4 Combining the Compensatory and Non Compensatory Part

In this section, the non compensational and compensational part of the aggregator are combined. The challenge is that the compensation factors a_i are real numbers out of the unit-interval $[0, 1]$. Further, they shall affect the result in accordance with their exact position in $[0, 1]$. Additionally, every fuzzy set to be processed might have a different compensation factor assigned.

Concerning these challenges the aggregator is derived stepwise:
(1) Divide the problem: Solve it individually for every abscissa value $x \in [0, 1]$.
(2) Compute for every $x \in [0, 1]$ the value of the compensational part f_1:

$$f_1(x) = \left(\sum_{i \in N_n \wedge \mu_i \neq \emptyset} \omega_i \mu_i(x) \right) \Big/ \left(\sum_{i \in N_n} \omega_i \right)$$

Example 1 *Figure 3 depicts the situation for the value x := 0. Presented are ordinate values for three fuzzy sets μ_1, μ_2, μ_3, together with the value $f_1(0)$, which is computed by:*

$$f_1(0) = \left(\sum_{i=1}^{3} \omega_i \mu_i(0) \right) / \left(\sum_{i=1}^{3} \omega_i \right).$$

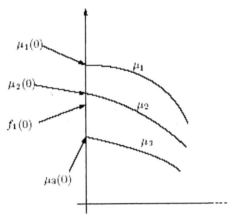

Fig. 3 Calculating $f_1(0)$

(3) For every μ_i, $i \in N_n$, compute the weighted sum of $\mu_i(x)$ and $f_1(x)$. Use the compensation factor a_i as the weight:

$$\left. \begin{aligned} \Delta_1(x) &:= a_1 f_1(x) + (1 - a_1)\mu_1(x) \\ &\vdots \\ \Delta_n(x) &:= a_n f_1(x) + (1 - a_n)\mu_n(x) \end{aligned} \right\} \forall x \in [0,1]$$

For a compensational value function μ_i it holds:

$$\Delta_i(x) = \mu_i(x) = \underbrace{a_i}_{=0} \cdot f_1(x) + \underbrace{(1 - a_i)}_{=1-0} \cdot \mu_i(x).$$

Analogue for a compensational one: $\Delta_i(x) = f_1(x)$. Hereby, it is secured that the exact position of the compensation factor a_i within the unit interval is considered.

Example 2 *In figure 4 the situation is depicted for a value function μ_i at*
$x = 0$. *The points $\mu_i(0)$ and $f_1(0)$ are shown. For $a_i = \dfrac{1}{4}$ the weighted sum*

$$\Delta_i(0) := \underbrace{a_i}_{=\frac{1}{4}} \cdot f_1(0) + \underbrace{(1-a_i)}_{=1-\frac{1}{4}} \cdot \mu_i(0)$$

lies between $f_1(0)$ and $\mu_i(0)$ at that point, which has from $f_1(0)$ the distance

$$\frac{3}{4} \cdot |f_1(0) - \mu_i(0)|$$

and from $\mu_i(0)$ the distance

$$\frac{1}{4} \cdot |f_1(0) - \mu_i(0)|.$$

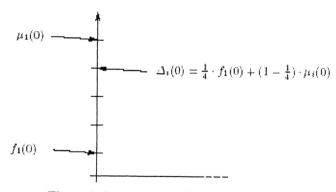

Figure4. Computing $\Delta(0)$ in Example 2

(4) For all $x \in [0, 1]$ collect all $\Delta_i(x)$, $i \in \{1, \ldots, n\}$ in a set M_x, and enhance that set by $f_1(x)$:

$$M_x := \{\Delta_1(x), \ldots, \Delta_n(x)\} \cup \{f_1(x)\} \quad \forall x \in [0,1]$$

The elements of this set represent the ordinate value of every value function μi at the position x in relation to the weighted average $f_1(x)$.

(5) Bring the non compensational part f_2 into play: As in section 4.3, it is not possible to simply define the final aggregator as the minimum of the sets M_x for all x. Analogue to the argumentation in section 4.3, for abscissa values $x \leq 0.5$ the maximum operator, for $x \geq 0.5$ the minimum operator is used. Both are combined with a linear combination, using the abscissa value x as parameter.

$$f(x) := (1-x) \cdot \max M_x + x \cdot \min M_x, \quad \forall x \in [0,1]$$

This is also the final version of the aggregator.

The following theorem presents a summarization.

Theorem 3 Definition of the Aggregator

Let $X := [0, 1]$. Let

$$(\mu_i, a_i, \omega_i), i \in N_n$$

be triples, consisting of a fuzzy set $\mu_i \in \{F_X, \varnothing\}$, a compensation factor $a_i \in \{X, \varnothing\}$ and an attribute weight $\omega_i \in N$.

Define the mapping

$$f : (\{F_X, \varnothing\}) \times \{X, \varnothing\} \times N)^n \to \{F_X, \varnothing\} \times \{X, \varnothing\}$$
$$\{(\mu_1, a_1, \omega_1), \ldots, (\mu_n, a_n, \omega_n)\} \mapsto (\mu, a)$$

with

$$\mu_1 = \ldots = \mu_n = \varnothing \Rightarrow \mu = \varnothing$$

and

$$a = \begin{cases} \varnothing, & a_i = \varnothing \forall i \in N_n, \\ \min\{a_i | i \in N_n \wedge a_i \neq \varnothing\} & \text{otherwise} \end{cases}$$

In the case:

$$\exists i \in N_n : \mu_i \neq \varnothing$$

define μ as follows: Let f_1 be as in Theorem 1 and set:

$$f_1(x) = f_1((\mu_1(x), \omega_1), \ldots, (\mu_n(x), \omega_n)), \quad \forall x \in X.$$

Define

$$M_x := \left\{ a_i f_1(x) + (1 - a_i) \mu_1(x) \big| i \in N_n \right\} \cup \left\{ f_1(x) \right\}, \quad \forall x \in X$$

Let f_2 as in theorem 2. Apply f_2 for all $x \in X$ to the sets M_x:

$$\mu(x) := (1 - x) \cdot \max M_x + x \cdot \min M_x, \quad \forall x \in X.$$

Then f is commutative, stable, idempotent, continuous, monotone, not injective, not strictly monotonic and fulfills requirement 9 for all $n \in N$.

Proof. Commutativity follows from the commutativity of addition, multiplication and the commutativity of the minimum resp. maximum operator.

Stability results from the fact, that no t-norms resp. t-conorms are used unequal to the minimum resp. maximum operator.
Idempotency: Consider an arbitrary triple

$$A := (\mu, a, \omega) \in F_x \times X \times N$$

and compute

$$f(\underbrace{A, ..., A}_{n-times})$$

Trivially it holds $f_1(x) = \mu(x)$, $x \in [0,1]$ and

$$M_x = \underbrace{\left\{ \mu(x), ..., \mu(x) \right\}}_{n+1-times}$$

Therefore $f(x) = \mu(x)$ for all $x \in X$. The case $A = (\emptyset, \emptyset, \omega)$ is trivial. Continuity follows from the fact that only continuous fuzzy sets are used to represent value and knowledge and that the parameterization of the minimum and maximum operator is conducted by linear combinations. Proving monotony is trivial.

The maximum and minimum operators are not injective, therefore also not f. Analogue monotony.

Requirement 9 is fulfilled, because the compensational part f_1 pervades requirement 9a and the non compensational fulfils requirement 9b.

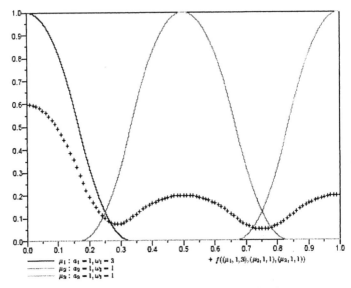

Figure5. Fuzzy Sets in Scenario 1

5.   EXAMPLES

This section demonstrates the properties of the aggregator by examples. The emphasis is on constructing extreme situations, in order to show, how the aggregator works. Starting point is that several fuzzy sets together with compensation factors and an attribute weight have been defined. The used fuzzy sets might represent value as well as knowledge functions.

5.1   Scenario 1

Figure 5 depicts the fuzzy sets used in scenario 1 (μ_1, μ_2, μ_3). All sets are marked compensational, as depicted in the caption ($a_1 = a_2 = a_3 = 1$). Hence, the compensational part of the aggregation is used and averaging is conducted. Because μ_1 is assigned weight 3 ($\omega_3 = 3$), this function effects the result more than the other two, which can best be seen in the subinterval [0, 0.2].

5.2   Scenario 2

Figure 6 shows the same fuzzy sets as in scenario 1, but now furnished with different compensation factors and weights. Fuzzy set μ_1, representing low value, is classified as "low compensational": $a_1 = 0.1$. It can be seen that in subinterval $T_1 = [0, 1 - \alpha]$ (the crossover point is set to $\alpha := 2/3$ for all scenarios) the aggregated membership function runs only slightly above the abscissa, i. e. The high value of μ_3 is no longer in effect. If it was set $a_1 := 0$, i. e. μ_1 is not compensational at all, the aggregated function would run congruently with μ_1.

Additionally, to give proof of the properties of the aggregator, the weights ω_2 and ω_3 are set to $\omega_2 = \omega_3 = 100$. Despite these high weights, it can be seen that the compensation factor $a_1 = 0.1$ is considered correctly by the aggregator.

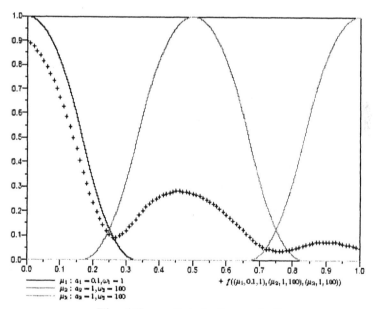

Fig. 6 Fuzzy Sets in Scenario 2

5.3   Scenario 3

Figure 7 demonstrates the effect of the aggregator, if used with a fuzzy set incorporating high value but a little compensation factor. Again, three fuzzy sets μ_1, μ_2, μ_3 are depicted, whereas μ_1 represents a moderate, μ_2 a moderate up to high, and μ_3 a high value. All sets are assigned with weight 1.

Value function μ_2 is assigned compensation factor 0. In the interval [0, 0.5] the compensational part of the aggregator, i. e. averaging, is used. The reason is that here negative values are represented and the graph of μ_1 is above that one of μ_2.

In the right part of the figure – subinterval [0.5, 1] – averaging is only conducted where the average value, i. e. $f_1(x)$, is greater than $\mu_2(x)$. Starting from $x = 0.875$ the ordinates of the non compensational function μ_2 run below the average, therefore the aggregated value function – due to applying the minimum of the non compensational part f_2 – runs (nearly) congruent with the graph of value function μ_2. The affect of requirement 9b can clearly be seen, which states that for a non compensational value the aggregated value must not be higher than this compensational one.

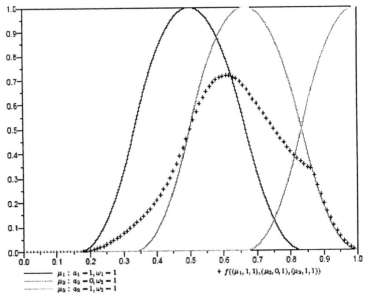

Fig. 7 Fuzzy Sets in Scenario 3

6. SUMMARY AND OUTLOOK

Verification, Validation and Accreditation (VV&A) deals with the credibility assessment of models and simulations. One possibility to support the VV&A process is the usage of decision analytical methods. Especially the concept of value tree analysis has been implemented and shows great benefit. Nevertheless, applying classical value tree analysis is not flawless, as it was shown by the identification of deficits in [4]. Therefore, the main proposal of [4] was the use of value tree analysis enhanced by fuzzy set theoretic methods.

The concept itself consists of the steps:
- establishing the value tree,
- assigning values to attributes,
- aggregating the attribute values, and
- interpreting the result.

As a first step, the definition of attribute values was dealt with in [5] and the paper at hand focused on the definition of the aggregating function.

After shortly having reviewed [4, 5] first the requirements and constraints the aggregating function must adhere to were defined. Also, mathematical counterparts of these requirements were identified in order to determine possible fuzzy set theoretic operators. A compensational as well as a non compensational part of the aggregator were defined and finally combined by a linear combination. In the last section the effect of the aggregator was illustrated by examples.

At the moment, the author considers also the step 3 and 4 in the above itemization as solved. It is planned to publish the results in the near future.

REFERENCE

[1]   Osman Balci, Michael L. Talbert, and Richard E. Nance. Application of the Analytic Hierarchy Process to Complex System Design Evalutation. Technical Report 24061-0106, Virginia Polytechnic Institute and State University, 1994.

[2]   Osman Balci, Robin J. Adams, David S. Myers, and Richard E. Nance. A Collaborative Evaluation Environment for Credibility Assessment of Modeling and Simulation Applications. In E. Y"ucesan, C.-H. Chen, J. L. Snowdon, and J. M. Charnes, editors, Proceedings of the 2002 Winter Simulation Conference, pages 214–220, 2002.

[3]   Robert O. Lewis. A Comprehensive VV&A Cost Estimating Tool. MORS SIMVAL, 1999.

[4]   Siegfried Pohl and Dirk Brade. Using Fuzzy Multi Attribute Decision Theory in the Accreditation of Models, Simulations, and Federation. Spring Simulation Interoperability Workshop, 05S-SIW-042, 2005.

[5]   Siegfried Pohl and Dirk Brade. Representing Results of V&V-Activities in Fuzzy Decision Analysis Value Trees. European Simulation Interoperability Workshop, 05E-SIW-012, 2005.

[6]   Paul Goodwin and George Wright. Decision Analysis for Management Judgement. John Wiley & Sons, 1991.

[7]   Helsinki University of Technology Systems Analysis Laboratory. Introduction to value tree analysis – http://www.mcda.hut.fi/value tree/.

[8]   Osman Balci. Verification, Validation, and Certification of Modeling and Simulation Applications. In S. Chick, P. J. Sanchez, D. Ferrin, and D. J. Morrice, editors, Proceedings of the 2003 Winter Simulation Conference, pages 150–158, 2003.

[9]   Osman Balci. A Methodology for Certification of Modeling and Simulation Applications. ACM Transactions on Modeling and Computer Simulation, 11(4):352–377, Oktober 2001.

[10]   Hans-J"urgen Zimmermann and Lothar Gutsche. Multi-Criteria Anlyse. Springer Verlag, 1991.

[11]   Ralph L. Keeney and Howard Raiffa. Decisions with Multiple Objectives: Preferences and Value Tradeoffs. John Wiley & Sons, 1976.

[12]   Didier Dubois and Henry Prade, editors. Fundamentals of Fuzzy Sets. Kluwer Academic Publishers, 2000.

[13] D. Dubois and H. Prade. Fuzzy Sets and Systems: Theory and Application. Academic Press, 1980.

[14] Adolf Grauel. Fuzzy-Logik: Einf'uhrung in die Grundlagen mit Anwendungen. Bibliographisches Institut & F. A. Brockhaus AG, 1995.

[15] Zadeh, L. A. The Concept of a Linguistic Variable and its Application to Approximate Reasoning. Technical Report Memorandum ERL-M 411, Berkeley, Oktober 1973.

[16] Zimmermann, Hans-J"urgen. Fuzzy Set Theory and its Applications. Kluwer Academic Publishers, 1991.

[17] R"udiger von Nitzsch. Entscheidung bei Zielkonflikten: Ein PC-gest"uztes Verfahren. Dissertation, T. H. Aachen, 1991.

[18] Shu-Jen Chen and Ching-Lai Hwang. Fuzzy Multiple Attribute Decision Making. Springer Verlag, 1992.

[19] Karl-Dieter Schwab. Ein auf dem Konzept der unscharfen Mengen basierendes Entscheidungsmodell bei mehrfacher Zielsetzung. Dissertation, T. H. Aachen, 1983.

CHAPTER 21

On The Use Of Simulation For The Improvement And Measurement Validation Of A Smart Antenna Prototype

Laura García, Ramón Martínez, Leandro de Haro, Miguel Calvo, Alberto Martínez, F. Javier García-Madrid.

Universidad Politécnica de Madrid. ETSI Telecomunicación.
C. Universitaria s/n. 28040 Madrid, Spain

Abstract. *The benefits of using smart antennas in a wireless mobile system have been thoroughly studied in recent years, showing an improvement in the antenna gain, a reduction of interfering power and, as a consequence, an increase in the system capacity. The real implementation and measurement of a smart antenna for the UMTS system are presented in this paper. Implementation impairments produce undesirable effects, such as frequency mismatch and coupling between RF channels. In order to minimize these effects, some simulations of the real system and the main non-ideal effects were carried out, so the analysis of these effects were affordable. Some solutions to the real implementation impairments were proposed and simulated, and the ones with better performance have been implemented and the obtained results are reported in this paper. Since traditional measurement methods are no longer suitable for the characterization of a smart antenna, a new method to measure the performance of the system is also introduced. The results show the feasibility of the system, which outperforms the performance of a conventional sectored antenna. An improvement of 6 dB in gain and 12.5 dB in carrier to interference ratio (CIR) was obtained with this prototype. The measurements were also validated by comparison with simulation results.*

Keywords-smart antenna prototype, implementation impairments, beamformer, antenna system simulation, smart antenna measurements.

1. INTRODUCTION

The use of smart antennas and their interference cancellation capabilities in mobile systems have been widely studied in recent years [1]. Smart antennas, and specifically adaptive antennas, not only can improve the antenna gain in the user direction but they can also adapt their radiation pattern in order to cancel interferences within the angular control range. This ability increases the carrier to interference plus noise ratio (SINR) for each user, which for CDMA systems implies an increase of cell or sector capacity in those cells where a smart antenna is used. This increase in capacity is higher in the case of canceling the interference produced by high bit rate users or by hot spots.

In spite of their great advantages, adaptive antennas have not been extensively deployed in W-CDMA cellular networks yet, due to their cost and complexity. Furthermore, few real implementations of adaptive antennas have been reported. Consequently, new methods of measurement and calibration, and also the effects of adaptive antennas implementation impairments are novel topics to consider. This paper summarizes the non-ideal behaviors observed in a real implementation of a W-CDMA adaptive antenna, presents the solutions proposed and implemented to overcome these effects, based on simulations of the antenna system, and the measured results.

The rest of the paper is organized as follows. Section II presents the general architecture of adaptive smart antennas and presents the peculiarities of our antenna system prototype called ADAM (Adaptive Antenna for Multioperator scenarios), [2], [3]. Section III discusses the use of simulation as a tool to design the real system implementation, to characterize the degradation of performance due to implementation, the use of simulation to design and evaluate solutions to mitigate implementation impairments and how we have used simulation to. characterize and validate measurements. Finally Section IV draws the main conclusions of this work.

2. ADAM SMART ANTENNA: GENERAL ARCHITECTURE

2.1 ADAM vs Conventional Smart Antennas

The general architecture of a WCDMA receiver adaptive antenna system is shown in Fig. 1. The antenna array is formed in this case by a set of either conventional sectored antennas. Their ports are connected to RF/IF chains and converted digitally to base band by suitable A/D converters. The signals for each of the different users are demodulated and the Pilot bits of the DPCCHs are used by the LMS beam forming algorithm to calculate the weights to combine, for each user, the signals from each antenna element.

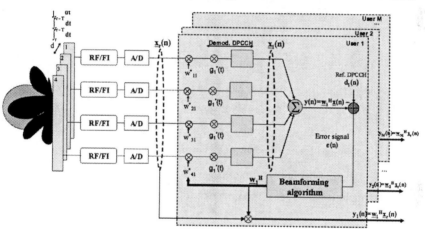

Fig. 1 Typical Architecture of a WCDMA Adaptive Antenna

2.2 General Architecture of ADAM Prototype

ADAM stands for ADaptive Antenna for Multioperator scenarios, a smart antenna prototype developed for UMTS. Unlike typical adaptive antenna systems that interconnects at base band level (manufacturer dependent) to the Node B, ADAM remodulates the signals and interfaces the Node B at RF level (standard Uu interface). ADAM can be easily connected to any standard Node B site even shared by several operators using equipment from different manufacturers. Those unique characteristics allow a straightforward deployment of ADAM over any existing UMTS cellular network replacing sectored antennas.

Fig. 2 shows the architecture of ADAM implementation [4]. To allow the "plug and play" functionality ADAM antenna is connected to Node B using the 3GPP standard Uu interface [5]. There is a direct connection between the smart antenna RF outputs and the base station RF inputs. Therefore demodulated UMTS signals, used to determine antenna weights, are remodulated again and upconverted before being sent to the base station in the uplink.

In the downlink, RF signals from the Node B are despread and demodulated, to separate contributions from different users and the common channels. Once the beamformed user signals are obtained, a remodulation and upconversion is performed for each array element.

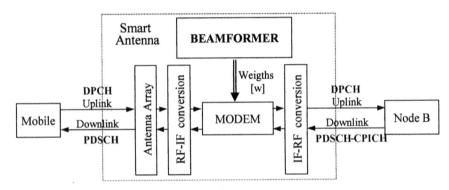

Fig. 2 Implementation Architecture of ADAM to be Deployed on a Standard Node-B.

ADAM has been implemented using a Software Defined Radio (SDR) architecture and it can be easily reconfigured and updated with more advanced versions. ADAM prototype is formed by four independent subsystems:

Antenna array: it uses four conventional UMTS sectored antennas to produce a 120° cellular coverage.

RF-IF sections: this stage performs the up- and down-conversion stages, using analog signals at 44 MHz IF. Wideband ADCs and DACs are used at IF instead of at baseband

MODEM: its main tasks include the set-up procedure, synchronization stages and modulation/demodulation of the DPCCH for each user in order to obtain the reference signal needed for the beamforming section.

Adaptive beamfomer: its main task is the calculation of the adaptive set of beamforming weights for each user. For this computation, the NLMS algorithm has been used.

Fig. 3 shows the antenna array subsystem (left) and the rack with boards of the other subsystems connected to the measurement equipment (right).

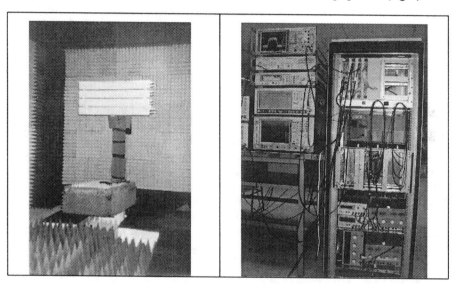

Fig. 3 Subsystems of ADAM Prototype. Antenna Array (Left) and RF-IF and SDR Modules (Right).

3.   USE OF SIMULATIONS TO IMPROVE REAL SYSTEMS

3.1   Motivation

Typically, simulations are used as a tool to study and analyse a wide range of systems and situations. By means of simulations, the parameters of the system can be properly chosen and the expected performance is evaluated. Regarding mobile communications and their related topics, such as smart antennas, simulations are generally used first in the design phase, so the system under study can be initially characterised. Specifically, the conventional simulations related to smart antennas in a system for mobile communications deal with the improvement that can be obtained by means of including smart antennas in the system, [1], [6], or even the design of several basic parameters, such as the number of elements in the array and the separation between them, the number of smart antennas to be included in the system, etc.

However, another aspect that can be considered in the use of simulations is their use as a tool to design and evaluate the real-time implementation of the system. The simulation of several aspects related to the implementation of the real system can be utilised to properly select the main parameters in both the overall system and the modules that constitute it. Moreover, a very attractive use of simulations in this topic, that has not been extensively considered yet, is the analysis of real implementation impairments, as well as the study of possible solutions to mitigate their effects.

When a system is initially designed, several simplifications and approximations are assumed, in order to make the design simpler. However, when the system is implemented by using real devices, several non-ideal behaviours may appear, which generate undesired effects and degradation of the system performance. In most cases the separation of these effects is not straightforward, which makes the understanding, analysis and solution of them a difficult task. The simulation of these effects, as well as other aspects of the real implemented system may help us to improve the performance and reduce the non-ideal effects. In the next sections several examples of simulations applied to the specific case of the ADAM prototype are shown.

3.2   Use of Simulation as a Tool to Design the Real System.

Apart from the conventional parameters that must be selected in a system, when a real implementation is carried out there are several features that must be chosen in the design, which are related to the use of real devices and platforms where the system is developed. In the case of modules implemented on software-radio platforms, for example digital signal processors, it is interesting to consider aspects such as the quantification error introduced by the use of an analogue to digital converter (ADC) with a certain number of bits, the type of arithmetic that will be used, etc. Also significant are the considerations to take into account when a real-time performance is necessary, as it is usually found in a prototype. In order to analyse all these aspects, the use of simulation becomes of great importance.

If ADAM prototype is considered, several aspects were selected in the design of the system. Regarding the digital processing modules, one of the main features to be studied and chosen is the type of arithmetic to be used and, consequently, the type of digital signal processors. In the case of ADAM prototype several considerations were taken into account. First of all, the system must work in real time. Therefore, fast operations and high computational capabilities were needed. Moreover, most of the operations in the digital modules are products and additions (in order to combine signals in the beamformer module, modulate and demodulate them in the MODEM module, etc). Taken into account these points, a fixed-point arithmetic was selected for ADAM prototype, since it fulfils high computational requirements, especially if linear operations are to be done. However, when using fixed-point arithmetic the accuracy in variables may be lower than the obtained when using floating-point arithmetic, since they are represented by integer values with a certain precision. Therefore, a not negligible quantification error may appear. An important design aspect in fixed-point case is the number of bits to be used in the representation of variables. In the specific case of the DSPs used in ADAM prototype, two possibilities were considered: 16-bit or 32-bit variables. In order to study which size fits better the system, simulations were carried out to compare fixed-point and floating-point accuracy. One of the critical modules in the digital processing part is the beamformer, due to accuracy requirements. Thus, the simulations were focused in this module. As an example, in Fig. 4 a comparison between the value of one of the variables in the adaptive algorithm used in the beamformer module implemented with 16-bit word fixed-point (label "C") and floating point (label "Matlab") arithmetic is depicted.

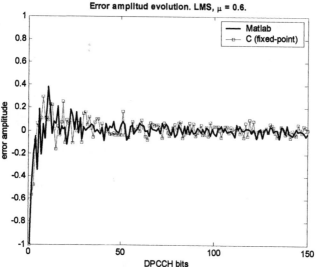

Fig. 4 Variable Error E(N). Comparison Between 16-Bit Fixed-Point and Floating Point.

As it can be observed, 16-bit fixed-point arithmetic allows obtaining enough accuracy, almost as good as the one obtained with floating point. Several variables and situations were analysed by means of simulations, and the computation of quantification error was carried out to study the possible word sizes. As a result, simulations showed that the 16-bit fixed-point implementation offers enough accuracy for our application.

3.3 Use of Simulation to Characterize Implementation Degradation.

Once a real system is implemented, there may appear several undesired effects that were not considered in the design step and produce a certain level of degradation in the performance of the system. Generally, the possible causes of degradation may be difficult to observe separately in the real system, thus making quite complex their characterization and solution. The simulation of these effects simplifies their analysis, since they can be studied separately. Moreover, simulations can be used to know which real implementation effects are more important in the degradation of the system, so they can be treated to reduce their undesirable effects. It is also of high importance to be able to quantify the maximum implementation impairment or deviation of system behavior compared to the ideal one, in order to guarantee a certain performance of the system.

Implementation impairments that were not considered in the first stages of the design can be crucial in the performance of the system. An example of this situation in ADAM prototype is the coupling and signal instability in RF-IF chains. In the ideal system considered for design steps, the RF-IF chains are assumed to be perfectly isolated, that is, no coupling appear among them, the phase lag and power losses in each chain is considered to be the same, and the signal in the chains present a perfect stability in time. However, these features cannot be fulfilled in a real implementation because the real system will introduce coupling between chains, a certain level of instability in RF-IF signal and different power losses and phase lag in each chain. What is more, to separate all these impairments and their effects may be a difficult task in the real system. In order to overcome this problem, several simulations were carried out to characterize their effects. It was observed that the main problems were the coupling between chains and instability in time, since the adaptive algorithm used in the beamforming module (an N-LMS conventional algorithm) is capable of compensating the differences in phase lag and power losses in the chains. In order to characterize the maximum allowable instability in the system several simulations were carried out, where variations in the signal were included, in both the phase and the amplitude. They were simulated as a Gaussian random variable, independent in each chain. The obtained radiation patter when two sources (one desired user and one interfering user) are considered in the system was studied in two situations: with high instability in time and with low instability in time. Fig. 5 shows the two situations, where several radiation patterns have been depicted simultaneously in order to compare the variation of the results.

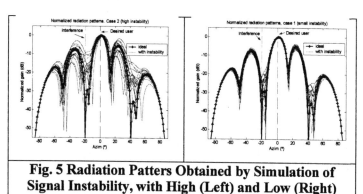

Fig. 5 Radiation Patters Obtained by Simulation of Signal Instability, with High (Left) and Low (Right) Variation.

As it can be noted, the instability in chains provoke an important degradation of the radiation patter, where the main effects are null filling and higher side lobes. After many simulations it was concluded that, in order to obtain an accuracy in the radiation pattern of at least -20 dB, the maximum allowable standard deviation in relative phase and amplitude was 5° in phase and 1 dB in amplitude. Measurements of the RF-IF chains in the ADAM prototype showed that the system fulfills these maximum levels of instability.

3.4 Use of Simulation to Design and Evaluate Solutions to Mitigate Real Implementation Impairments.

In the previous section, simulations were used to characterize implementation impairments in the system. Once the main effects are known we can try to mitigate some of them, by including new modules or modifying the existing ones in the digital system. Since several possibilities can be considered, it may be interesting to simulate them in a real system simulator before implementing them, since the real implementation in a DSP platform usually implies a more complex design and takes more time than the simulation in a PC, especially when fixed-point and real-time aspects are to be considered. Several possibilities can be analysed via simulation and then only the one with the best performance can be implemented and integrated in the system. This way the simulations allow reducing the time to solve the implementation impairments.

In ADAM prototype, one of the implementation impairments in the digital modules, whose effect is significant, is the frequency mismatch. Since the IF frequency for the input signal in the ADC is not exactly the expected value, the digital downconverter (which only consider a fixed IF frequency) offers as output a signal with a certain phase error, due to the frequencies differences. In the Fig. 6 this idea is depicted.

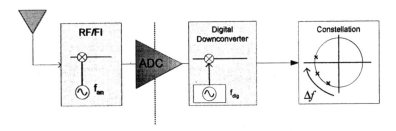

Fig. 6 Scheme of Receiver Chain: The Frequency Mismatch Produces a Phase Error.

As a consequence of the frequency mismatch, $\Delta f(t)$, a phase error appears, that can be modelled as:

$$\Delta \theta(t) = 2\pi \, \Delta f(t) \cdot t$$

The phase error was studied in the DPCCH demodulated bits for the desired user. Since these bits are used as a time reference for the adaptive algorithm in the beamformer, it is important to correct the possible frequency and phase error. Note that considering the expression above, if the frequency mismatch is approximately a constant value compared to the rest of the variables (as it is in our system), the phase error should be linearly dependent with time. However, when considering several users in the simulations, it was observed that the demodulated bits not only suffer a constant phase error, but also a small deviation from the expected value, even when the noise level is very small. This effect is mainly due to the residual signal of interfering users, which is not totally eliminated by the despreading operation in the demodulation of desired signal. To take into account this effect, a new term can be included in the above expression, so the phase error can be expressed as:

$$\Delta \theta(t) = 2\pi \, \Delta f(t) \cdot t + \Delta \varepsilon_{interf}(t)$$

Note that the residual signal of interfering users is a complex value and its effect in $\Delta\theta(t)$ is not a constant value. In Fig. 7 the constellation of the demodulated signal with and without this effect is depicted.

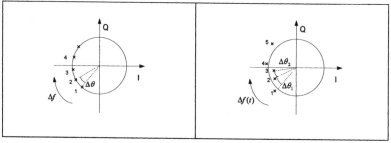

Fig. 7 Demodulated DPCCH Bits, with Constant ΔF, with No Interferences Considered (Left) and Considering an Interfering User (Right).

If the users signals are supposed to have zero mean, the residual interfering signal will have a zero mean effect as well, which means that the mean value of $\Delta\theta$ computed in a certain number of samples will be approximately the same in both cases.

In order to track the phase variation in the demodulated bits, two schemes were considered:

- A bit-by-bit track module, which computed the phase deviation between demodulated bits (taking into account the possible sign change depending on the transmitted bit) and carried out on-time compensation.
- A block track module, which computed a mean value of the phase deviation and carried out a "late" compensation, that is, compensate the phase error in the current frame with the mean value computed in the previous frame.
-

Both possibilities were analysed with two methods: by direct simulation of the whole system and by processing samples taken from the real system, in an "off-line" way. The study showed that the second scheme offers a better performance, regarding the capability of adjusting the radiation pattern to obtain a null in the interference direction of arrival.

The phase track had to be implemented in the fixed-point platforms. The computation of the actual and desired phase values requires the calculation of trigonometric functions, which in a fixed-point DSP is not efficient. In order to obtain a real-time implementation, a look-up table with quantized values of phase and trigonometric functions was used, so the operations that imply a higher computational load are avoided. But again the use of quantification requires to select the look-up table size. Therefore some simulations were carried out to be able to choose a proper number of quantized values.

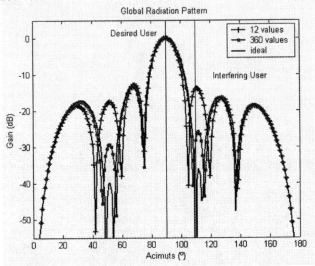

Fig. 8 Radiation Pattern Simulated for Different Quantized Phase Values in the Phase Track Module.

In Fig. 8 the radiation pattern obtained by simulating the system with different number of phase values in the look-up table is shown. It can be observed that with only 12 possible values of phase (when values can be between 0° and 360°) the accuracy is not enough for the beamforming algorithm to be able to put a null in the interference direction, although the maximum gain is properly pointing at the desired user. This is due to the demodulation effect, which reduces the information about the interfering user, so with only 12 values of possible angles the adaptive algorithm loses track of it. When using a 360-value quantification, the obtained radiation pattern shows a very low side lobe in the interfering direction, so that this number of quantified phase values can be considered accurate enough for our application.

3.5 Use of Simulation to Characterize and Validate Measurements

Once the different antenna components have been integrated and tested the complete system has been measured in normal operation conditions. Two scenarios have been considered to make the final measurements. The first one deals with the measurement of ADAM performance in a controlled scenario inside an anechoic chamber. Measurements have been carried out in a chamber where the far-field condition is satisfied ($d>2R^2/\lambda\sim5.2$ m). A positioning device is used to roll the smart antenna over the azimuth domain.

To evaluate the correct performance of the smart antenna, traditional measurement method for characterizing the radiation features of an antenna, based on the use of probe signals (tones), is no longer applicable since the antenna works with wide band UMTS signals. The measurement system has been modified to record the received power in the 5 MHz used carrier band versus positioner angle variation.

One UMTS user and one interfering source are allocated changing the relative angular position between them to demonstrate beamforming capability. The up-link pattern is measured by fixing the obtained beamformer weights for a given relative position between desired and interfering users and then rotating the antenna end recording the received power. For the downlink the same uplink weights are used but corrected to account for the difference in frequency and calibration of the corresponding RF/IF chains. Fig. 9 shows the measurement scheme used in the anechoic chamber.

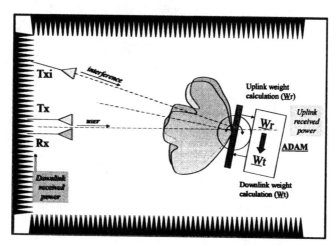

Fig. 9 Scheme for Array Pattern Measurement in Anechoic Chamber.

The first step that must be performed is the calibration of the receiver chains. Due to array imperfections, mutual coupling, differences in the RF-IF chains, the signal vector that is fed into the adaptive beamforming module differs from the one received in the array from the propagation scenario. The calibration is performed in the absence of interfering signals and the selected calibration scheme used is shown in

Fig. 10. Amplitude and phase differences between receiver chains can be obtained comparing the received signal with the reference signal vector. That is, if the user reaches the array through the broadside direction, the downconverted signal vector x' must be equal in amplitude and phase to x, x=x'. However, due to the calibration errors, this condition is not satisfied.

The objective of the calibration unit is to correct the received signal vector $\underline{x}'$, so that $\underline{x}' \approx \underline{\tilde{x}}$ is satisfied. The calibration can be carried out positioning the user in any other known direction of arrival (DOA), but for simplicity it has been placed in the broadside (in this particular case, all the components in $\underline{x}'$ should be equal in amplitude and phase in absence of receiver chain imperfections).

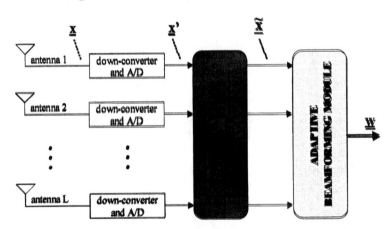

Fig. 10 Calibration scheme used in the uplink for correcting coupling between the antenna elements and amplitude and phase differences in the chains.

The measured adaptive radiation patterns can be compared with those obtained using a simulator, that includes most of the studied effects and considers a radiation pattern model of the antenna elements of the type $\cos^q$. This type of pattern is considered to be the one of the sector antenna. As an example of comparison the Fig. 11 shows the uplink patterns simulated, measured and that of the array element (sector) when the desired user is at $0°$ and interfering at $18°$. The higher side lobes at $\pm 60°$ obtained in the measurements indicate that the element pattern is broader than the model and the attenuation of the grating lobes is lower than predicted. Otherwise the cancellation of the interferer is well accomplished.

In normal operation the weights are continuously changing when the antenna is rotating. In this case the antenna pattern can not be measured and the performance of the antenna is characterized by the increase in gain and carrier to interference ratio (CIR) with respect to the sectored one. Those parameters have been simulated and directly measured in the anechoic chamber. The measured increase in gain was of 6 dB as simulated. The measured increase in CIR was of 12.5 dB that compares with the simulated value of 14.5 dB or 16 dB indicated in the measured pattern in Fig. 11.

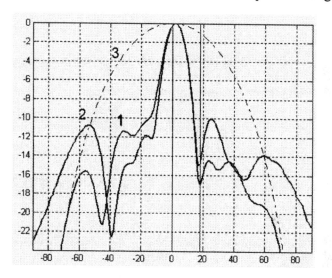

Fig. 11 Simulated (1), Measured (2) and Element (3) Patterns When Desired User is at 0° and Interfering at 18°.

4. CONCLUSIONS

A novel prototype of smart antenna applied to W-CDMA systems has been presented, focusing on the implementation of the digital modules. The proposed scheme allows to simplify the smart antenna system, reducing the complexity and thus allowing a feasible architecture.

The simulation of several aspects of the real implementation, such as the quantification error due to the use of fixed-point arithmetic, has been used to improve the design of the real system. The realized prototype was measured and several implementation impairments were characterised via simulation of parts of the system and their non-ideal effects. The simulation of these effects has simplified the analysis and posterior correction of them. Several examples have been shown, such as the design of a phase track module to correct the frequency mismatch in the real system, as well as the performance improvement obtained. As another example of results from the simulations carried out, the maximum allowed values in the implementation impairments were obtained.

The measurement of the main performance parameters of the antenna is described and the results from those measurements are compared with those predicted by simulations.

Acknowledgment

The authors wish to thank SICE and MCYT (TIC2002-01569) for the support in the development of ADAM project.

REFERENCES

[1] L.C. Godara, "Applications of Antenna Arrays to Mobile Communications, Part I: Performance, Improvement, Feasibility and System Considerations", Proc. of the IEEE, vol. 85, no. 7, pp. 1031-1060, July 1997.

[2] M. Sierra, M. Calvo, L. de Haro et al., "Modular Smart Antenna Multi-Standard for Multi-Operator Cellular Communications Scenarios", Patent no. P200102780, Spain, 2001 (patent pending).

[3] R. Martinez-Rodriguez Osorio, L. García-García, A. Martínez-Ollero, F.J. García-Madrid Velázquez, L. De Haro-Ariet, M. Calvo-Ramón, "ADAM: A Realistic Implementation for a W-CDMA Smart antenna", Special Issue of Eurasip Journal on Applied Signal Processing on "Advances in Smart Antennas" (3rd Quarter, 2004).

[4] M. Sierra, M. Calvo, L. De Haro, R. Martinez, L. Garcia, A. Martinez, J. Garcia-Madrid, J.L. Masa , J.M. Serna: "Integration and Measurements of an UMTS Smart Antenna". IEEE Antennas and Propagation 2004 Symposium. Monterrey, june 2004.

[5] Physical layer–General Description (FDD). 3GPP TS 25.201 v5.2.0 (2002-09).

[6] M. Sierra-Pérez, M. Calvo, L. de Haro, R. Martínez, M. Sierra-Castañer, L. García, A. Martínez, F.J. García-Madrid, J.L. Masa, J.M. Serna, "Integration, measurements and calibration of a UMTS smart antenna", 15th IEEE International Symposium on Personal, Indoor and Mobile Radio Communications, Barcelona, Spain, September 2004.

CHAPTER 22

The Need for Credible Modeling and Simulation in the Context of the Network Based Defense

Dirk Brade

Kungl Tekniska Högskolan
Department for Electronics and Computer Science
10044 Stockholm

Abstract. *In concert with numerous other western world nations, in Sweden military doctrine is changing focus from a platform-oriented towards a network-based approach. Military platforms become participants in "the network", share information, and provide services that allow them to sense, decide, and act beyond their individual capabilities in a cooperative and distributed manner. The information exchange occurs through an information infrastructure, the so called "infostructure", which constitutes a communication network spanned by the participating nodes. Although technologically not far beyond the edge of today's state of the art, are functional and technological requirements on the infostructure high: Mobile units exposed to lethal threats and electronic countermeasures shall connect and disconnect rapidly and reliably to build networks in an adhoc manner, ensuring secure and fault tolerant communication with sufficiently high quality of service. Nodes in the infostructure consist of heterogeneous systems provided by numerous nations, providing various types of end-user and communication services.*

Various alternative approaches can be taken for the creation of the infostructure, which must be expected to become a highly complex system, challenging to control and administrate. Modeling and Simulation are considered to be the mandatory enablers to optimize its design, maintenance, use, and improvement. This concept paper describes a vision of the infostructure, operational conditions, and desired properties, and identifies challenges which must be met on order to early avoid dead-end roads. Finally, it outlines application areas for communication related Modeling and Simulation to support the transition to and continuous improvement of a Network Based Defense.

1.   INTRODUCTION

Western-worldwide, the defense is in transformation. The trend for Future Combat Systems (FCS) drifts away from the integrated monolithic platform to specialized components, which can be quickly assembled to combat systems with capabilities fine-tuned to the current situation's demands. The individual platform shall unfold its full power as integral part of "the network", sharing information with other "nodes" in this network, invoking "end-user services" offered by others, and providing end-user services to others [Alberts, Garstka, Stein, 1999]. In a vision of the Network Based Defense (NBD), nodes constitute of a range of subsystems, spanning from the individual human on the ground, who contributes to the network by providing services according to his perception, reasoning, and engagement capabilities, to a satellite in space. Not only in Operations Other Than War (OOTW) also other nodes not maintained or owned by the military, but by civil organizations such as ambulance, police, or fire brigade may join the network, who provide their own ideally compatible communication infrastructure for Integrated Disaster Management [Meissner et. al., 2002].

1.1   What is the Network-Based Defense About?

In a nutshell, the Network-Based Defense (NBD) is about increasing the flexibility and the agility of the defense, to enable it to deal more efficiently and effectively with traditional threats (e.g., symmetric warfare, but also civil disasters such as train accidents or floods), but also to cope with new threats (e.g., extreme terrorism, invoking the application of weapons of mass destruction). Creating the network and teaching the participants in taking benefit of it aims on the improvement of response times and the optimization of resource management. Generally speaking, collaboration shall be fostered and cooperation improved, to create a true joined coalition environment, including civil and military units of numerous nations, forces, and military services.

1.2 NBD Information Concepts

[Alberts, Garstka, Stein, 1999] brings one of the main concepts of the NBD with the proverb "mass effects, not forces" to the point. The infostructure enables to "move information, not people" and enhances thus the elementary activities "sense, decide, act". Independently of military leadership level, soldiers, vehicles, weapon systems, and other pieces of equipment and their operators are considered to be end-systems and, as such, nodes in a network. A node can use the sensors of other nodes to sense information that it otherwise would not be able to retrieve, may access intellectual, authoritative, or computational capability it otherwise would not have to decide, and may utilize a remote effector to take some action through the network it otherwise could not take [Alvå and Palmquist, 2003].

- A sensor node feeds information sensed in the battle space into the net. Any other authorized relevant node should be able to access the sensor information, if desired. Sensor nodes provide "sense services". An envisioned application area for remote sense services is the creation of the "single integrated situation picture", which can be broadcasted and "continuously" updated for interested, authorized receivers.
- A decision node retrieves information from the net and places additional modified "value-added" information into the net, including control commands for the sensor nodes or actor nodes. Decision nodes provide "decision services". They are supposed to increase efficiency and effectiveness of reporting, controlling, commanding, and collaborative decision making.
- An actor node receives instructions from the net and creates an effect in the battle space. Actor nodes provide "action services". To provide this action may require a human operator at this node, but this is not mandatory as demonstrated by current achievements with, e.g., unmanned vehicles and other robots.

From a communications perspective it is highly relevant that everyone who is a part of the network gets the information or communication service required with sufficient quality in time (if authorized & concerned). For this purpose a communication infrastructure is required, which meets the application area's needs for information sharing and exchange. In Sweden not only the military's needs should be taken into account, but also those of the "Krisberedskapsmyndigheten" (KBM), who plan for the country's integrated disaster management [KBM, 2005].

According to new doctrine, defense capabilities shall be provided based on services, and the technical solution favored by the Swedish Defense (FM) are so called "end-user services" (Verksamhetstödjande Tjänster) [Jönsson, 2002]. These shall directly support the users of the infostructure to sense, decide, and act, without revealing to them, how a service actually is implemented. (E.g., when requesting enemy position data for a given area, the service to provide this information can be implemented by a human observer, by a UAV, or by a satellite, but this remains hidden to the service requester.)

The infostructure shall support all three end-user service classes by transmitting information collected by remote sensor nodes, reporting to and getting orders from a decision node, and transmitting the request for a remote action. From the communication point of view, the infostructure is intended to serve as a distributor of data, information, and knowledge, which filters and forwards data into local networks, and stores, checks, and propagates information with is relevant for other nodes according to their authorization. Communication services include email, voice, messaging, video, and as typical new information service desirable for the (self-) coordination of several units the so-called "single integrated situation picture".

1.3 The Current Situation

The technology currently available promises technological solutions, which meet the needs of the customer already in close future. To build the infostructure one definitely does not start from scratch: There are numerous technology solutions and existing infrastructures such as (wired and wireless) telecommunication networks, data networks, or the digital radio (professional mobile radio, PMR) currently in use, subjected to field trials, or under development and exploration. Science, theory, and practice are matured concerning network architectures (e.g., ISO OSI layer model, Internet Protocol). Wireless standards exist for various ranges, speeds, and bandwidths, including the IEEE 802.11 series for WLAN, UTMS, GPRS, TETRA, TETRAPOL, GSM, BlueTooth, IrDA, and others – quite different standards for different purposes, each with advantages, but also limitations. When it comes to the content to be transferred, there are military standards and data models, such as Link 16, JC3IEDM, or ATCCIS, which all are more or less appropriate to describe particular information items. How to invoke remote services is not at all new in the area of distributed systems, just to mention "Remote Procedure Calls (RPC)", or web-services, building und service oriented architectures enabled by XML, SOAP, and UDDI.

Briefly, there is a bunch of technologies and partial solutions already available. In Sweden, several demonstrators for network-based concepts are quite far advanced [Flink, 2004], and with the lessons learned from demonstrator development, the picture of the infostructure needs to be re-adjusted. A clear statement needs to be made of what is expected from the infostructure in mid-term future; already Seneca knew, "if a man does not know what port he is steering, no wind is favorable". The challenge for the infostructure is not any more to find *a* technical solution, but to find a good one.

2. NBD NEEDS STATEMENT

When going back to the roots, the high level objective of building the infostructure is to facilitate secure information sharing and exchange for globally distributed participants under a variety of operational conditions to provide the end-user services with sufficiently high quality for distributed and collaborative sensing, deciding, and acting. In principle, this is nothing exciting new in the community of teletraffic researchers and providers, but rather among the issues addressed since quite a while [Ince, 2002]. However, the operational conditions under which the infostructure shall be accessed create new, challenging demands.

2.1 Operational Conditions

The infostructure may span from the head quarters in a nation's capital, through space, to the soldiers under fire at the frontline, or rescue teams in the basement of a skyscraper close to collapse, and needs to provide the end-user services in geographically distributed locations under a variety of conditions. In some controlled, friendly areas it can be assumed that there is a stable pre-installed communication infrastructure available that can be re-used, but it might be insecure. In a hostile foreign or devastated environment, the infostructure must be self-maintaining, i.e., it cannot be assumed that carrier signals for wireless communication or pre-installed hardware are available. The communication network must be built quickly – and in the most dynamic locations in an ad-hoc manner. Many of the nodes close to the forward line of own troups will be exposed to lethal threats. The enemy will try to suppress communication by systematic elimination of communication key nodes and links, wireless communication by electronic counter measures (e.g., jamming), and to intrude the system with highly skilled professional hackers. To summarize:

- The geographical distribution of end-users will range from close collocation in a single building to global distribution. Global distribution is well experienced by telecommunication and the internet (which does not mean that no challenges remain), and the development of the required technology commercially driven. However, in contrast to most users of commercial networks, users of the infostructure also need to hook up to the network in sparsely populated and rural places with weak communication infrastructure.
- Mobility of end-users and access points: End-user movement will range from stationary in a single given location to mobile with supersonic speed. This condition applies in similar form to commercial networks, where mobile end-users, which are actually online during movement, will also make up a significant share of network users.
- Count of Mobile End-Users: The count of mobile nodes in the defense network is upwards open, ranging from a small group of rescue team members to all units in a theater scenario. This the defense network has in common with commercial networks, which currently work and also will do so in future with huge numbers of participants.
- The climatic environment of end-users ranges from air-conditioned offices to the extreme outdoors such as jungle, desert, or arctic environments. It must be assumed that commercial development will continue to mainly aim at indoor environments, with a few products for outdoor use to fill niches.
- The pre-installed infrastructure available for communication may reach from a fully functional digital communications network in a first world urban area to none at all in a crisis or disaster area. An operational requirement of the NCD not shared with commercial networking is the ability to install in a short time the communication infrastructure required by the own staff.
- The acceptable signal emission (influencing the stealth of a node) ranges from very low (e.g., for a submarine on a reconnaissance mission) to very high (e.g., the head quarters in the national capital). To keep those nodes in the defense network informed, for

which the communication must be minimized, is both a conceptual and a technical challenge (quite) unique to the NCD.

- Front nodes in the defense network will be exposed to jamming or other electrionic suppression measures. Communication devices should remain as functional as possible under unfriendly conditions. It cannot be assumed that the commercial market will provide devices that meet the military's needs concerning robustness against counter measures.

- The exposure to massive professional cyber-attacks varies from extremely low to extremely high. A defense network exclusively dedicated to national security will be attacked by many professional, well-educated specialists with significantly higher resources eceeding by far the intensity of attacks against today's internet or other communication networks.

- The physical and mental state of the end-users also must be taken into account, because healthy and fresh staff can take much more benefit from a complex multi-functional man-machine-interface than tired, exhausted, or even wounded soldiers or rescue personell.

2.2 Desired Properties

From the needs statement and the operational conditions a set of desired properties can be derived and directly compared to the current properties of the internet.

- Throughput and capacity: The throughput and capacity of a connection must be sufficiently high to meet QoS requirements, which depend on the connected participants and their demand for information. A general statement that always high bandwidth and low delay are required does not hold: Text-based communication can be achieved with a quite low bandwidth, but whenever audio and video material is transmitted, high capacity communication is required. Looking ahead, there should be enough buffer space to support future, more bandwidth-intensive information transfers, but communication discipline promises to achieve meaningful information exchange within the infostructure with a throughput and capacity which is in general lower than those required by the community of internet users, who are confronted with thousands of spam mails, bit-torrents, and entertainment video streams.

- Security: The infostructure will be subjected to professional intrusion attempts, including all classical forms of network attacks (such as denial of service, phishing, hacking, cracking, bribery of operators), as well as attacks that are less common in peacetime working environments (e.g., physically enforced access to a network node, hostage situations). Only authorized nodes shall access the information intended for them, without an unauthorized party getting them also. Compared to the internet, the needs of the infostructure concerning privacy and integrity are significantly higher.
- Reliability and fault tolerance: The communication must be robust against electronic warfare measures like jamming or the systematic elimination of communication key nodes. It must remain operational, even if numerous nodes or links fail to function. The infostructure's requirements on reliability and fault tolerance are higher than on the internet – although down times of connections and servers are often not tolerated, the infrastructure is rarely exposed to physical threats.
- Flexibility and Mobility: Stationary units are extremely vulnerable to attack. Nodes will need to disconnect from the infostructure to avoid detection and to reconnect later, maybe in a different location (e.g., during reconnaissance operations). Data must be intelligently cached that it is available to those who connect later, even if the source of the data has disconnected. Infostructure nodes shall be able to quickly join and leave (in combat the transition from a sending/receiving state to silent/ stealthy state must be assumed to be rapid and unannounced), without negatively affecting the network functionality. Demands on flexibility and mobility of network nodes are higher than for the internet.
- Interoperability: In coalition warfare or cooperation with civil organizations for Integrated Disaster Management different parties will work together to achieve a common objective, which can be supported by information exchange and sharing. Interoperability needs are equally high as for the internet, although – due to the existence of strong national interests – most likely harder to achieve.
- Robustness: Although the users of the infostructure are expected to be trained specialists, the infostructure's man-machine-interface (MMI) should be highly robust with respect to end-user errors. The infostructure must be useful in crisis situations for exhausted, tired, or wounded end-users under stress, too, and whenever a user

operation error might results in a security breach, robustness needs are higher than in the internet.

- Safety: Some communication devices hold a risk for the health of the human end-user, and this risk must be reduced to an acceptable minimum. However, the criticality of this constraint is rather low in comparison to safety requirements imposed on those systems of the infostructure that actually create an effect in the battle space. Safety needs are considered to be equal to the internet's infrastructure and in exceptional crisis situations even lower.

To which degree the above properties can be established depends on the architecture of the infostructure, the quality of the design and implementation of its subsystems, and its configuration in a range of more or less unfriendly environments.

3.   TOPOLOGY, ARCHITECTURE, AND CHALLENGES

The infostructure will consist of components to mainly satisfy two roles with different properties and requirements, which can be distinguished as:

- Interface nodes, which serve as access points for the end-users from and to the battle space (e.g., end-user workstations, sensor interfaces, notebooks, PDAs, HUDs, or interactive maps). They must be built and configured in such a manner that the end-user gets the desired functionality from the infostructure, under consideration of his training and education, mental and physical state, and pressure created by the environment.
- Networking nodes and links, i.e. components exclusively dedicated to information transfer without direct interfaces to the end-user, which nevertheless need to be maintained by infostructure operators/administrators to ensure infostructure functionality (e.g., fibers, routers, switches), constitute internal nodes and links. They must be built, connected, and configured in such a manner that the end-user requirements on availability, reliability, security, or connectivity are not violated.

3.1 Network Topology and Layer Model

The requirement to reuse currently available and mastered scalable concepts and technology for information transfer implies also the topologies of the infostructure, resulting in a "network-of-networks" (NoN) similar to the internet with higher demands on mobility and security due to the extended operational conditions, as discussed in section 0. A NoN can enable point-to-point connections between the participants of different subnets of the infostructure, while maintaining the ability to deal with a large number of end-nodes. Different communication technologies can be used in the various subnets, which are connected to lower, higher, and neighboring networks by gateways. To achieve a degree of fault-tolerance higher than in the internet, gateways should be redundant; otherwise in a NoN topology the single, non-redundant gateway constitutes a bottle neck, which could disable any communication to the outside of the (sub-) network.

Similar to the adoption of the NoN-Topology, also the architectural concepts for the technical realization of the infostructure discussed in the Swedish defense are mainly borrowed from the architecture of the internet, where extensive experience with a NoN is available. End-user services which support distributed and remote sensing, deciding, and acting, will rely on infrastructure services for their execution, which, e.g., encrypt and decrypt, cache, re-route, or perform other fundamental tasks for their completion. Although the architecture details are not yet stabilized, demonstrators are under construction and give opportunity for evaluation of architecture proposals, which all have in common that there will be "layers". On a quite abstract level, [Carling et al, 2003] envision a layer architecture as depicted in Fig.1.

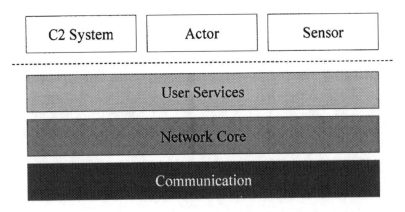

Fig. 1 Network Structure from [Carling et al, 2003]

- End-User Service Layer: This layer holds the end-user services of the three main categories sense, decide, and act (compare [Jönsson, 2002]). To avoid even higher complexity of the infostructure, all data fusion or transformation services, ontology's, and other "helper applications" should be provided as services from the outside of the infostructure. Drawing a parallel to the ISO-OSI layer model (for a clear introduction refer to, e.g., [Tanenbaum, 1992]), this corresponds roughly to the layers 5 – 7. The definition of end-user services should be driven by the end-user.

- Network Core Layer: On this layer those services are located, which deal with node identification, addressing, routing, distribution (single, multi-cast, broad-cast, dynamic reconfiguration, continuous QoS control/reassessment, robustness/fault tolerance). In addition and conformance to current wireless communication standards, here also data security, including encryption and sender/transmitter/receiver authorization should be addressed (including a vulnerability assessment based on the situations of all involved participants), and intelligent caching for joining and disconnecting nodes be managed. A parallel can be drawn to ISO-OSI layers 2 – 4. Here the focus of examination lies on different possible software solutions, including both architecture and realization. The definition of services provided by the network core layer should be driven by experts for networking and communication.

- Communication layer: This layer provides the technical (physical) hardware. It deals with, e.g., the generation, transfer, and reception of binary signals. A parallel can be drawn to ISO-OSI layers 0 and 1. Various combinations of hardware solutions must be compared and evaluated for efficiency and effectiveness.

For the sake of modularity, flexible implementation, maintenance and mastering of complexity, here again the rule of transparency must be strictly obeyed: for a service of a higher layer it must be transparent how a lower layer service implements the requested service – a condition that is not necessarily met by all proposals around in the defense.

3.2 Technical Challenges

Given the topology of a network of networks and given a layered architecture such as outlined in section 0, there is due to the extended operational conditions and desired properties of the infostructure a range of challenges to overcome. Below these challenges are addressed by architectural layer.

3.2.1   Communication Layer

The lowest layer of the infostructure consists of those senders, transmitters, and receivers that enable connectivity, including nodes (e.g., gateways, routers, relays/repeaters, or terminals) and wired and unwired links (e.g., radio, laser, hydro-acoustic, fibre) with the main objective to transfer bits. In contrast to communication links in commercial teletraffic networks, both links and nodes of the communication network will not only be exposed to heavy weather conditions (which massively influence the throughput and capacity of a link [Grushevsky, McCann, and Welsh, 2005]), but are expected to be exposed to electronic counter measures, deliberately created by someone with the intent to block communication. Jamming attacks on wireless links give error-detecting/correcting data encoding methods significantly increased importance.

Numerous international standards for wireless communication exist, including for short distances the IEEE 802.11 series (for WLAN), Bluetooth, or IrDA (see also section 0). Those standards should be reused for the infostructure whenever possible, not only under consideration of the effort that the development of own proprietary communication formats would imply, but also under aspects of interoperability with other forces and civilian services. An undesirable situation (which should be avoided for the infostructure) occurred with respect to the professional digital radio networks used by the public authorities of several European nations, where France and Czechia decided to use the FDMA-based TETRAPOL [Tetrapol, 2005], while Finland [VIRVE, 2005], the Netherlands, Belgium, and Austria build on the TDMA-based ETSI TETRA [ETSI TETRA, 2005] [c't, 2005].

A special role beyond the needs of the "traditional" internet must be allocated to data encoding, which helps to compress information to be transferred, but also to detect and correct bit errors that occur during transfer. In the internet, communication links are quite reliable, but also there it is hardly possible to transfer un-encoded bit-streams without errors. Considering the high importance of error-prone wireless communication links in the NBD, the requirement for senders in hostile territory to minimize their signal emission (stealth), and the impact of electronic warfare measures on the probability for erroneous bits, compression, error-detection, and error-correction seem to be of even higher importance for the NBD than for today's digital communication structures. Which degree of data encoding is reasonable certainly is situation dependent; to optimize data encoding for the infostructure, knowledge about most appropriate encoding strategies should be acquired.

Weight, size, and robustness of the deployed technical devices may span a wide range: Broad-band communication might occur among head quarters and major command posts via fiber or satellite, among (communication) vehicles by laser link, using UAVs not only as sensors, but also as hubs or repeaters to increase laser range, while the devices that create a soldier's body area network must be light, small, and resistant to physical stress. As long as open standards are applied, at least for portable devices COTS products can be used, but considering requirements on robustness, signal emission, and encryption, at least modifications of these devices are likely to become necessary.

To avoid a dead-end road concerning communication technology, two technological branches, mobile telephony and wireless internet technology, need to continue to merge to enable connectivity as imagined for the infostructure. Questions that should be answered for the military customer include appropriate technologies for different ranges, speeds, terrain, and the most appropriate common communication standard.

3.2.2   Network Core Layer

Under the assumption that the lowest connectivity layer enables directed transfer or broadcast of bits from one node to the other, for the transfer of (bit-encoded) information many concepts and methodological solutions can be adopted. Challenges that need to be addressed include classical cyber attacks (DoS, eavesdropping, manipulation, replay), which must be assumed to be more massive and more professional than those targeted at today's internet, but also the physical take-over of portable communication devices not only at the frontline. In addition to all the technological difficulties unsolved at the moment for mobile ad-hoc networks, also the secure reuse of existing infrastructures (e.g., IP networks) deserves special attention.

3.2.2.1 Addressing, Routing, and Caching in Mobile Cell Networks

For the infostructure it must be assumed that not only the nodes, but also the network cells (known from mobile telephone networks) are mobile and that the carrier signal in the cells is not constantly active, thus, a routing path may become invalid, as soon as the first data packages are sent. The receiver may be connected then via another path, or be completely disconnected from the net. Besides the capability to dynamically reconfigure routing paths, the infostructure also must be able to intelligently store information that cannot be directly delivered to their intended receivers at appropriate cache nodes. This can be started with a "mail box approach" and update notifications, but more appropriate seems to be the development of intelligent caching policies, with would both protect information sources (they transmit the information once and go offline then again, others access the cached information) and helps to reduce network traffic. Open questions associated with addressing, routing, and caching in the infostructure include how to deal with rapid changes in the topology and how to cache data in such a manner that end-nodes with low signal emission requirements can provide and receive up-to-date information. The needs for data access and end-user behavior in a NCD environment need to be studied before the design phase is finished.

Ad-hoc Networking: Another critical issue that currently is intensively exploited by automotive industry and mobile communication enterprises is ad-hoc networking. In contrast to the "traditional" internet approach where a user connects to and disconnects from a limited number of access points using a quite slow connection protocol, in network centric environments nodes might need to disconnect rapidly from one domain and connect to another domain just seconds later. A helicopter might quickly join a peer-to-peer network of an infantry group, transferring ground target position data, and then immediately disconnect again. The high dynamics of the relative positions of the network nodes pose a currently technically unsolved challenge [Grönquist 2003]. Cells, known from mobile telephony as quite static elements, are likely to become mobile, too. Open questions include how to deal with the dynamic, continuously changing topology, and how do highly flexible routing algorithms look like. However, these questions will be addressed (and eventually answered) also outside the NCD community.

Authorization, Encryption and Decryption: Generally speaking, the security of the internet is low, which is not due to the lack of methodological and theoretic security concepts, but originates from the mistakes that occur during the technical implementation of these concepts under economic constraints, errors during the configuration of the end-nodes, and last but not least user mistakes. Authorization, encryption, and decryption are in a presumably hostile environment where a highly educated and technically competent opponent will try to take over, manipulate, distort, block, deny, or corrupt the communication flow of extreme importance. Under the constraints that allies shall be able to share the infostructure (at least to a certain degree), that login shall occur rapidly without (much) human interaction, that formal logout may be completely skipped, and that end-nodes must be expected to fall into the enemy's hands, especially the question whether a node really is authorized to access some sensitive information does not become easier to answer. Different strengths of encryption will be used for different types of information and different addressees. The open questions are mainly related to the identification and authorization in ad-hoc networks. Although encryption is also requested on the larger commercial market, encryption within the info structure must meet the high military needs.

3.2.2   End-User Service Layer

The highest layer builds on the foundation laid by the lower layers, i.e. the mechanisms to transfer the required information from a source node to a (set of) target node(s). On the highest layer the services are situated that are intended to directly support the end-users to fulfill their duties. End-user services are available at a variety of end-user devices, including notebooks, goggles with compass, distance laser, and integrated displays, soldier-carried GPS position receivers, or field telephones, sensors of an UAVs, but also remote controlled demolition charges, or artillery. [Alvå and Palmquist, 2003] envisions a new type of ground unit, the "information vehicle", which serves both as gateway between the platoon's LAN and the division's WAN and as user interface displaying incoming information and providing the ability to uplink own information. End-user devices must be sufficiently robust, and they require an ergonomic man-machine-interface, that can still be operated by exhausted, tired, or maybe wounded users. Technical solutions to describe, register, and invoke services, internet research made available several technologies with matured to a quite reliable state. However, service composition cannot yet be completely automated, which requires human operators in the loop [Cabral et. al., 2004].

Concerning information representation in an interchangeable manner, good results were achieved in the C4ISR community, information fusion, and artificial intelligence. Information that is relevant as result of a sensing action and as input to decision making and action includes, e.g., terrain information, position data, movement information, combat values, or, most difficult, statement about the own and estimated enemy intentions. These information items must be represented appropriately to facilitate their transfer and processing by other participants in the infostructure. Compatible representations of the battle space are required, encouraging the development and use of common (or at least compatible) mission space models and data models, such as the JC3IEDM. An open question that urgently should be answered is, which services does the end-user want, and which information shall be shared to make this service possible? These questions cannot be reasonably answered outside the NCD community.

4. SIMULATION SUPPORT

Being a distributed, heterogeneous multi-purpose communication system with countless different methodological and technological building blocks with a hardly exhaustively predictable interplay, the infostructure certainly deserves the title "complex system". It must not only be technologically mastered, which is a non-trivial issue [Lüthi, Brade, and Lehmann, 2001], but also needs to be properly handled by its users. Not only [Alberts and Hayes, 2003] points out that the transition from the current way of command and control to the network-based information sharing and communication also requires training for the end-users.

During the transformation towards a NBD, M&S can play an important role to cost-efficiently support the design, exploration, improvement, use, and maintenance of the infostructure for combat, OOTW, or disaster management. Simulation support should be made available and used for:

- Analysis and optimization of the infostructure, with a horizon ranging from long-term strategic examinations to faster-than-real-time "online" simulation of sub-networks. Here the main objective is to support the process of optimizing the desired properties discussed in section 0, and finding answers to those open questions identified in section 0;
- Training, for both the end-users and the administrators of the infostructure, enabling the end-users to get familiar with the options that the invocation of end-user services offers (even beyond the originally planed options), and the administrators to control and maintain it in critical situations; and
- Test and Evaluation of materiel, addressing the certification and smooth integration of upgraded or new infostructure components,

and the assessment of their capabilities enabled by the integrated system of networked systems.

During simulation in the above disciplines of special relevance for armed forces [NATO, 1998], now in the context of military operations research additional attention needs to be paid to communication and networking aspects, such as load balancing, bottleneck analysis, connectivity, signal-to-noise ratios, fault insertion, or network scalablity. The repositories of simulation models of the armed forces need to be extended by the new chances, but also the new risks, provided to the end-user and with models for the simulation of communication networks.

4.1 Analysis and Optimization

To reveal capability gaps and to close them in the short term (concerning, e.g., deployment of forces, mission), medium term (concerning, e.g., acquisition or reconfiguration of minor new systems), or long term (concerning, e.g., structural changes of the whole communication architecture), systematic monitoring and reflection of the own capabilities is required. Communication weaknesses and problems must be detected as early as possible to identify the real needs of the end-users. Potential solutions need to be compared, the best solution should be implemented, and then this implementation should be continuously monitored and evaluated to find the next weaknesses to overcome during another iteration of the cycle. Modeling and Simulation of the infostructure in the context of its operational environment are enablers to gain and keep the overview over problems and solution alternatives. Efforts addressing high level design (concept, system architecture), subsystem design (subnets, nodes, links), planning (operation/mission preparation), and infostructure maintenance during operation (runtime administration, online verification, validation, and exploration) are going on, but it seems that the co-ordination among them could be improved.

Changes to the infostructure can be either demand-driven, as determined by defense analysts ("what do we need?"), or technology-driven, as proposed by providers of technology solutions ("what is possible?"). The properties that should be observed and evaluated during simulation can be directly derived from the desired infostructure properties identified in section 0, and consist of a combination of traditional observables from both domains "military operations research" and "communication network research". The problem to solve can be considered either as a multidimensional optimization task or as optimization under constraints. From the communications perspective, appropriate metrics are required to measure in the simulation of the combat operation or disaster management effort throughput and capacity, reliability and fault tolerance, robustness, security, mobility, interoperability, safety, and utility of the modeled infostructure and its components at various time horizons: On the one hand, during an ongoing military or civilian operation with a given set of communication systems involved, potential situations that are likely to negatively impact communication during the close future (minutes) must be detected and solved before a problem actually appears. On the other hand we also need to analyze and optimize the infostructure's underlying concepts and architecture, to make sure that in the future (decade) information superiority can be maintained. To summarize, simulation objectives (ordered by time looked ahead into the future) include:

- Architecture design and evolution (long time horizon): Especially now during the transition from a platform-oriented to a network-based defense, a flexible, robust, and extensible architecture for the defense network should be developed. Various architectural frameworks are being discussed at the moment, and without in-depth analysis it is hard to tell which proposal is superior. For this analysis activity the freedom of the simulationists is great, numerous concepts under discussion should be explored, and the constraints on the (simulated) architectural framework are just limited concerning already existing solutions. Lessons learned from the M&S of other (e.g., commercial) communication networks should be considered.

- Sub-system exploration: Constructive simulation of the battle space, including conventional symmetric warfare as well as asymmetric warfare or non-article 5 operations is already used to identify capability gaps, which could be closed using a network based defense structure. New needs for communication and information exchange have been identified and made explicit, but the impact of their realization on future scenarios requires more examination. NBD concepts and methods can be evaluated in such constructive simulations, where NBD entities are involved. All modifications and changes must remain within the architectural framework, but there is a lot of freedom with respect to the solutions on node and link level.

- Operation/mission preparation: If an end-user requests a service and another (team of) end-users is supposed to provide it, besides the infostructure required to communicate the request some other end-users (humans or machines) are required that ultimately do what is necessary to provide the service. In a current configuration of the infostructure may exist bottlenecks such as overloaded resources, non-redundant communication hubs, nodes may be taken over by the enemy or simply go down to a lack of electricity, and also other complications may occur. The robustness of the architecture against such types of complications can be examined by M&S, where various configurations of the infostructure build from various subsets of a set of available components [Alfredson, Lindoff, Rencrantz, 2004] can be compared.

- Runtime administration (real time and faster than real time): It must be assumed that during operation there will be rapid changes in the topology of the infostructure, both planned as well as unexpected, imposed by external circumstances. To maintain control of the infostructure, it is advisable to constantly monitor it and to (quasi) continuously update a model, which comprehensively communicates the current state of the infostructure to its administrators. Furthermore, simulating the infostructure "online" will allow the execution of quick "what-if" analysis concerning rapid reconfiguration of the network, intrusion and network vulnerability, or network load distribution. Online simulation takes place in parallel to the real system's operation.

4.2 Training and Education

Numerous human beings will be confronted with the infostructure, and most can be allocated either to the groups "end-users" or "administrators". Such as in other learning areas, simulation aided training can efficiently and effectively be used to prepare the trainee for real world situations.

- End-users should be familiarized with the infostructure, learn how to benefit from the new possibilities, and get acquainted with the new risks. End-user services need to be embedded into combat and crisis management simulation (including their reliability, availability, security). Who the end-users are depends on the situation in which the infostructure is used: During a military operation military end-users of various ranks will benefit from the infostructure; during disasters fire fighters and ambulance staff may exchange information over it. Those end-users should practice in sufficiently realistic environments and situations, including the network based information exchange and sharing. The simulation support in computer aided exercises needs to be updated to reflect the new communication possibilities, but also the new dangers that come with the communication network. According to [Hill, 2005] network centric communication is today not adequately represented in such exercises.

- Infostructure administrators are responsible to keep the infostructure operational. Although a high degree of automation is required to handle rapid changes in the infostructure's topology, human administrators need to monitor the communication traffic to be able to quickly react on undesired developments. They need to learn, e.g., how to detect and deflect cyber attacks, how to deal with subsystem failure and system degradation, how to reconfigure the network to optimize data flow, how to kick flooding or otherwise suspicious senders or receivers, or how to position communication nodes, which serve as hubs or routers, without exposing them to lethal threats. Simulators must be made available to prepare the infostructure administrators also for those rare events that are likely to results in a breakdown of communication means.

4.3 Test and Evaluation

For the defense material administrations that are supposed to accept or certify new defense systems prior to their transition to use, a new challenge arises in form of the need to focus test and evaluation (T&E) not any longer on the individual system only, but also on the system of networked systems created by the interplay of infostructure components and end-users connected. This does not mean that the T&E of the individual system looses importance, each node or link in the infostructure still needs to meet its requirements, but that the efficiency and effectiveness of testing needs to be increased. Subsystems in the NBD will have to satisfy an interface specification, which must be tested in a well structured compliance test. However, considering the rapid re-configurability of the infostructure and the unpredictability of the types of systems within future configurations of the NBD, it will be hard, if not impossible, to capture all possible communication with the new component in a single standardized set of I/O sequences that the subsystem need to satisfy. Because it is expensive to create a rich real test environment for new infostructure components, virtual testing (hardware in the loop simulation) of new prototypes or COTS products has the potential to cost-effectively complement traditional testing [Lagerström, 2005]. In such a virtual testing environment, the device under test hooks up to the virtual environment, is stimulated by simulation generated environment information, feeds back its own response to the simulation and causes some effect in the virtual environment. During the virtual test, the NBD entity under test (EUT) is stimulated with simulated service requests from other simulated NBD entities the EUT is intended to cooperate with, subjected to simulated data attacks. This virtual "NBD entity in the loop" test can be used to complement the mandatory NBD compliance check. Compounds of real and simulated NBD components could be tested, to complement the results gained from "traditional" test and evaluation.

5. SUMMARY AND CONCLUSIONS

This paper outlines challenges that can be addressed using Modeling and Simulation (M&S) concerning concept, architecture, and technological solutions that accompany the transition towards a Network Based Defense, including the planning, design, realization, and introduction of an appropriate integrated communication infrastructure, the so-called "infostructure". The concepts and tools that matured in the past in the teletraffic community and military operations research promise in concert with already existing technology solutions to provide a stable foundation for a first generation of infostructure components. An initial fully functional infostructure can be „assembled" from state of the art knowledge and technology, but without a clear needs statement concerning NBD communication objectives and requirements, there is the great danger to miss an appropriate, already existing solution. It will be hard to avoid dead end roads, which are opened to those lost in the sheer amount of potential solutions offered by technical opportunity and not driven by need. A first attempt is made in this paper to summarize operational constraints for the infostructure and to separate out those, which should be individually readdressed by defense researchers, because they hardly apply to commercial networking efforts and are unlikely to be investigated in depth in non-defense efforts. Based on the assumption that the high level objective of building the infostructure still is to enable secure information sharing and exchange for globally distributed participants under a variety of tough operational conditions to provide the end-user services with sufficiently high quality for distributed and collaborative sensing, deciding, and acting, and based on the identified operational constraints, a list of desired properties is developed. This list includs requirements on throughput and capacity, security, reliability and fault tolerance, flexibility and mobility, interoperability, robustness, and safety.

The use of Modeling and Simulation is motivated by the likely complexity of the infostructure. Already during its early design stages a way must be found to master the complexity of the potential interactions in the defense network, and this control must be maintained during use and evolutionary improvement of the infostructure. Modeling and simulation turned out to be powerful enablers for virtual testing, design, maintenance, improvement, and training in numerous military and non-military application domains, and promise to also be useful in the context of the Network Based Defense. During the design phase of the infostructure, it is advisable to combine the knowledge and experience from mainly two areas of interest, which are military operations research and teletraffic research into one integrated effort. Half-hearted approaches with a lack of knowledge in any of those two areas bear the risk to lead to invalid results and should be avoided. The paper sketches opportunities for application of M&S in analysis and optimization, training, and test and evaluation in the context of the Network Based Defense, and aims to inspire more related in-depth work.

Conclusions and recommendations from the work documented in this paper include that

(1) Governmental agencies and representatives should include M&S studies on the infostructure in close- and mid-term budget planning to (co-) sponsor cooperation between experts in telecommunication and military operations research or disaster management, respectively.

(2) Defense analysts should update and extend their simulation models by the new possibilities for information sharing and exchange, and effects of the invocation of remote end-user services in the end-user's local environment.

(3) Training simulators should soon reflect new man-machine-interfaces of end-user devices, services, and their behavior, to enable the trainees to familiarize themselves with the now chances, but also the risks associated with being a part of the network.

(4) To enable the smooth introduction of new and upgraded infostructure components into the existing network, Material Acquisition Administrations should prepare Virtual Proving

Grounds for the test and evaluation, verification, validation, and exploration of infostructure (sub-) networks and components.

Acknowledgements

The study work from which the presented results originate was sponsored by the Swedish Defense Material Administration FMV (Försvarets Materielverk).

ACRONYMS

ATCCIS	Army Tactical Command and Control Information System
COTS	Commercial of the Shelf
DoS	Denial of Service
EUT	Entity Under Test
FCS	Future Combat System
FDMA	Frequency Division Multiple Access
FM	Försvarsmakten, Swedish Defense
FMV	Försvaretsmaterielverk, Defense Material Administration
GPRS	General Packet Radio Service
GSM	Global System for Mobile Communications
IEEE	Institute of Electrical and Electronic Engineers, Incorporated
IrDA	Infrared Data Association
JC3IEDM	Joint Communications, Command, and Control Information Exchange Data Model
KBM	Krisberedskapsmyndigheten, Emergency Management Agency
M&S	Modeling and Simulation
MMI	Man Machine Interface
NATO	North Atlantic Treaty Organization
NBD	Network Based Defense
NoN	Network of Networks
OOTW	Operations Other Than War
PMR	Professional Mobile Radio
RPC	Remote Procedure Call
SOAP	Simple Object Access Protocol
TDMA	Time Division Multiple Access

TETRA	Terrestrial Trunked Radio
UDDI	Universal Description, Discovery and Integration
UTMS	Universal Mobile Telecommunications System
WLAN	Wireless Local Area Network
XML	Extensible Markup Language

REFERENCES

[1] Alberts, D.S., J.J. Garstka, F.P. Stein. 1999. Network Centric Warfare, Developing and Leveraging Information Superiority. CCRP Publication Series.

[2] Alfredson, J., K. Lindoff, C. Rencrantz. 2004. Behovsanalys av SitSyst-byggaren. FOI Metodrapport, ISSN 1650-1942

[3] Alvå, P. and U. Palmquist. 2003. Stridsfältnära Ledningsstöd i NBF: Scenariobeskrivning. Underslagsrapport, FOI-R—1030—SE, ISSN 1650-1942.

[4] c't. 2005. EADS übernimmt Tetra-Sparte von Nokia. c't-Magazin 2005, Heft 9

[5] Cabral, L., J. Domingue, E. Motta, T.Payne, F. Hakimpour. 2004. Approaches to Semantic Web Services: An Overview and Comparisons. First European Semantic Web Symposium, Heraklion, Greece

[6] Carling, C., R. Enander, J. Lindström, U. Palmquist, J. Tofte. 2003. FoRMA 2003: Nätverkstrukturer. Underlagsrapport, FOI-R—1037—SE. Försvarsanalys, ISSN 1650-1942

[7] ETSI TETRA. 2005. Website http://portal.etsi.org/tetra/, accessed in Apr 2005.

[8] Flink, H. 2004. Metodbeskrivning för Integration i NBF. Version för Demo 04-06, version 1.0A

[9] Grönquist, J. 2003. Distributed STDMA Algorithms for Ad Hoc Networks. Scientific Report, FOI-R—0159—SE.

[10] Grushevsky, Y., J.C. McCann, R. Welsh. 2005. The moving rain model for KU band communication. In Proceedings of the Spring Simulation Multiconference, San Diego, CA.

[11] Hill, F. 2005. The Challenge of Simulation Support for Network Centric Exercises. European Simulation Interoperability Workshop, Toulouse, organized by the Simulation Interoperability Standardization Organization (SISO).

[12] Ince, A.N (editor). 2002. Modeling and Simulation environment for Satellite and Terrestrial Communication Networks. Proceedings of the European COST Telecommunications Symposium. Kluwer Academic Publishers. ISBN 0-7923-7547-5

[13] Jönsson, P.G. 2002. FMA_AR_Tjänstebegreppet. Rapport Funktion 09100:54976/02, v2.0-M5.

[14] KBM, 2005. http://www.krisberedskapsmyndigheten.se/

[15] Lagerström, H, D. Brade, H. Flink. 2005. Verification, Validation and Exploration in Network Centric, Service oriented Defense Systems. FMV Rapport VO VoV 21 121:44034/05, Ve rsion 1.0

[16] Lüthi, J., D. Brade, A. Lehmann. 2001. A Modeling Concept for the Mastering of Systems Complexity. In: Modelling and Simulation 2001, 15th European Simulation Multiconference 2001, ESM'2001, Eds. Kerckhoffs, Snorek, Prague, Czech Republic, June 6-9, 2001, Society for Computer Simulation, Delft, The Netherlands.

[17] Meissner, A., T. Luckenbach, T. Risse, T. Kirste, and H. Kirchner. 2002. Design Challenges for an Integrated Disaster Management Communication and Information System. The First IEEE Workshop on Disaster Recovery Networks (DIREN 2002), New York City.

[18] North Atlantic Treaty Organization. 1998. NATO Modelling and Simulation Master Plan. Document AC/323 (SGMS)D/2, Version 1.0.

[19] Tanenbaum, A.S. 1992. Modern Operating Systems. Prentice-Hall Inc.

[20] Tetrapol. 2005. Website http://www.tetrapol.com, accessed in Apr 2005.

[21] VIRVE, http://www.virve.com/ . Accessed in September 2005

CHAPTER 23

An Analysis Tool for Markovian Traffic Model Validation

Rachid El Abdouni Khayari, Axel Lehmann, Markus Siegle

Universitat der Bundeswehr München, Deutschland
Institut für Technische Informatik,
WernerHeisenberg Weg 39, 85577 Neubiberg, Deutschland

Abstract. *Even though the problem of developing adequate models for self-similar data traffic is relatively old, it still remains unresolved. One of the limitations of the existing models is their lack of general applicability. Many traffic models re validated using a single trace only, such that the developed model may well fit the traffic of this trace, but not necessary of another one. Therefore, it is mandatory to use more than one trace for model validation, and in order to fit the collected traffic, specific descriptors of the traffic features have to be chosen (e.g. Hurst parameter, index of disperson of counts, etc.). The used traces should stem from different areas, reflecting various user behaviour. In this paper, we present a tool box which helps modelers to analyse measured traces and to validate their developed Markovian traffic models.*

Keywords: Markovian modeling, workload analysis, self-similarity, heavy-tailed distributions, synthetic load generation.

1. INTRODUCTION

Historically, traffic modeling has its origin in conventional telephony. Traditionally, Poisson processes have been used for modeling arrival processes in communication systems, or more generally, in queueing systems. Furthermore, Markovian assumptions about arrival patterns as well as resource holding times have been made. However, more recent measurement studies of network traffic from local area networks [14], multimedia and video traffic [1, 4, 10], signaling traffic [7] and high-speed networks [3,9] have shown that the traffic has slowly decaying variances. In many studies, the presence of long-term correlations, burstiness, self-similarity, fractality and long-range dependencies in network traffic has been shown. The property of self-similarity has been associated with the heavy-tailedness of the involved distributions, e.g. for the object-size distributions or inter-arrival times. Furthermore, it has been realized that traffic with self-similar properties has serious implications on the system performance, and that ignoring these properties can lead to underestimation of performance measures [5, 14, 22]. Consequently, designing and dimensioning of systems by only taking into account the mean load requirements is not acceptable, since it fails to consider the behavioral dynamics associated with burstiness, self-similarity and long-range dependency [15]

Even though the problem of developing adequate models for self-similar data traffic is relatively old, it still remains unresolved. Until now, adequate and universally applicable models describing real traffic behavior do not exist. Many different models have been proposed in the literature, some of which work very well for particular situations, but the development of a stochastic model which is robust enough to be widely applicable remains a big challenge.

Most of the existing traffic models have been validated by the use of a single network trace only, which means that only the specific features of that particular trace could be considered. A test for the general validity of the developed models has never been carried out. We have shown in [12] that such an examination makes sense in the context of traffic modeling. In this study, we found that even when traffic models appear to perform well in a specific situation, that is not necessary sufficient to affirm the model's usability. In fact, existing traffic models suffer from many other constraints which, to a greater or lesser extent, affect their quality [12]. Most of the models presented in the literature are not accompanied by any analysis tool which would allow their reuse in practical traffic engineering. Therefore, we have developed a tool box which helps modelers to analyse measured traces and to validate their developed Markovian models. The tool aims to achieve the following goals:

— **Reliable validation:** Many traffic models are validated using only a single trace. Our tool enables its user to employ more than one trace for model validation. It offers an extensible data base of recorded traces, which enables the modeler to carry out reliable tests over a broad spectrum of different traffic situations. The traces in the data base should stem from different areas, reflecting various kinds of behavior.

— **Traffic descriptors:** According to [11], a commonly accepted, accurate and compact traffic characterization is not available up to now. However, in order to fit the collected traffic, some specific descriptors for representing the traffic features have to be chosen. The most important traffic parameters to be fitted are the mean arrival rate and its variance, the lag-1 correlation, the Hurst parameter and the index of dispersion of counts. For a given measured or generated trace, our tool computes these traffic indices.

— **Complexity vs. accuracy:** The basic criterion for the suitability of a model is its ability to reproduce the observed traffic. By considering more traffic descriptors during the fitting process, the approximation will typically improve, but the model complexity and therefore the time complexity of the fitting algorithm will increase. As a rule of thumb, the developed model should be accurate enough to reproduce the observed traffic by using only a reasonable number of traffic descriptors. Our tool supports the choice of an appropriate model by offering several basic types of models together with the associated fitting and validation algorithms.

— **Asymptotic behavior:** The proposed traffic models have important asymptotic properties, but all tests are of course based on finite data sets [11] which constitutes a contradiction in terms. The correctness of the use of an asymptotic model to describe finite systems has never been analyzed.

— **The proof of generalization:** As it has been seen in this Chapter, the proof for the suitability, or non-suitability, of the developed model can be very useful. In fact, such proofs are in general very difficult, and are therefore not very widerspread. Indeed, it sometimes suffices to make a small test to verify, if the suggested model has some limitations (see [12]).

The rest of the paper is organized as follows. In Section 2, we first introduce some notation and definitions before we describe our tool framework in Section 3. Experimentation results are then presented in Section 4. Finally, the paper is concluded by a summary in Section 5.

2. FORMAL DEFINITIONS

2.1   Self-Similarity

Stationarity
A stochastic process $X = (X_t, t \geq 0)$, with $t \in T$ (the index set) is called weakly stationary if
1. its expectation is constant over time, i.e., $E[X_t] = \mu$, for all $t \in T$, and
2. its covariance function γ is shift-invariant, i.e.,

$$\gamma\left(X_{t_1+s}, X_{t_2+s}\right) = \gamma\left(X_{t_1}, X_{t_2}\right), \text{ for all } s, t_1, t_2 \in T, s > 0.$$

Aggregated process
The aggregated process $X^{(m)}$ is obtained from an original process X by "averaging" over non-overlapping blocks of size m, that is:

$$X_k^{(m)} = \frac{1}{m}(X_{km-m+1} + ... + X_{km}), \quad k=1, 2, ... \tag{1}$$

Note that $X^{(m)} = X_k^{(m)}$ is weakly stationary if X is weakly stationary.

Exact self-similarity
A stochastic process $X = (X_t \geq 0)$ is called exactly self-similar with Hurst parameter H if

$$X =_d m^{1-H} X^{(m)}, \quad \text{for all } m = 1, 2, ..., \tag{2}$$

where, $X^{(m)}$ describes the m-aggregated process of X.

This definition implies that the aggregated process $X^{(m)}$ is related to X via a simple scaling relationship involving H in the sense of finite-dimensional distributions (denoted by $=_d$) [18, Section 1.4.1.2]

Second-order self-similarity

A process X is called exactly second-order self-similar if the aggregated processes $X^{(m)}$ has the same correlation structure as X, that is,

$$r^{(m)}(k) = r(k), \qquad \text{for all } m = 1, 2,..., \text{ and } k = 1, 2,...,$$

where $r^{(m)}(k)$ denotes the autocorrelation function at lag k of the aggregated process $X_k^{(m)}$ and $r(k)$ denotes the autocorrelation function at lag k of the original process X.

Asymptotically second-order self-similarity

A process is called asymptotically second-order self-similar if

$$r^{(m)}(k) \sim r(k), \qquad m \to \infty$$

Self-similar processes have the so-called property of *long-range dependency*, i.e., their autocorrelation function decays hyperbolically. This implies that $\sum_k r(k) \to \infty$. In contrast, *short-range dependency* implies an exponentially decaying autocorrelation function for which $\sum_k r(k) < \infty$.

The Hurst parameter H defines *the degree of self-similarity* and expresses the rate of decay of the autocorrelation function. From (2) we obtain (for details, see [18, Section 1.4.1.2])

$$\text{var}\left[X^{(m)}\right] \sim \alpha m^{-\beta}, \qquad \beta = 2 - 2H \quad 0 \leq \beta \leq 1. \qquad (3)$$

For the Hurst parameter H, it holds:

$$\frac{1}{2} \leq H \leq 1 \qquad (4)$$

thereby, the degree of self-similarity increases with the value of the Hurst parameter. For $H = \frac{1}{2}$, the process is short-range dependent and completely uncorrected, for instance Poisson processes. For the case where $H \to 1$, the process is strongly correlated.

Variance-time plot method:

The variance-time plot method estimates the Hurst parameter H from a graph of $var[X^{(m)}]$ versus m, plotted on a log-log scale. An example of such a variance-time plot is given in Figure 4, which will be discussed later. From (3), we derive that

$$\log(\text{var}[X^{(m)}]) \sim \log(\alpha) - \beta \log(m),\tag{5}$$

so that β emerges as the negative gradient. Using a linear regression technique, we estimate β and hence $H = 1 - \beta/2$.

2.2   Heavy-Tailedness

Self-similarity in network traffic has been explained by the fact that many of the involved distributions, e.g., of file sizes, are heavy-tailed. In a HTD, the complementary cumulative distribution function F^c decays more slowly than exponentially, i.e., $e^{\gamma x}F^c(x) \rightarrow \infty$ as $x \rightarrow \infty$ for all $\gamma > 0$. For a random variable X, distributed according to some HTD, we typically have:

$$P[X > x] \sim x^{-\alpha}, \quad x \longrightarrow \infty \quad 0 < \alpha < 2.\tag{6}$$

Note that the term "$x \rightarrow \infty$" should be read as "for very large x" in case of measurements.

The degree of the heavy-tailedness is given by the value of the shape parameter a which can be determined by plotting the complementary cumulative distribution $F^c(x) = 1 - F(x) = P[X > x]$ on a log-log scale. The slope of the plot, found, for instance, via a linear regression, then gives the value of a.

2.3   Stochastic Modeling Of Traffic Processes: MMPP, MAP And PSST

The Markov-modulated Poisson Process (MMPP) is a doubly stochastic process of an m-state irreducible Markov process M, and a Poisson process whose arrival rate λ_j depends on the current state $j, 1 \leq j \leq m$. When the Markov chain is in state j, arrivals occur according to a Poisson process with rate λ_j. Thus, the MMPP is parameterized by the m-state CTMC (continuous time Markov chain) with infinitesimal generator $\mathbf{Q}$ and the m Poisson arrival rates $\lambda_1, \lambda_2,..., \lambda_m$:

$$Q = \begin{pmatrix} q_{11} & q_{12} & \cdots & q_{1m} \\ q_{21} & q_{22} & \cdots & q_{2m} \\ \vdots & \vdots & \ddots & \vdots \\ q_{m1} & q_{m2} & \cdots & q_{mm} \end{pmatrix}$$

$$q_{ii} = -\sum_{j=1, j \neq i}^{m} q_{ij}$$

$$\Lambda = \mathrm{diag}(\lambda_1, \lambda_2, \ldots, \lambda_m)$$

For the above given MMPP, the steady-state vector π can be calculated by solving the following system of equations:

$$\pi Q = 0, \pi \cdot \underline{e} = 1,$$

where $\underline{e}$ is the 1-column vector of length m, $\underline{e} = (1,1, \ldots, 1)^T$. The mean process arrival rate λ can then be computed as:

$$\lambda = \pi \Lambda \underline{e}$$

The inter-arrival time disribution function of the interval-stationary process is given by [8,21]

$$F(t) = \int_0^t e^{[(Q-\Lambda)x]} dx \Lambda \tag{7}$$

$$= \left(I - e^{(Q-\Lambda)t}\right)(\Lambda - Q)^{-1} \Lambda$$

$$= \left(I - e^{(Q-\Lambda)t}\right) F(\infty)$$

where $F(\infty) = (\Lambda - Q)^{-1} \Lambda$ is a stochastic matrix representing the transition probabilities of the Markov chain embedded at arrival epochs [8]. The stationary vector of $F(\infty)$, denoted by $\underline{p}$, is given by [8,13]

$$\underline{p} = \frac{1}{\pi \Lambda \underline{e}} . \pi \Lambda. \tag{8}$$

The vector $\underline{p}$ is chosen as the initial probability vector of the MMPP. By substituting (8) in (7), we obtain for the inter-arrival time distribution the following result:

$$F(t) = 1 - \underline{p} . e^{(Q-\Lambda)t} \underline{e}. \tag{9}$$

From (9), the k-th moment of the inter-arrival time can be computed as:

$$E[T^k] = k! \, \underline{p} (\Lambda - Q)^{-(k+1)} \Lambda \underline{e} \tag{10}$$

Now, consider the time T; between two arrivals; the i-th and the $(i + 1)$-th in a MMPP. The autocorrelation function $r(k)$ for T_i and T_{k+1}, with $k \geq 1$ is given by [8]

$$r(k) = E[\ (T_i - E[T_i] \ ) \ (T_{k+1} - E[T_{k+1}] \ )]$$
$$= \underline{p}(\Lambda - Q)^{-2} \Lambda \left\{ \left[(\Lambda - Q)^{-1} \Lambda \right]^{k-1} - \underline{e}\underline{p} \right\} (\Lambda - Q)^{-2} \Lambda\underline{e}. \tag{11}$$

Furthermore, the limiting index of dispersion can be computed as [8,21]

$$I = \lim_{t \to \infty} \frac{Var[N(t)]}{E[N(t)]} \tag{12}$$

$$= 1 + 2 \left(\underline{\pi}\Lambda\underline{e} - \frac{1}{\underline{\pi}\Lambda\underline{e}} \underline{\pi}\Lambda(Q + \underline{e}.\underline{\pi})^{-1} \Lambda\underline{e} \right), \tag{13}$$

where $N(t)$ defines the counting process of the MMPP, that is the number of arrivals in $(0, t]$

Markovian arrival processes (MAPs)

MAPs are a special class of Markovian renewal processes that form a generalization of the MMPPs. Their particularity is that the inter-arrival times are phase-type distributed, with a simple constraint: at the absorption instants, arrivals still occur, and thereafter, the process starts with a new phase-type distribution. The MAP is very simple and very interesting at the same time; it includes Poisson processes, phase-types renewal processes (and others) as special cases, and is dense in the space of point processes. The superposition of independent MAP traffic streams results in a MAP with the cross product state space of the component state spaces [21]. The splitting of a MAP stream again results in MAP streams.

Pseudo Self-Similar Traffic model (PSST) [19, 20]

The PSST model attempts to characterise traffic self-similarity by the use of a discrete-time Markov modulated Bernoulli process (MMBP), i.e., the discrete-time analog of a Markov modulated Poisson process. The modulating Markov chain has n states, numbered 0 through $n - 1$. Its one-step transition probability matrix is given as:

$$A = \begin{pmatrix} \Sigma_0 & 1/a & 1/a^2 & \cdots & 1/a^n \\ q/a & \Sigma_1 & 0 & \cdots & 0 \\ (q/a)^2 & 0 & \Sigma_3 & \cdots & 0 \\ \vdots & \vdots & \vdots & \ddots & \vdots \\ (q/a)^{n-1} & 0 & 0 & \cdots & \Sigma_{n-1} \end{pmatrix},$$

With $\Sigma_0 = 1 - \dfrac{1}{a} - \dfrac{1}{a^2} - \cdots - \dfrac{1}{a^{n-1}}$ and $\Sigma_i = 1 - \left(\dfrac{q}{a}\right)^i$, for $i=1,\ldots,n-1$

At every discrete time step, a state transition, possibly a self-loop, takes place in the modulating chain. Only upon entry in state 0 a packet arrival takes place. As can be observed, the PSST model is completely specified by the three parameters q, n and a. This makes the model attractive, as it requires only three parameters to be set, e.g., based on some fitting procedure.

Notice that the parameters a and q need to fulfill certain conditions so that A is indeed a stochastic matrix describing a discrete-time Markov chain: $q, a > 0$, $q \leq a$ and a such that $0 \leq A_{0,0} \leq 1$.

In the sequel, we denote with $A_{i,j}^k$ the entry in row i and column j of A^k. We furthermore define $N = (N_t, t \in \mathbb{N})$ as the discrete-time stochastic process describing the number of arrivals over time, as described by the MMBP.

Moments. Using the notation and terminology of the MMPP cookbook [8], we can derive the following results for the first and second moment of the number of arrivals N in an interval of length 1, i.e., per discrete time step:

$$E[N] = \underline{\pi} \Lambda \underline{e} \quad \text{and} \quad E[N^2] = \underline{\pi} \Lambda^2 \underline{e} \tag{14}$$

where

- $\underline{\pi}$ is the steady-state solution of the ergodic DTMC given by A, that is,
- $\underline{\pi} = \underline{\pi} A$ and $\sum_i \pi_i = 1$; it can easily be shown (by substitution) that

$$\underline{\pi} = (\pi_0, \ldots, \pi_{n-1}) = \frac{1 - 1/q}{1 - 1/q^n} \left(1, \frac{1}{q}, \ldots, \frac{1}{q^{n-1}}\right);$$

- $\underline{e} = (1,1,\ldots,1)^T$, a column vector of just 1's;
- and the n x n-matrix Λ has the simple form

$$\Lambda = \begin{pmatrix} 1 & 0 & \cdots \\ 0 & 0 & \cdots \\ \vdots & \vdots & \ddots \end{pmatrix}.$$

Notice that (14) is in a form typical for Markov modulated arrival processes [8] However, given the above explicit expressions for $\underline{\pi}$ and Λ, we can generalise and reduce (14) as follows. The k-th moment of the number of packet arrivals per unit time is given as:

$$E[N^k] = E[N] = \frac{1 - 1/q}{1 - 1/q^n} = \pi_0. \tag{15}$$

Note that the first moment of N can also easily be derived in the following way: an arrival takes place with probability 1, whenever the current state i is occupied (with happens, on the long run, with probability π_i) and the next state is state 0 (which happens with transition probability $A_{i,0}$). Hence, we have $E[N] = \sum_{i=0}^{n-1} \pi_i A_{i,0}$ which, after simple manipulations, indeed yields π_0.

Aggregated process. The m-aggregated process $N^{(m)} = (N_t^{(m)}, t \geq 0)$ is introduced, defined as the average number of arrivals over m successive intervals (preceding t)

$$N_t^{(m)} = \frac{1}{m}(N_{t-m+1} + N_{t-m+2} + \ldots + N_t), t > m.$$

Since N is second-order stationary, we obtain for the first moment of $N_t^{(m)}$

$$E[N_t^{(m)}] = E[N^{(m)}] = E[N].$$

For the variance of $N_t^{(m)}$, we follow the definition, that is, $\mathrm{var}[N_t^{(m)}] = \mathrm{var}[N^{(m)}] = E[(N^{(m)})^2] - E[N^{(m)}]^2$, which can be reduced to [19,20]

$$\mathrm{var}[N^{(m)}] = \frac{1}{m}E[N^2] - E[N]^2 + \frac{2}{m^2}\sum_{i=1}^{m-1}(m-i)\underline{\pi}\Lambda A^i \Lambda \underline{e}. \tag{16}$$

The matrix $\Lambda A^i \Lambda$ is an $n \times n$ matrix consisting completely of zeroes except for the non-zero entry $A'_{0,0}$ in the upper left corner. An explicit expression for the autocorrelation of the m-aggregated process at lag k, that is, $r^{(m)}(k)$, cannot be easily obtained.

Computation of the parameters n, q and a

In this section we assume that the expectation $E[N]$, the variance $\mathrm{var}[N^{(m)}]$ and the Hurst parameter H of the process under study are known, that is, they have been obtained from a trace using some estimation procedure. Given these (required) workload parameters, we describe how the model parameters n, q and a for the PSST can be computed. Note that the iterative recipe given below has been proposed in [20]; its simplicity makes it attractive to use. It is not the aim of the current paper to improve on this scheme.

Computation of n: The value for n is chosen by experience. It is suggested that values around $n = 6$ give good results in most cases [19, 20]. We used similar values in our experiments [2, 16, 17].

Computation of q: The Newton iterative method is used to solve the nonlinear equation (15) in order to compute from a known estimate for $E[N]$ and given n.

Computation of a: Assume that the Hurst parameter H of the measured workload has been estimated from the log-log plot of $\mathrm{var}[N^{(m)}]$ against m, for instance using a least-squares fitting procedure. From (16) we see that $\mathrm{var}[N^{(m)}]$ depends, via the entry $A^{i}_{0,0}$ in the summation, on the actual value of a. Thus, implicitly a function $V(a)$ is defined that yields, for given a, the function of $\mathrm{var}[N^{(m)}]$ against m. Hence, for a starting value $\acute{a}$, we can esti mate the negative gradient of $\log \mathrm{var}\,[N^{(m)}]$ against $\log m$, giving β and H (an estimate for H). If $\hat{H}$ differs from the measured value for H, we compute a next estimate for a, using an interval splitting procedure, and iterate until we have achieved the desired accuracy. We do not address the issue of uniqueness of the found value for a.

3.   DESCRIPTION OF THE TOOL FRAMEWORK

Figure 1 gives an overview of our tool. On the left hand side top, there is the data base of recorded traffic traces. For any trace from this data base, all important traffic statistics, such as moments and quantiles of the inter-arrival time (or request size), autocorrelation coefficients, Hurst parameter, etc., can be computed by the tool. Furthermore, the selected trace can be used as input for a queueing model to be simulated (see below).

On the right hand side top, different Markovian traffic models are available for modelling real traffic. The user may choose between MMPP, PSST or general MAP processes. Various fitting algorithms which we have implemented are connected to the tool. With the help of these fitting algorithms, an analytical distribution can be fitted to a given measured trace. The results of the fitting, especially the indices of the goodness of fit, are displayed to the user.

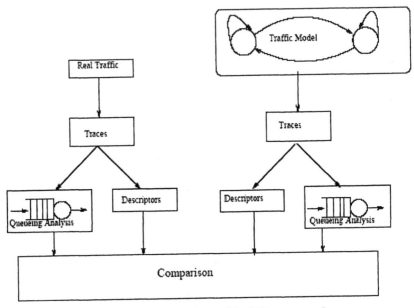

Fig. 1 Overview of Our Tool Framework

Our tool is then capable of generating a synthetic trace from the fitted analytic model. The generation is based on a pseudo random number generator, and the user may specify the length of the generated trace. The synthetic traces can then be analysed for their traffic characteristics. Furthermore, they can be used as input for a queueing model simulation, in order to be able to compare the queueing behaviour of the original measured trace and the synthetic trace which was generated from the fitted analytical model. So, in order to test the goodness of the analytic models, the original trace and the synthetic trace derived from the analytical model may both be used in simulations of a queueing system. The tool supports trace-driven discrete event simulation of */M/1, */D/1 and M/*/1 queueing systems, where the * stands for either a measured or synthetic trace.

3.1   Traces Considered

For the purpose of this paper, we consider a selected set of four traces collected from the access log files of web/proxy servers. The traces are called DEC96, RWTH00, RWTH04 and UniBw04, and their basic data is summarised in Table 1. The DEC96 trace consists of the logged requests of the web proxy of the Digital Equipment Company [6] . The RWTH00 and RWTH04 traces contain the internet traffic of the Aachen University of Technology as recorded by the computing centre of that University during some weeks in the years 2000 and 2004. The UniBW04 trace contains the traffic directed to the web proxy of the University of the Federal Armed Forces Munich as recorded during 10 days in June of 2004. In this paper, we focus on the measured interarrival times. Similar experiments can be achieved for the object sizes of the requested documents.

Table 1. Statistics of Traces Considered (Inter-Arrival Times)

trace	meas. interv	# data sets	min [s]	max [s]	mean [s]	ariance [s²]	cv²
EC96	09 05 – 09 12 1996	7.127.800	0.0	13.682	0.085	0.018	2.53
achenOO	02 17 – 03 02 2000	9.000.000	0.0	608.936	0.136	0.599	32.45
achen04	06 22 – 07 11 2004	37.253.065	0.0	58.11	0.044	0.038	9.73
UniBw04	06 16 – 06 25 2004	38.008.629	0.0	337.044	0.0125	0.005	32.97

3.2 Basic Trace Analysis

Figure 2 shows the number of requestst per hour over the measurement period for the selected traces. In all four cases the periodical fluctuation with daily peaks can be observed. Among the four traces, UniBw04 exhibits the highest traffic intensity and the highest coefficient of variation. The DEC96 trace stands out as the one with by far the smallest coefficient of variation (which, however, is still 2.53 times larger than the one of a Poisson process).

Figure 3 shows the empirical complementary distribution function of the inter-arrival times in the four selected traces on a log-log scale, as displayed to the user by our tool. The variance time plots of the four traces are shown in Figure 4. The parameter m which denotes the number of values aggregated was varied between 1 and 30,000. On the y-axis, the corresponding variance $Var(N_m)$ is given. As described above, N is a discrete time stochastic process describing the number of arrival over the time. From these results, our tool estimates the Hurst parameter of the underlying distribution by linear regression. This yields the following results:

$$H_{Aachen00} = 0.938, \quad H_{Aachen04} = 0.919, \quad H_{DEC96} = 0.953, \quad H_{UniBw04} = 0.944$$

According to these results, the counting processes for all considered traces are strongly self-correlated, i.e. they exhibit a second-order self-similarity.

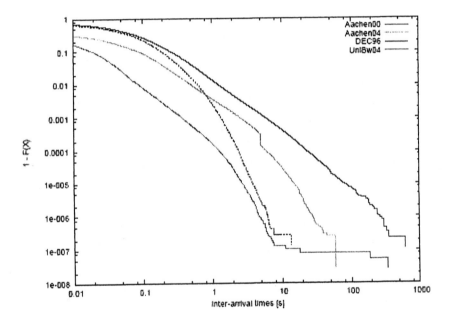

Fig. 3   Complementary distribution functions of inter-arrival time

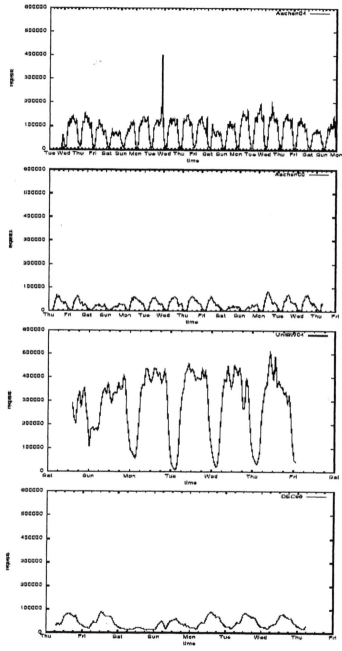

**Fig. 2 Number of Requests per Hour for the Traces Aachen04,
Aachen00, Unibw04 and DEC96**

4. EXPERIMENTAL RESULTS

We now present the results of some experiments which we conducted with the help of our tool framework.

4.1 Simulation Of Trace/D/1 System

First we used the UniBw04 trace as input for a simulation of the */D/1 queueing system. Fig 5 shows the behaviour of the queue length for two different service times, yielding the utilizations $\rho = 0.3$ and $\rho = 0.9$. Some statistical results of these simulations are presented in Table 2. For utilization $\rho = 0.9$, the mean waiting time is 148 seconds (!).

Table 2. Simulation Results for Unibw04/D/L System.

	queue length			
ρ	min	max	mean	mean waiting time [s]
0.3	0.0	79	0.267	0.003
0.9	0.0	567746	109258	1468

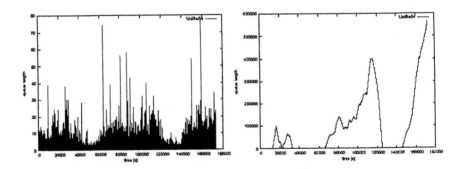

Fig. 5 Simulation Results of Unibw04/D/L System: Behaviour of queue length for utilizations $\rho = 0,3$ and $\rho = 0,9$.

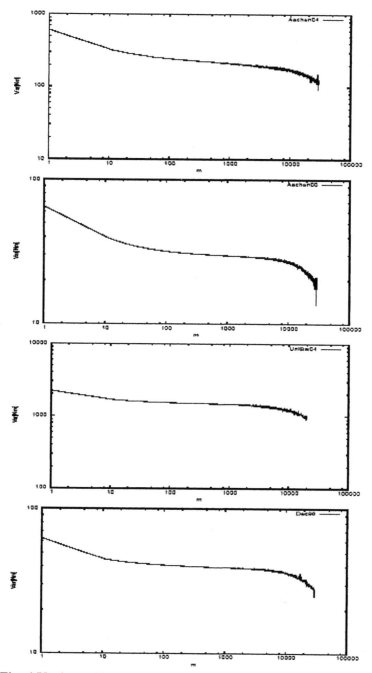

Fig. 4 Variance Time Plots for the Four Traces, $M = 1,... 30000$

4.2 PSST Model Validation

Table 3 shows the fitting results for the Aachen00 and UniBw04 traces. Comparing to Table 1, one notices immediately that the mean inter-arrival times are matched well whereas the variance (and therefore the squared coefficient of variation) are not fitted well at all by this model. Figure 6 depicts the complementary distribution function of the inter-arrival times as modelled by the PSST model, i.e. the figure shows the inter-arrival time distributions for synthetic traces generated from the PSST model. Figure 6 should be compared to Fig. 3 (the reader should mind the different scaling of the axes). We observe that the distributions from the synthetic traces exhibit a "wavy" character which is not present in the original traces.

Table 3. Inter-Arrival Times for PSST Models of Aachen00 and Unibw04.

Trace	min [s]	max [s]	mean [s]	variance [s^2]	cv^2
$PSST_{Aachen00}$	0	11601	0.1348	145.1145	7980
$PSST_{UniBw04}$	0	745	0.0123	0.0889	581

We decided to employ this PSST model for simulating the */D/1 queuing system, even though the fitting results are not very good. Figure 7 shows the complementary distribution function of the queue length as obtained from simulations where the input stream was given by

    (i)   the original measured traces,

    (ii)   (ii) synthetic traces generated from the fitted PSST models.

As one can observe from Figure 7, the PSST model results are not satisfying in both cases (for both, the lightly and the heavily loaded server), since the differences between the curves are very large.

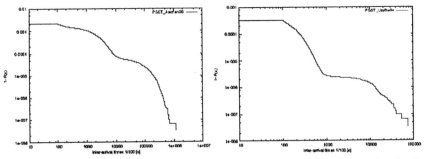

Fig. 6   Complementary Distribution Function for the Inter-Arrival Times for the Model

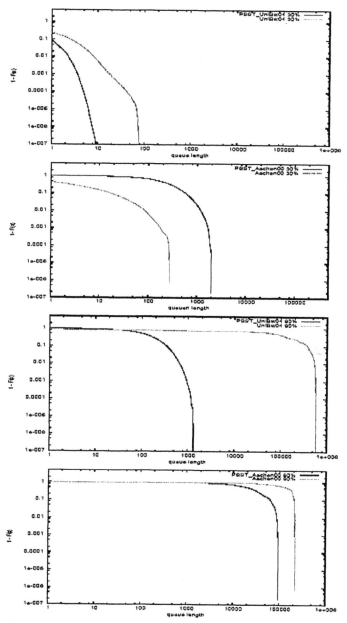

Fig. 7 Comparison of the complementary distribution function of the queue length for */D/1 simulations of the original traces and the synthetic traces derived from the PSST Models. Utiliations $\rho = 0,3$ and $\rho = 0,9$.

5.   CONCLUSIONS AND FURTHER WORK

In this paper afther recapitulating some basics of traffic modeling, we presented a tool framework for Markovian traffic model validation. It is our objective to give modelers the opportunity to validate their developed models against a large number of collected traces and also to compare their models to existing ones. Therefore, we are currently extending our tool in order to incorporate more models. Furthermore, it is important to reach a consensus on the used traces for model validation. We think that it is mandatory to have a kind of database of traces with different charcteristics which should serve as a basis of mutual model validations.

REFERENCES

[1] J. Beran, R. Sherman, M. S. Taqqu, and W. Willinger. Long-Range Dependence in Variable-Bit-Rate Video Traffic. *IEEE Transactions on Communications,* 43(2-4):1566-1579, February/March/April 1995.

[2] D. Brocker. Messung und Modellierung komplexer Verkehrsstrukturen in Hochgeschwindigkeitsnetzen. Master's thesis, RWTH-Aachen, Germany, 1998.

[3] T. Chiotis, F. Staatelopoulos, and B. Maglaris. Traffic Source Models for Realistic ATM Performance Modelling. In *Proc. 5th IFIP Workshop on Performance Modelling and Evaluation of ATM Networks,* Ilkley, UK, July 1997.

[4] T. Chiotis, C. Stathis, and B. Maglaris. The Impact of Self-Similarity on the Statistical Multiplexing of MPG Video Data. In *Proc. 6th IFIP Workshop on Performance Modelling and Evaluation of ATM Networks,* Ilkley, UK, July 1998.

[5] M. E. Crovella and, Bestavros Self-Similarity in World Wide Web Traffic: Evidence and Possible Causes. *IEEE/ACM Transactions on Networking,* 5(6)835-846, 1997.

[6] Digital Equipment Cooperation. Digital's Web Proxy Traces ftp//ftp.digitalcom/pub/D/traces/proxy.

[7] D. E. Duffy, A. A. McIntosh, M. Rosenstein, and W. Willinger. Statistical analysis of CCS/SS7 traffic data from working sub-networks. *IEEE Journal on Selected Areas in Communications,* 12(3):544-551, 1994.

[8] W. Fischer and K. Meier-Hellstern. The Markov-modulated Poisson Process (MMPP) Cookbook. *Performance Evaluation,* 18:149-171, 1992.

[9] H. J. Fowler and W. E Leland. Local area network traffic characteristics, with implications for broadband network congestion management. *IEEE Journal on Selected Areas in Communications,* 9(7):1139-1149, 1991.

[10] M.W. Garrett and W. Willinger. Analysis, Modeling and Generation of Self imilr VBR Video Traffic.   In *Proceedings of ACM SIGCOMM '94,* volume 24, ages 269-280, London, October 1994.

[11] A. Horvath and T. Telek. Markovain Modeling of Real Traffic: Heuristic Phase Type and MA   Fitting of Heavy Tailed and Fractal Like Samples. In M. C. Calzarossa and S. Tucci, editors, *Performance Evaluation of Complex Systems: Techniques and Tools,* pages 405-434. Lecture Notes in   Computer   Science 2459, 2002.

[12] R. El Abdouni Khaari, R. Sadre, B. Haverkort, and A. Ost. The pseudo-self-similar traffic model application and validation. *Performance Evaluation,* 56(1-4):3-22, March 2004.

[13] D. Koenig and V. Schmidt. Extended and Conditional Versions of Property. *Advanced Application Probability,* 22:510-512, 1990.

[14] W. E. Leland, M. S. Taqqu, W. Willinger, and D. V. Wilson. On the self-similar nature of Ethernet traffic. In *Proc. ACM SIGCOMM '93,* volume 23 of *Computer Communications Review,* pages 183-193, October 1993.

[15] S. Q. Li.   Queue Response to Input Correlation Functions: Continuous Spectral analysis. *IEEE/ACM Transactions on Networking,* l(6)678-692, December 1993.

[16] A. Ost. *Model-Based Performance Evaluation of Complex Communication Systems with Matrix-Geometric Methods* PhD thesis, RWTH-Aachen, Springer Verlag, Germany, 2001.

[17] A. Ost and B. R. Haverkort. Modeling and Evaluation of Pseudo Self-Similar Traffic with Infinite-State Stochastic Petri Nets. In M. Ajmone Marsan, J. Quemada T. Robles and M. Silva, editors, *Formal Methods and Telecommunications,* pages 120-136. Zaragossa University Press, 1999.

[18] K. Park and W. Willinger. *Self-Similar Network Traffic and Performance Evaluation.* John Wiley & Sons, 2000.

[19] S. Robert. *Modelisation Markovienne du trafic dans les reseaux de communication.* PhD thesis EPFL, Lausanne, Switzerland, 1996.

[20] S. Robert and J.Y. Le Boudec. New models for pseudo self-similar traffic. *Performance Evaluation,* 30:57-68, 1997.

[21] R. Sadre and B. Haverkort.   Caracterising Traffic Streams in Networks of MAP|MAP|1 Queues. In B. R. Haverkort, editor, *Modeling and Evaluation of Computer and Communication Systems,* pages 195-208. VDE Verlag, 2001.

[22] W. Willinger, M. Taqqu, and A. Erramilli. A Bibliographical Guide to SelfSimilar Traffic and Performance Modeling for Modern High-Speed Networks. In F. P. Kelly, Zachary, and I. Ziedins, editors, *Stochastic Networks: Theory and Applications* ages 339-366. Oxford, UK, 1996.

CHAPTER 24

Some General Terminal and Network Teletraffic Equations for Virtual Circuit Switching Systems

Stoyan A. Poryazov[1], Emiliya T. Saranova[1,2]

[1] Institute of Mathematics and Informatics,
Bulgarian Academy of Sciences.

[2] College of Telecommunications and Post, Sofia, Bulgaria.

Abstract. *A virtual circuit is a connection between two users, devices, or terminals that functions as if it were a direct connection. Forwarding, switching, and/or routing over directly-connected circuits takes place at intermediate devices within the virtual circuit, but the details are hidden from the end points. Virtual circuits are widely used in many telecommunications and networking architectures; viz.: in wireline (PSTN, ISDN, BISDN) and wireless (GSM) telephony; in packet-switched networks at various layers (e.g., transport layer connections); in Multiprotocol Label Switched (MPLS) networks where tunnels are a type of virtual circuit; in Asynchronous Transfer Mode (ATM) networks where permanent and switched virtual circuits are fundamental components; in optical networks where wavelength routing and burst switching are abstract forms of provisioned virtual circuits; and in emerging multi-service networks where managed Ethernet virtual circuits are used to bridge legacy technologies like frame relay and ATM.*

The names of the service stages, branches, and exits used herein to describe and model virtual circuits are borrowed directly from telephony and teletraffic engineering. For example, 'blocking' is a branch point in virtual circuit provisioning with counterparts in telephony, ATM, MPLS, Ethernet, and optical networks. The terminology is a convenient way to define a conceptual teletraffic model for these networks. Most of the service stages, branches and exits defined herein have direct counterparts in the data and/or control planes of multi-service networks.

In this paper, we define a conceptual model that is sufficiently general for a broad family of virtual circuit switching systems, and we use the model to derive expressions for traffic intensity, blocking probability, and quality of service (QoS) dimensioning of resources. The results are applicable to a number of virtual circuit switching systems and paradigms, and can be used as a basis for terminal and network teletraffic engineering tools for multi-service networks.

1. INTRODUCTION

Fourth-generation telecommunication networks are being designed and successfully deployed, and experts agree that network and terminal teletraffic theory is an important tool for modeling, simulating, and analyzing traffic in emerging multi-service telecommunication networks.

In this paper, we present a refined approach for network and terminal teletraffic modelling of (virtual) circuit switching systems. As noted in the Abstract, the virtual circuit paradigm is used in various single- and multi-service networks, and we believe that much of the methodology defined and developed herein will be useful for traffic modelling in these networks. We also believe that teletraffic theory for very complex mobile heterogeneous multimedia telecommunications networks depends on describing the simplest case.

Therefore, we propose a detailed conceptual traffic model of a (virtual) circuit switching telecommunication network, like the PSTN and GSM networks, and which includes users' behaviour. We use the reference model to describe an analytical macro-state model of the system in stationary state, with: BPP (Bernoulli–Poisson–Pascal) input flows; repeated calls; a limited number of homogeneous terminals; and losses due to abandoned and interrupted dialing, blocked and interrupted switching, unavailable terminals, blocked and abandoned ringing, and abandoned communication. As noted in the Abstract, these service stages, branches, and exits have direct counterparts outside of telephony – viz., provisioning, releasing, and otherwise managing virtual circuits in optical, ATM, Ethernet, and multi-service networks.

We begin by defining useful terms and formulating the necessary mathematical assumptions. Next, we assert and prove a system of mathematical expressions, and present qualitative results.

The network dimensioning task is formulated on the basis of preassigned values of various quality of service (QoS) parameters; QoS is a fundamental component of emerging multi-service networks. The results lead to a straightforward dimensioning of network resources, based on these QoS requirements.

The general equations (described in the paper) may reduce the amount of stored measurements data. They may also be used as a basis for developing Internet-based network and terminal teletraffic engineering tools for modeling and simulation.

The approach is directly applicable to every (virtual) circuit switching telecommunication system (e.g., GSM and PSTN), and may help considerably in modeling traffic for ISDN, BISDN, and most core and access networks. For packet switching systems (as in the Internet), the approach may be used as a basis for comparing virtual circuit paradigms with competing methods.

2. CONCEPTUAL MODEL

We describe two types of virtual devices: base devices, and systems comprising base devices.

2.1 Base Virtual Devices and Their Parameters

We illustrate the conceptual model by using the base virtual device type names and graphic notation shown in Fig. 1. For every device we propose the following notation for its parameters: F = intensity of the flow [calls/sec.], P = probability of directing the calls of an external flow to the device considered, T = mean service time, in the device, of a served call [sec.], Y = intensity of the device traffic [Erl], and N = number of service places (lines, servers) in the virtual device (i.e., the capacity of the device). In the normalized models used in this paper (Poryazov 2001), every virtual device except the switch has no more than one entrance and/or one exit. Switches have one entrance and two exits. To characterize the intensity of the flow, we use the following standard notation: $inc.F$ for incoming flow, *dem.F*, *ofd.F* and *rep.F* for demand, offered and repeated flows respectively (ITU E.600). The same characterization is used for traffic intensity (*Y*).

2.2 Virtual Base Device Terminology

In the conceptual model, each virtual device has a unique name. The device names are constructed according to their position in the model.

The model is partitioned into service stages (**d**ialing, **s**witching, **r**inging and **c**ommunication).

Every service stage has branches (**e**nter, **a**bandoned, **b**locked, **i**nterrupted, **n**ot available, **c**arried), corresponding to all the possible cases of the end-of-calls' service in the branch considered.

Every branch has two exits (**r**epeated, **t**erminated) which show what happens with the calls after they leave the telecommunication system. Users may make a new bid (repeated call), or stop attempts (terminated call).

In constructing the virtual device name, the corresponding bold first letters of the names of stages, branches, and exits are used in the following order:

Virtual Device Name = <BRANCH EXIT><BRANCH><STAGE>

A parameter's name (for a virtual device) is a concatenation of the parameter's name letter and the virtual device name. For example, "*Yid*" means "traffic intensity in the interrupted dialing case"; "*Fid*" means "flow (calls) intensity in the interrupted dialing case"; "*Pid*" means "probability of interrupted dialing"; *Tid* = "mean duration of interrupted dialing"; "*Frid*" = "intensity of repeated flow calls caused by interrupted dialing".

2.3 Call Paths

The network under consideration corresponds to the reference configuration "terminal - subscriber switch - terminal" (Iversen 2003). We ignore the signalling network.

In this paper "call" means "call attempt" or "bid" according to (ITU E.600). (This abstraction also corresponds to 'calls' in ATM networks, provisioning requests in optical networks, etc.) Figure 1 shows the call paths generated from (and occupying) the A-terminals in the proposed network traffic model and its environment. *Fo* is the intent intensity of calls for one idle terminal; *M* is a constant, characterizing the BPP flow of demand calls (*dem.Fa*). If *M* = -1, the intensity of demand flow corresponds to the Bernoulli (Engset) distribution, if M = 0, to the Poisson (Erlang), and if *M* = +1, to the Pascal (Negative Binomial) distribution. In our analytical model, every value of *M* in the interval [-1, +1] is allowed. The BPP-traffic model is applicable, but in the numerical examples presented herein, *M* = 0 because the conclusions are independent of the input flow model (Iversen 2003; Iversen's *Handbook* is a basis for comparison of the main ideas discussed in this paper.)

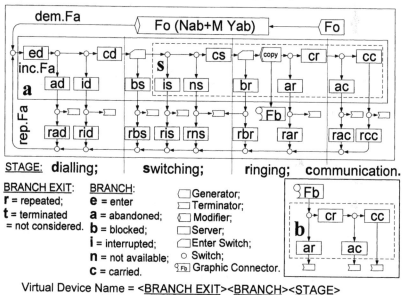

Fig. 1 Conceptual model of a telecommunication system and its environment, including: The paths of the calls, occupying A-Terminals (A-Device), switching system (S-Device) and B-Terminals (B-Device); base virtual device types, with their names and graphic notation.

2.4 Systems Comprising Virtual Devices, and Terminology

The following virtual devices (comprising several base virtual devices as shown in Fig. 1) are considered:

a = virtual device that comprises all the A-terminals (calling) in the system (shown with a continuous-line box). The devices outside the A-device belong to the network environment. The calls in the environment do not occupy network devices, but they form the incoming flows to the network;

b = virtual device that comprises all the B-terminals (called) in the system (box with dashed line). The paths of the calls occupying B-terminal and corresponding virtual devices are shown;

ab = a device comprising all the terminals (calling and called) in the system (not shown in Fig. 1);

s = virtual device corresponding to the switching system. It is shown with a dashed line box into the A-device. Ns stand for the capacity (number of internal switching lines) of the switching system.

The flow of calls (B-calls), with intensity Fb, occupying the B-terminals, comes from the Copy device. This corresponds to the fact that at the beginning of the ringing stage, a second (B) terminal in the system becomes busy. Another reason for this conceptual modelling trick is that the paths of the A and B-calls are different in the telecommunication system's environment after releasing the terminals; (compare the environments of a and b - devices in Fig. 1).

There are two virtual devices of type Enter Switch (Fig. 1) before the Blocked Switching (bs) and Blocked Ringing (br) devices. These devices deflect calls if there is no free line in the switching system, or if the intended B-terminal is busy, respectively. The corresponding transition probabilities depend on the macrostate of the system (Yab).

The macrostate of a (virtual) device (including the overall network, when considered as a device) is defined as the mean number of simultaneously served calls in this device, in the observed time interval (similar to "mean traffic intensity" in (ITU E.600)).

3. FORMULATION OF GENERAL TELETRAFFIC TASKS

3.1 Full Parameter Set

In our conceptual model, we have at least 35 virtual devices (31 base and four (a, b, ab and s) comprising devices). These devices are important because values of their parameters are specific to the characteristics and state of the modeled telecommunication system. Each device has five parameters (P, F, T, Y, N), so the total number of parameters is 175.

3.2 System Tuples

Definition 3.2.1. A **system tuple** is a finite set of distinguishable (by name and/or position) parameter values, which fulfills simultaneously the three following requirements:

1. All parameters (parameter set), evaluated by the system tuple, correspond to one considered (observed, modeled) system;

2. All the values of a system tuple correspond to one and the same time interval of measurements or considerations;

3. The beginning and duration of this time interval are elements of the system's tuple set.

Remarks 3.2.1: 1. Our definition of system tuple is based on the tuple definitions in Computer Science (Relational Data Bases) and Mathematics. It is adjusted for real systems' measurements, modelling and simulation; 2. The duration of the time interval, recommended by ITU-T, is from 15 minutes to one hour; 3. In the stationary state of the system, the beginning and duration of the time interval may be omitted; 4. In this paper we consider system tuples only, and use "tuple" instead "system tuple"; 5. Every subset of a system tuple is called a **"sub-tuple"**.

3.3 Base Parameter Set

There are many obvious dependencies in a system tuple, corresponding to the Full Parameter Set of the Conceptual Model. For example, the sum of probabilities of outgoing transitions in every virtual switch device has value one; in stationary state, Little's formula ($Y = F T$) is in force for every virtual device; we assume most devices have infinite capacity. As a result, there are sets of base parameters (sub-tuples) with the following property: If we know the values of the base parameters, we may calculate the values of all other parameters of the same system tuple. Several different base parameter sets may exist. After careful analysis and some assumptions (see below), we have chosen a base parameter set with 41 parameters. We call this the *base tuple*. Note that the base tuple is a sub-tuple of a system tuple.

3.4 Classification of Origination Parameters

The parameters of the base set may be classified in five groups:

1. Human Behaviour Parameters, 21: *Fo, Nab, Prad, Tid, Prid, Pris, Tis, Pns , Tns , Prns, Tbs, Prbs, Tbr, Prbr, Par, Tar, Prar, , Tcr, Prac, Tcc, Prcc*;

2. Technical Characteristics Parameters, 4: *Pid, Pis, Tcs, Ns*;

3. Mix Factors Parameters, 6: *Ted, Pad, Tad, Tcd, Pac, Tac*;

4. Modeler Chosen Values Parameter, 1: *M*;

5. Parameters Derived from the previous four groups, 9: *Yab, Fa, dem.Fa, rep.Fa, Pbs, Pbr, ofr.Fs, Ts, ofr.Ys.*

3.5 Stationary Teletraffic Tasks

For a short observation interval, we usually consider processes in this system as being in a stationary state and described with a system tuple. Some values of the system tuple may be known (measured or stipulated), while others are not known. The proposed classification allows definition of different types of teletraffic tasks. Depending on the task specificity, Mix Factors Parameters may be considered as belonging to the Human Behaviour or Technical Characteristics Parameters groups. Excluding *M*, we have three groups of parameter types and their corresponding stationary task types:

1. **System State Task** finds values of the 5th group of parameters if the values of the base parameters from the same base tuple are known. Note that *Yab* is the macrostate of the system and the values of *Pbs* and *Pbr* belong to Quality of Service (QoS) parameters. A decision of this task is described in Sections 4, 5 and 6;

2. **Technical Characteristics Task** finds values of the 4th group of parameters if the values of the base parameters from the same base tuple are known. Note that *Pid* and *Pis* are usually caused by technical failures, *Tcs* is limited, and *Ns* is a main network dimensioning parameter. The Network Dimensioning Task (NDT) is to find *Ns* if the target values of *Pbs* and *Pbr* are known. In Section 7, we formulate the NDT, describe necessary conditions for analytical solution, and find a solution.

3. **Human Behaviour Task** finds values of the 1st group of parameters if the values of the rest of the base parameters from the same base tuple are known. This task is difficult due the relatively large number of unknown values. There are some results for finding important parameters with respect to the number of active terminals *Nab* (Poryazov, Saranova, 2002).

Other criteria for parameter classification and corresponding teletraffic tasks within a system tuple are theoretically and practically interesting. For example the **Inconvenient Measurements Task** finds values of parameters that are difficult to measure – e.g., intensity of repeated attempts flow – if the values of easily-measured parameters are known.

3.6 Dynamic Teletraffic Tasks

The system's dynamics may be represented by a series of system tuples. There is a difference between long- and short-term dynamics. In long-term considerations, <u>all</u> the system parameters may have variable values. In short term analyses, <u>some</u> of the parameters may have constant values.

3.7 Classification of Static and Dynamic Parameters

We propose a short-term classification of the chosen base parameter set having 31 static and 10 dynamic parameters.

For static parameters, we assume that their values do not depend on the state of the system, nor on the intensity of the input flow. They may depend on other factors, e.g. time of day, season, human temperament, Telecom Administration, Gross Domestic Product, and so on; but for the observed and modeled time interval we consider them as constants.

The 31 static parameters are: M; Nab; Ns; $Ted = 2$ sec.; $Pad = 9\%$; $Tad = 5$ sec.; $Prad = 95\%$; $Pid = 1\%$; $Tid = 11$ sec.; $Prid = 10\%$; $Tcd = 12$ sec.; $Tbs = 5$ sec.; $Prbs = 82\%$; $Pis = 0\%$; $Tis = 5$ sec.; $Pris = 80\%$; $Pns = 1\%$;

Tns = 6 sec.; *Tcs* = 0.5 sec.; *Prns* = 1%; *Tbr* = 5 sec.; *Prbr* = 80%; *Par* = 15%; *Tar* = 45 sec.; *Prar* = 75%; *Tcr* = 10 sec.; *Pac* = 20%; *Tac* = 13 sec.; *Prac* = 90%; *Tcc* = 180 sec.; *Prcc* = 1%. The values of the static parameters correspond to the classic PSTN, and are illustrative only.

The 10 dynamic parameters, with mutually dependent values are: *Fo Yab, Fa, dem.Fa, rep.Fa, Pbs, Pbr, ofd.Fs, Ts, ofd.Ys*.

Network and Terminal Teletraffic Dynamics is an area for future research.

4. FINDING TERMINAL TELETRAFFIC PARAMETERS

4.1 Task Formulation: We consider the full telecommunication system conceptual model as depicted in Fig. 1 and described in Section 2. Parameters with known values include all the P (probability for call direction) and T (holding time) parameters of the base virtual devices, plus values of the intensity of incoming call flows (*Fa*). Parameters with unknown values include those of the comprising devices, except *Fa* and *Nab* (see assumption A-2 below). The task is to analytically determine the unknown parameter values of the devices *a* (A-terminals), *b* (busy B-terminals), and *ab* (all the simultaneously busy terminals).

4.2 Main Assumptions

To create a simple analytical model, we make the following ten assumptions (A-1 – A-10):

A-1. (Closed System Structure) We consider a closed telecommunication system with the functional structure shown in Fig. 1;

A-2. (Device Capacity) All base virtual devices in the model have unlimited capacity. Comprising devices are limited: *ab*-devices contain all the active *Nab* $\in [2, \infty)$ terminals; switching system (*s*) has a capacity of *Ns* calls (every internal switching line may carry only one call); every terminal has a capacity of one call, common for both incoming and outgoing calls;

A-3. (A-Terminal Occupation) Every call from the incoming flow in the telecommunication system (*inc.Fa*) falls only on a free terminal. This terminal becomes a busy A-terminal;

A-4. (Stationarity) The system is in a stationary state. This means that in every virtual device in the model (including comprising devices like the switching system), the intensity of input flow $F(0, t)$, call holding time $T(0, t)$ and traffic intensity $Y(0, t)$ in the observed interval $(0,t)$ converge to the corresponding finite numbers F, T and Y as $t \to \infty$. In this case we apply Little's Theorem (1961), and for every device: $Y = FT$;

A-5. (Calls' Capacity) Every call occupies one place in a base virtual device, independent of the other devices (e.g., a call may occupy one internal switching line if it finds one free, independent of the state of the intended B-terminal (busy or free));

A-6. (Environment) The calls in the communication systems' environment (outside the blocks a and b in Fig. 1) do not occupy any telecommunication systems' device and therefore do not create a communication system load. (For example, unsuccessful calls waiting for the next attempt are in "the head of" the user only. The calls and devices in this environment form the intent and repeated call flows.) Calls leave the environment (and the model) when they enter a Terminator virtual device;

A-7. (Parameters' Independence) We consider probabilities for direction of calls to, and holding times in, the base virtual devices as independent of each other, and from intensity $Fa = inc.Fa$ of the incoming flow of calls. Values of these parameters are determined by users' behavior and the technical characteristics of the communication system. (Obviously, this is not applicable to the devices of type Enter Switch, corresponding to Pbs and Pbr - see 2.4.);

A-8. (Randomness) All variables in the analytical model may be random, and we work with their mean values (following Little's Theorem).

A-9. (B-Terminal Occupation) Probabilities of direction of calls to, and duration of occupation, of devices ar, cr, ac and cc are the same for A and B-calls;

A-10. (Channel Switching) Every call simultaneously occupies places in all of the base virtual devices in the telecommunication system (comprised of devices a or b) that it passed through, including the base device it is in at the moment of observation. Every call releases all of its occupied places in all base virtual devices of the communication system at the instant it leaves the comprising devices a or b.

4.3 Mathematical Expressions

Theorem 4.3.1. The intensity of traffic (traffic) of all the terminals (Yab) is the sum of traffic of the A (Ya) and B-terminals (Yb):

$$Yab = Ya + Yb. \qquad (4.3.1)$$

Proof: i) There are no terminals outside Fig. 1 because the modeled system is closed (assumptions A-1); ii) Every terminal, at a particular instant of time, may be free or busy. If it is busy, it may be calling (A) or called (B), but not simultaneously both because every terminal has a capacity of one call (A-2). Obviously from i) and ii) follows (1).

Absolute Terminal Traffic Limitations. Since the number of terminals is limited to Nab (A-2), and there is no negative occupancy, we have the following obvious terminal traffic limitation in the system:

$$0 \leq Yab \leq Nab. \qquad (4.3.2)$$

Theorem 4.3.2. The calls flow intensity occupying all terminals (Fab) is the sum of intensities of calls flow occupying A-terminals (Fa) and calls flow occupying B-terminals (Fb):

$$Fab = Fa + Fb. \qquad (4.3.3)$$

This follows assumption A-1 and our conceptual model (section 2.4) in which $Fb = inc.Fb$ is a different flow from Fa and corresponds to cases when successful calls, after dialing and switching, occupy B-terminals. From assumption A-3 it follows that $Fa = inc.Fa$ (there are no blocked calls, directed to the communication system).

Theorem 4.3.3. The traffic intensity of B-terminals (Yb) may be calculated from the equation

$$Yb = FbTb, \qquad (4.3.4)$$

where Tb is the mean holding time of calls in a B-terminal, and:

$$Fb = Fa(1 - Pad)(1 - Pid)(1 - Pbs)(1 - Pis)(1 - Pns)(1 - Pbr) = Fa\,Eb, \qquad (4.3.5)$$

$$Tb = ParTar + (1 - Par)[Tcr + PacTac + (1 - Pac)Tcc]. \qquad (4.3.6)$$

Proof: Equation (4) is Little's formula for device b in stationary state (A-4).

Equation (4.3.5) expresses that A-calls have to avoid the six modeled losses before occupying the intent B-terminals, with mean intensity of calls

Fb. Equation (4.3.5) is a direct corollary from: Fig. 1 (box for *B-device*, in the lower right); closed system structure (A-1); calls' capacity (A-5); excluding calls in the environment (A-6); parameters' independence (A-7); randomness (A-8); and the B-terminal occupation assumption (A-9).

Remark 4.3-1: *Fb* is the intensity of call attempts that receive an answer signal (after occupying B-terminals). For those, the ITU E.600 expressions are: "completed call attempt" and "effective call attempt". Because of this, we call the quantity *Eb* in (4.3.5) *"B-efficiency"*.

We derive the expression (4.3.6) for B-terminals holding time (*Tb*) as follows. From Fig. 1, parameters' independence (A-7), and channel switching (A-10), it follows that *Yb* is a sum of the traffic of the base blocks comprised in it; The assumption for B-terminal occupation (A-9) implies that *Yar, Ycr, Yac* and *Ycc* have the same traffic intensities for A and B-terminals, so:

$$Yb = Yar + Ycr + Yac + Ycc. \tag{4.3.7}$$

On the other hand, we may also express the traffic intensities using Little's formula by presenting every flow intensity in the base devices as a function of *Fb*:

$$Yar = Far\ Tar = Fb\ Par\ Tar; \tag{4.3.8}$$

$$Ycr = Fcr\ Tcr = Fb\ (1-Par)\ Tcr; \tag{4.3.9}$$

$$Yac = Fac\ Tac = Fb(1-Par)Pac\ Tac; \tag{4.3.10}$$

$$Ycc = Fcc\ Tcc = Fb(1-Par)(1-Pac)\ Tcc. \tag{4.3.11}$$

After substituting (4.3.8), (4.3.9), (4.3.10) and (4.3.11) in (4.3.7), taking into consideration A-9, and using (4.3.4), we arrive at (4.3.6).

Remark 4.3-2: A-9 is used because it is close to the reality of digital switching systems and simplifies the expressions, but it is not a part of the described modelling approach, nor is it linked with other assumptions and thus may be ignored if necessary. This contingency is accounted for in the parameter notation system.

Theorem 4.3.4. A-terminals' traffic intensity (*Ya*) is determining from Little's formula

$$Ya = FaTa \tag{4.3.12},$$

where *Ta* is:

$$Ta = Ted + PadTad + (1\text{-}Pad)(PidTid + (1\text{-}Pid)$$
$$(Tcd + PbsTbs + (1\text{-}Pbs)(PisTis + (1\text{-}Pis)$$
$$(PnsTns + (1\text{-}Pns)(Tcs + PbrTbr + (1\text{-}Pbr)Tb))))). \quad (4.3.13)$$

The proof of Theorem 4.3.4 is very similar to the proof of Theorem 4.3.3, but includes more base devices as shown in Fig. 1.

Theorem 4.3.5. Distinguishing static and dynamic parameters, we have:

$$Ya = Fa[Sa_1 - Sa_2(1 - Pbs)Pbr - Sa_3Pbs] , \quad (4.3.14)$$

where:

$Sa_1 = Ted + PadTad + (1 - Pad)[PidTid + (1 - Pid)[Tcd + PisTis +$
$\quad (1 - Pis)[PnsTns + (1 - Pns)[Tcs + Tb]]]],$
$Sa_2 = (1 - Pad)(1 - Pid)(1 - Pis)(1 - Pns)[Tb - Tbr],$
$Sa_3 = (1 - Pad)(1 - Pid)[PisTis + (1 - Pis)[PnsTns + (1 - Pns)[Tcs + Tb]]] -$
$(1 - Pad)(1 - Pid)Tbs.$

$$(4.3.15)$$

Proof: Equations (4.3.14) and (4.3.15) are the result of simple mathematical transformations of (4.3.13) after applying Little's formula $Ya = Fa\,Ta$.

Theorem 4.3.6. Traffic of all the simultaneously busy terminals (Yab) may be determined from equation (4.3.16) as a function of Fa and other parameters:

$$Yab = Fa\,\{Ted + PadTad + (1\text{-}Pad)[PidTid + (1\text{-}Pid)[Tcd + PbsTbs +$$
$$(1\text{-}Pbs)[PisTis + (1\text{-}Pis)\,[PnsTns + (1\text{-}Pns)[Tcs + PbrTbr + 2(1\text{-}Pbr)Tb]]]]\},$$
$$(4.3.16)$$

or equation (4.3.17), after distinguishing static and dynamic parameters in it:

$$Yab = Fa[S_1 - S_2(1 - Pbs)\,Pbr - S_3\,Pbs], \quad (4.3.17)$$

where:

$$S_1 = Ted + PadTad + (1 - Pad)[PidTid + (1 - Pid)[Tcd + PisTis +$$
$$(1 - Pis)[PnsTns + (1 - Pns)[Tcs + 2Tb]]]],$$
$$S_2 = (1 - Pad)(1 - Pid)(1 - Pis)(1 - Pns)[2Tb - Tbr],$$
$$S_3 = (1 - Pad)(1 - Pid)[PisTis + (1 - Pis)[PnsTns + (1 - Pns)$$
$$[Tcs + 2Tb]]] - (1 - Pad)(1 - Pid)Tbs.$$

$$(4.3.18)$$

Proof: Adding equations (4.3.4) and (4.3.13), using (4.3.1), and performing elementary mathematical transformations yields (4.3.16), and from it (4.3.17) and (4.3.18).

Remark 4.3.3. The expression in brackets in (4.3.17) is the mean duration of A and B-terminal occupation, corresponding to one A-call. Note that Sa_1, Sa_2 and Sa_3 differ from the corresponding S_1, S_2 and S_3 only in factor 2 before the B-terminal holding time Tb.

Theorem 4.3.7. In equations (4.3.18), $0 < S_1$ always. In normal circumstances, $0 < S_2$ and $0 < S_3$.

Proof: All possible holding times are positive, because they correspond to real actions. Since probabilities are always in the interval $[0,1]$, obviously $0 < S_1$;

If there are successful communications (normal circumstances) in the system, then the following four probabilities are less than 1: $Pad < 1$, $Pid < 1$, $Pis < 1$ and $Pns < 1$ (see Fig. 1.), and therefore $0 < S_2$. If one or more of these probabilities has value 1, the communications are impossible and $0 = S_2$ (see (4.3.18).). Note that $Tb > 0$ always, and in normal circumstances $[2Tb - Tbr] > 0$ (see (4.3.6)) because Tbr is the duration of listening for a busy tone (with a mean value of approximately 5 seconds, and this duration is usually limited in digital switching centers), $Tar \approx 45$ sec. (usually limited to 60 sec.), $Tcr \approx 10$ sec., $Tac \approx 13$ sec., $Tbr = 5$ sec. and $Tcc \approx 180$ sec.

Theorem 4.3.8. The mean occupation time (Tab) of all the simultaneously busy terminals may be determined from equation (4.3.19) as a function of Ta, Tb and Eb.

Proof: Is obvious from formula $Yab = FabTab$ after replacing Yab with (4.3.1), Fab with (4.3.3) and Fb with (4.3.5), yielding:

$$Tab = \frac{Yab}{Fab} = \frac{Ya+Yb}{Fa+Fb} = \frac{FaTa+FbTb}{Fa+Fb} = \frac{FaTa+FaEbTb}{Fa+FaEb} = \frac{Ta+EbTb}{1+Eb}.$$
$$(4.3.19)$$

Theorem 4.3.9. The mean occupation time (Tab) of all the simultaneously busy terminals may be determined from equation (4.3.20) as a function of $S1, S2, S3, Pbr$, Pbs and Eb.

Proof: From the formula $Yab = FabTab$, after replacing Yab with (4.3.17), Fab with (4.3.3) and Fb with (4.3.5), we have:

$$Tab = \frac{Yab}{Fab} = \frac{Fa[S_1 - S_2(1-Pbs)\,Pbr - S_3\,Pbs]}{Fa + FaEb} = \frac{S_1 - S_2(1-Pbs)\,Pbr - S_3\,Pbs}{1+Eb}.$$
$$(4.3.20)$$

Theorem 4.3.10. The mean occupation time of A-terminals (Ta) can be determined from equation (4.3.21) as a function of $S1, S2, S3, Pbr$, Pbs and Eb.

Proof: From the formula $Ya = FaTa$, after replacing Ya with (4.3.14), Fa with (4.3.3) and Fb with (4.3.5), we have:

$$Ta = \frac{Fa\,[Sa_1 - Sa_2(1-Pbs)Pbr - Sa_3Pbs]}{Fa} = Sa_1 - Sa_2(1-Pbs)Pbr - Sa_3Pbs$$
$$. \qquad (4.3.21)$$

Theorem 4.3.11. The mean occupation time of A-terminals (Ta) and of all the simultaneously busy terminals (AB-terminals) (Tab) depend on the intensity of the input flow (Fa) through the probabilities of blocked switching (Pbs) and blocked ringing (Pbr); and their absolute maxima ($max.Ta$ and $max.Tab$) correspond to the values $Pbs = 0$ and $Pbr = 0$.

Proof: Dependencies are presented in equations (4.3.20).and (4.3.21). Following Theorem 4.3.7, their absolute maxima are reached when $Pbr = 0$ and $Pbs = 0$:

$$max.Ta = Sa_1 \qquad\qquad (4.3.22)$$

$$max.Tab = \frac{S_1}{1+Eb}. \qquad\qquad (4.3.23)$$

5. DETERMINING INPUT FLOW INTENSITY

We must also determine the mean intensity of the input flow to the telecommunication system. This is the flow occupying the calling (A) terminals Fa. From the ITU E.600 definitions and Fig. 1, it is obvious that the intensity of incoming flows is the sum of the intensities of primary (demand) calls ($dem.Fa$) and repeated attempts ($rep.Fa$):

$$Fa = dem.Fa + rep.Fa. \tag{5.1}$$

From the definition of a BBP-flow we have (see Section 2.3.):

$$dem.Fa = Fo\ (Nab + M\ Yab), \tag{5.2}$$

Definition 5.1. Probabilities for repeated attempts ($Prxx$) have mean (expected) values for all attempts and calls in the stationary state. For example, the probability of the repetition of an attempt after abandoned ringing ($Prar$) may be expressed as a ratio of the number of all repeated attempts after abandoned ringing $Zrar(0,t)$ observed in the interval $(0,t)$ to the number of all abandoned ringing cases $Zar(0,t)$:

$$Prar = \lim_{t\to\infty} Prar(0,t) = \lim_{t\to\infty} \frac{Zrar(0,t)}{Zar(0,t)} . \tag{5.3}$$

Theorem 5.1. The intensity of calls of repeated attempts $rep.Fa$ may be determined from the expression (5.4).

$$
\begin{aligned}
rep.Fa = \ &Fa\ \{PadPrad + (1\text{-}Pad)[PidPrid + (1\text{-}Pid)[PbsPrbs + (1\text{-}Pbs) \\
&[PisPris + (1\text{-}Pis)[PnsPrns + (1\text{-}Pns)[PbrPrbr + (1\text{-}Pbr) \\
&[ParPrar + (1\text{-}Par)[PacPrac + (1\text{-}Pac)Prcc]]]]]]]\}.
\end{aligned}
\tag{5.4}
$$

Proof: As seen in Fig. 1 and assumption A-1, $rep.Fa$ is the sum of intensities of repeated attempt flows in all branches:

$$rep.Fa = Frad + Frid + Frbs + Fris + Frns + Frbr + Frar + Frac + Frcc. \tag{5.5}$$

The intensities of repeated attempt flows in all branches may be easily expressed as functions of Fa, following the structure of Fig.1:

$$Frad = Fa\ Pad\ Prad; \tag{5.6}$$

$$Frid = Fa\ (1 - Pad)\ Pid\ Prid; \tag{5.7}$$

$$Frbs = Fa\ (1 - Pad)(1 - Pid)\ Pbs\ Prbs; \tag{5.8}$$

$$Fris = Fa\ (1 - Pad)(1 - Pid)(1 - Pbs)Pis\ Pris; \tag{5.9}$$

$$Frns = Fa\ (1 - Pad)(1 - Pid)(1 - Pbs)(1 - Pis)Pns\ Prns; \tag{5.10}$$

$$Frbr = Fa\ (1 - Pad)(1 - Pid)(1 - Pbs)(1 - Pis)(1 - Pns)Pbr\ Prbr; \tag{5.11}$$

$$Frar = Fa\ (1 - Pad)(1 - Pid)(1 - Pbs)(1 - Pis)(1 - Pns)(1 - Pbr)\ Par\ Prar; \tag{5.12}$$

$$Frac = Fa\ (1 - Pad)(1 - Pid)(1 - Pbs)(1 - Pis)(1 - Pns)(1 - Pbr)(1 - Par)\ Pac\ Prac; \tag{5.13}$$

$$Frcc = Fa\ (1 - Pad)(1 - Pid)(1 - Pbs)(1 - Pis)(1 - Pns)(1 - Pbr)(1 - Par)(1 - Pac)\ Prcc. \tag{5.14}$$

Adding equations from (5.6) to (5.14) and taking into account (5.5), we obtain (5.4).

Theorem 5.2. Distinguishing static and dynamic parameters in (5.4), following elementary algebraic operations, yields *rep.Fa* as a simple function of *Fa, Pbr* and *Pbs*:

$$rep.Fa = Fa\ [R_1 - R_2\ Pbr\ (1 - Pbs) - R_3\ Pbs], \tag{5.15}$$

where:

$$Q = ParPrar + (1 - Par)[PacPrac + (1 - Pac)Prcc]$$
$$R_1 = Pad\ Prad + (1 - Pad)(Pid\ Prid + (1 - Pid)\ Pis\ Pris + (1 - Pis)(Pns\ Prns + (1 - Pns)\ Q));$$
$$R_2 = (1 - Pad)(1 - Pid)(1 - Pis)(1 - Pns)(Prbr - Q);$$
$$R_3 = (1 - Pad)(1 - Pid)\{PisPris + (1 - Pis)[PnsPrms + (1 - Pns)Q] - Prbs\}. \tag{5.16}$$

Theorem 5.3. Equation (5.17) gives the dependencies between Fa and: the intensity of the input flow from one idle terminal (Fo); Pbr; Pbs; and the flow model chosen (M):

$$Fa = \frac{Fo\,Nab}{[1 - FoS_1M - R_1] + [FoS_2M + R_2](1 - Pbs)Pbr + [FoS_3M + R_3]\,Pbs}.$$

$$(5.17)$$

Proof: Equation (5.17) is a direct corollary from equations (5.1), (5.2) and (5.15).

After (5.17), we have Fa, Ya (4.3.14), Yab (4.3.17), Tab (4.3.20), Ta (4.3.21) and other terminal teletraffic characteristics represented as functions of static parameters Pbr and Pbs. Our next tasks are deriving the probabilities of finding a B-terminal busy (Pbr), and for switching system blocking (Pbs).

6. FINDING THE BLOCKING PROBABILITIES

6.1 Determining the Probability of Finding a B-terminal Busy

Parameters characterizing terminal traffic are used in teletraffic engineering for all telecommunication networks. One of the most important parameters is the probability of finding the called (B-terminal) busy. This parameter was studied in some of the earliest papers in teletraffic theory (Johannesen 1908). We describe a solution for the case of (virtual) channel systems, such as the PSTN and GSM, and more generally, for virtual channel/circuit paradigms in multi-service networks.

6.1.1 Task Formulation: We use the same conceptual model and assumptions, but our task is to analytically determine the unknown value of the mean probability of calls in the stationary state that find the intent B-terminals busy (Pbr).

6.1.2 Main Assumptions

To solve this task, we need to add the following four assumptions (A-11 – A-14) to our existing assumptions:

A-11. (Terminals' Homogeneity) All terminals are homogeneous; e.g. all relevant characteristics are equal for every terminal;

A-12. (A-Calls Directions) Every A-terminal uniformly directs all its calls only to the other terminals, and not to itself;

A-13. (B-flow ordinariness) The flow directed to B-terminals (*Fb*) is ordinary. (The importance of A-13 is limited only to the case when two or more calls may simultaneously reach a free B-terminal. A-13 may be assumed from results in (Burk 1956) and (Vere-Jones 1968);

A-14. (B-Blocking Probability for Repeated attempts) The mean probability (*Pbr*) of a call finding the same B-terminal busy at the first attempt and at all following repeated attempts is one and the same.

6.1.3 Mathematical Model

Theorem 6.1.3.1. The probability of finding the B-terminal busy (*Pbr*) if $1 \leq Yab \leq Nab$, is:

$$Pbr = \frac{Yab - 1}{Nab - 1} \qquad\qquad (6.1.3.1)$$

Proof: According to assumption A-1, all calls are directed to the terminals inside the system. Following A-11, all the terminals have equal probability to be called, but A-12 excludes the calling A-terminal. A similar assumption is made by [Ionin 1978] in a simulation model. Consequently, every call is directed with equal probability to *Nab-1* terminals, with *Yab*-1 of them busy. From A-13 it follows that two or more calls cannot come simultaneously to one terminal, and hence the probabilities of finding B-terminals busy are proportional to the number of asked busy terminals (*Yab*-1).

If call number i finds $Yab(i)$ busy terminals, then the probability of finding B-terminals busy (*Pbr(i)*) under our assumptions is:

$$Pbr(i) = \frac{Yab(i) - 1}{Nab - 1} \qquad\qquad (6.1.3.2)$$

If we consider n calls reaching B-terminals, then the mean probability *Pbr* of finding B-terminals busy is:

$$Pbr = \frac{1}{n}\sum_{i=1}^{n}Pbr(i) = \frac{1}{n}\sum_{i=1}^{n}\frac{Yab(i)-1}{Nab-1} = \frac{1}{(Nab-1)}\frac{\sum_{i=1}^{n}(Yab(i)-1)}{n}.$$

$$(6.1.3.3)$$

From A-4, it follows that the system is in a stationary state. Therefore, the mean value of the intensity of the terminal traffic exists and is equal to Yab. In the other words:

$$Yab = \lim_{n\to\infty}\frac{\sum_{i=1}^{n}Yab(i)}{n}. \qquad (6.1.3.4)$$

From (6.1.3.3) as $n \to \infty$ and (6.1.3.4), using the (strong) law of large numbers, we obtain (6.1.3.1).

Remark (6.1.3.1). The proof is based on assumption A-11 (terminals' homogeneity) which corresponds to the flow of primary (demand) calls. When we consider the repeated calls flow, assumption A-11 is not in force because repeated call attempts are directed to the same intent B-terminal as the primary attempt, and therefore terminals are not equal for repeated calls. In the case of a system with repeated calls, we use assumption A-14 as well. A-14 is the only assumption in this paper that causes a systematic error. In (Poryazov 1992) it is shown on the basis of simulation results (Todorov, Poryazov 1985) that the relative error does not exceed 5% of Pbr in a reasonable traffic load interval.

Theorem 6.1.3.2. A threshold ($thr.Fa > 0$) of the intensity of the input flow Fa exists; hence, busy terminals exist in the interval $Fa \in [0, thr.Fa]$ but there are no losses due to finding the B-terminal busy. In this case:

$$thr.Fa = \frac{1}{S_1} \qquad (6.1.3.5)$$

Proof: We can present (4.3.17) in the form:

$$Fa = \frac{Yab}{S_1 - S_2(1-Pbs)\,Pbr - S_3\,Pbs} \qquad (6.1.3.6)$$

If we replace Pbr with 0 in (6.1.3.1) we obtain $Yab=1$, and from (6.1.3.6):

$$Fa = \frac{1}{S_1 - S_3 \, Pbs}.$$ (6.1.3.8)

Equation (6.1.3.8) is derived without assuming any other dependence between Fa and Pbr, and therefore is true in every case when $Pbr = 0$.

If there are successful conversations, then there must be at least one internal switching line in the system. We assume that enough equivalent switching lines exist in the system to serve traffic of less than one Erlang without blocking (from (4.3.1) follows $Ya \leq Yab$), and therefore $Pbs = 0$, and from (6.1.3.8) follows (6.1.3.5). Since $S_1 > 0$ (Theorem 4.3.7), if follows from (6.1.3.1) that $thr.Fa > 0$.

Comment: The fact that we have no losses in the interval $Fa \in [0, \, thr.Fa]$ due to finding the B-terminal busy must be understand in the asymptotic case as $t \to \infty$. In the other words, losses may exist, but:

$$Pbr = \lim_{t \to \infty} \frac{Zbr.a(0,t)}{Zbr.a(0,t) + Z.b(0,t)} = 0,$$ (6.1.3.9)

where $Zbr.a(0,t)$ denotes the number of calls finding the B-terminal busy in the interval of observation $(0,t)$, and $Z.b(0,t)$ is the number of calls successfully seizing B-terminals in the same time interval.

Definition 6.1.3.1. Based on Theorem (6.1.3.1) and Theorem (6.1.3.2), we may define Pbr as a function of Yab and Nab:

$$Pbr = \frac{Yab\text{-}1}{Nab\text{-}1} \quad in \; case \; of \; 1 \leq Yab \leq Nab,$$

$$Pbr = 0 \quad in \; case \; of \; 0 \leq Yab < 1.$$

(6.1.3.10)

This definition is used in a very simple teletraffic model (Poryazov 1991) without detailed proof.

6.2 Determining the Probability of Switching System Blocking

6.2.1 Task Formulation: We will use the same conceptual model and assumptions, but our task is to find analytically the unknown value of the mean blocking probability (Pbs) due to insufficient equivalent switching lines (Ns).

6.2.2 Mathematical Model

Theorem 6.2.2.1. The mean holding time of the switching system (Ts) may be expressed through equation (6.2.1.1):

$$Ts = [PisTis + (1-Pis)[PnsTns +$$
$$(1-Pns)[Tcs + PbrTbr + (1-Pbr)Tb]]] = S_{1Z} - S_{2Z} \; Pbr$$

$$(6.2.2.1)$$

Proof: According to the Channel Switching assumption (A-10), we have to consider switching system holding time as a function of the holding times of calls in the switching system itself, as well as the holding times of calls receiving a busy tone (Pbr) and occupying B-terminals (Tb) (see dashed box in Fig. 1.). Following the approach used in Theorem 4.3.3., we obtain (6.2.2.1) directly from Fig. 1.

Theorem 6.2.2. The intensity of calls offered to the switching system (ofr.Fs) may be expressed through equation (6.2.2.2):

$$ofr.Fs = Fa \; (1-Pad)(1-Pid)$$
$$(6.2.2.2)$$

Proof: Following the ITU E.600 definition of offered traffic, we obtain (6.2.2.2) directly from Fig. 1.

Theorem 6.2.3. The probability of switch blocking (Pbs) is determined from equations (6.2.2.3) and (6.2.2.4).

$$ofr.Ys = ofr.Fs \; Ts. \qquad (6.2.2.3)$$

$$Pbs = Erl\_b \; (Ns, ofr.Ys). \qquad (6.2.2.4)$$

Proof: Equation (6.2.2.4) expresses usage of the Erlang-B formula for determining the blocking probability in the switching system on the basis of the number of internal switching lines (Ns) and offered traffic $ofr.Ys$.

Equation (6.2.2.3) needs more attention because of the ITU E.600 definition of offered traffic: viz., "The traffic that would be carried by an infinitely large pool of resources." The problem is that Ts depends on Pbr (6.2.2.1), and through it, on the macrostate of the system Yab (6.1.3.10); however, Yab depends on Pbs (4.3.17). If we strictly follow the ITU E.600 definition, then we must use in (6.2.2.3) the value $ofr.Ts$ corresponding to the *Yab* in the case when *Pbs* = 0. Despite the ITU E.600 definition, we use the value of *Ts* (6.2.2.1) in the context of a system of equations, simultaneously determining values of all unknown dynamic parameters.

Remark 6.2.2.1. In (Poryazov 2005) we have shown that traffic offered to the switching system *ofd.Ys* is higher in the case of nonzero blocking (Pbs ≠ 0) in comparison with the case of zero blocking (Pbs = 0) for the same *Yab*. The relative difference between values of *ofd.Fs* in both cases may exceed 112%. This is caused by the influence of the lack of resources on the processes in the communication system, which is not foreseen in the ITU-T definition.

6.3 The Full System of Equations for Dynamic Variables

We consider the values of 10 basic dynamic parameters, which are mutually dependent: *Fo, Yab, Fa, dem.Fa, rep.Fa, Pbs, Pbr, ofd.Fs, Ts, ofd.Ys*. For these parameters, and under the assumptions we have made, we have a system of 9 equations with 6 generalized static parameters ($S_1, S_2, S_3, R_1, R_2, R_3$): Equation (4.3.17) for *Yab*; (5.1) for *Fa*; (5.2) for *dem.Fa*; (5.15) for *rep.Fa*; (6.1.3.10) for *Pbr*; (6.2.2.1) for *Ts*; (6.2.2.2) for *ofr.Fs*; (6.2.2.2) for *ofr.Ys*; (6.2.2.4) for *Pbs*. We have no equation in which *Fo* is present on the left side; it appears on the right side only in (5.2).

If we choose the intensity of the input calls flow *Fo* as the independent input variable, the system of equations has 9 equations and 9 output parameters with unknown values (the main output variables). Since this system of equations is derived directly from the conceptual model and the assumptions made, we have shown network and terminal teletraffic to be a function of parameters describing human behaviour and the technical characteristics of the communication system.

7. FINDING THE REQUISITE NUMBER OF EQUIVALENT INTERNAL SWITCHING LINES

Sometimes it is necessary (for network planning and control) to determine network dimensioning and the level of quality of service (QoS) in advance (ITU E.800).

For our conceptual model and corresponding analytical models, we next formulate a network dimensioning task (NDT), and discuss the solution of the NDT and the necessary conditions for an analytical solution. We propose an algorithm to calculate the number of switching lines and corresponding values of traffic parameters to simplify the management and control of QoS.

7.1 Formulation Of A Network Dimensioning Task (NDT):

1. Determine the number of switching lines required to 'dimension' a network when the desired level of QoS is administratively determined in advance, and the values of known parameters are provided or calculated in some way.

2. Determine the values of the unknown parameters that describe the system state – e.g., parameters describing the macrostate of the system (through the value of Yab); capacity of the system (the number of active terminals Nab); intensity of demanded and repeated call attempts ($dem.Fa$ and $rep.Fa$); offered traffic intensity ($ofr.Ys$); etc.

Parameters in the network dimensioning task:
Administratively preassigned parameters: $adm.Pbs$, $adm.Pbr$ and M
$$(7.1.1)$$
Known values of the parameters: Fo, Tb, S_1, S_2, S_3, S_{1z}, S_{2z}, R_1, R_2 and R_3 $$(7.1.2).$$

Aim: To determine

The number of necessary switching lines Ns;

Unknown parameters: Nab, Yab, Fa, $dem.Fa$, $rep.Fa$, $ofr.Fs$, Ts, $ofr.Ys$ $$(7.1.3).$$

7.2 Solving the NDT

The traffic intensity Yab characterizes the macrostate of the system.

Theorem 7.1: If $Pbr \neq 0$ and $Fo \neq 0$, then Yab in the NDT is expressed through (7.2.4).

Proof: Considering the system equations (5.2) - (6.1.3.10), when $Pbr \neq 0$ and $Fo \neq 0$ it follows that:

$$demF_a = \frac{F_0}{Pbr}\left[Pbr - 1 + (Mpr + 1)Yab\right] \tag{7.2.1}$$

From (5.1) and (5.15), it follows that:

$$demFa = Fa\{1 - R_1 + R_2Pbr + (R_3 - R_2Pbr)Pbs\} \tag{7.2.2}$$

Then (7.2.1), (7.2.2) and (4.3.17) yield:

$$Yab = \frac{Fo\,(1 - Pbr)\{S_1 - S_2Pbr - (S_3 - S_4Pbr).Pbs\}}{Fo\,(1 + M.Pbr)\{S_1 - S_2Pbr - (S_3 - S_2Pbr).Pbs\} - Pbr.\{1 - R_1 + R_2Pbr + (R_3 - R_2Pbr).Pbs\}} \tag{7.2.3}$$

The simplified expression is:

$$Yab = \frac{Fo\,(S_1 - S_3\,Pbs) - (Fo(S_1 - S_3\,Pbs) + Fo\,S_2\,(1 - Pbs))Pbr + Fo\,S_2\,(1 - Pbs)Pbr^2}{Fo\,(S_1 - S_3\,Pbs) - (Fo\,M\,(S_1 - S_3\,Pbs) + Fo\,S_2\,(1 - Pbs) - 1 + R_1 - R_3\,Pbs)Pbr + (1 - Pbs)(Fo\,MS_2 + R_2)Pbr^2}$$
$$\tag{7.2.4.}$$

If $Fo = 0$, then this is a case where $Fa = 0$, $dem.Fa = 0$, and $rep.Fa = 0$.

Therefore, when $Pbr \neq 0$ in the NDT, traffic intensity Yab is expressed on the basis of administratively determined values of parameters Pbs, Pbr, M, and the known parameters (7.1.2). The other system parameters in the NDT depend on the system state (and respectively on Yab).

Theorem 7.2: If

$$Pbr \neq 0 \text{ and } Pbs \neq \frac{S_1 - S_2Pbr}{S_3 - S_2Pbr}$$

in the NDT, then every unknown parameter (7.1.3) can be expressed (evaluated) through the Yab and the known parameters (7.1.2) analytically.

Proof: Using (4.3.17), if

$(S_1 - S_2\,Pbr) - (S_3 - S_2\,Pbr)\,Pbs \neq 0$ it follows that:

$$Fa = \frac{Yab}{S_1 - S_2Pbr - (S_3 - S_2Pbr)Pbs} \tag{7.2.6.}$$

For *dem.Fa* from (5.2) and (6.10) we obtain (7.2.1).

From (6.10) it follows that:

$$Nab = 1 + \frac{Yab - 1}{Pbr} \qquad (7.2.7).$$

$$rep.Fa = \frac{Yab\{R_1 - R_2\,Pbr - (R_3 - R_2\,Pbr)\,Pbs\}}{S_1 - S_2\,Pbr - (S_3 - S_2\,Pbr)\,Pbs} \qquad (7.2.8).$$

which is the result of (5.15) and (7.2.6).

From (6.2.2.2) and (7.2.6) it follows that:

The parameter Ts can be calculated from (6.2.2.1). From (7.2.9) and

$$ofr.Fs = \frac{Yab\,(1 - Pad)(1 - Pid)}{S_1 - S_2\,Pbr - (S_3 - S_2\,Pbr)\,Pbs} \qquad (7.2.9).$$

(6.2.2.1) it follows that:

$$ofr.Ys = \frac{(1 - Pad)(1 - Pid)(S_{1z} - S_{2z}\,Pbr)\,Yab}{S_1 - S_2\,Pbr - (S_3 - S_2\,Pbr)\,Pbs} \qquad (7.2.10).$$

Therefore, the values of the unknown parameters (7.1.3) in the NDT can be expressed and calculated by the conditions of Theorem 7.1 and Theorem 7.2.

For network dimensioning, when the level of QoS is determined administratively in advance (e.g., the blocking probability Pbs), then the Erlang B formula is used:

$$Pbs = Erl\_b(Ns, ofr.Ys) \qquad (7.2.11).$$

$Pbs = Pbs(Ns, ofr.Ys)$ is a function of the number of switching lines Ns and the values of *ofr.Ys*, which are calculated under the conditions of Theorem 7.1 and Theorem 7.2.

Remark: From (7.2.10) for *adm.Pbs* and *adm.Pbr*, the corresponding value of *ofr.Ys* is fixed. Then $Pbs = Pbs(Ns, ofr.Ys)$ is a function only of Ns.

Theorem 7.3: The function $Pbs = Pbs(Ns, ofr.Ys)$, defined through (7.2 11) in the NDT is strictly monotone decreasing, when *ofr.Ys* > 0.

Proof: It can be shown that $Pbs(Ns+1, ofr.Ys) < Pbs(Ns, ofr.Ys)$. Using the recursively-expressed Erlang B formula [Iversen, 2004] and $crr.Ys=ofr.Ys(1-Pbs)$ [ITU E.501], it follows that:

$Erl\_b(Ns, ofr.Ys) > 0$ when $ofr.Ys > 0$, $ofr.Ys\, Erl\_b(Ns, ofr.Ys)+Ns+1 > 0$,

$$Pbs(Ns+1, ofr.Ys) - Pbs(Ns, ofr.Ys) = Erl\_b(Ns+1, ofr.Ys) - Erl\_b(Ns, ofr.Ys) =$$

$$= Erl\_b(Ns, ofr.Ys)\frac{ofr.Ys[1 - Erl\_b(Ns, ofr.Ys)] - (Ns+1)}{ofr.Ys\, Erl\_b(Ns, ofr.Ys) + (Ns+1)} =$$

$$Erl\_b(Ns, ofr.Ys)\frac{crr.Ys - (Ns+1)}{ofr.Ys\, Erl\_b(Ns, ofr.Ys) + (Ns+1)}.$$

and $crr.Ys \leq Ns$, and hence $crr.Ys - (Ns+1) < 0$.

Therefore, $Pbs(Ns+1, ofr.Ys) - Pbs(Ns, ofr.Ys) < 0$, and the function $Pbs = Pbs(Ns, ofr.Ys)$, defined through (7.2.11), is strictly monotone decreasing according Ns, when $ofr.Ys > 0$.

7.3 Analytical Solution

Theorem 7.4: There is only one solution of the NDT through the equation

$$Erl\_b(Ns, ofr.Ys) = adm.Pbs \qquad\qquad (7.3.1),$$

according to the number of switching lines Ns and $of.rYs$ with fixed value. The values of QoS blocking probabilities $Adm.Pbs \in [0;1]$ and $adm.Pbr \in (0; 1]$ are determined administratively and in advance.

Proof: Existence: It was proved in Theorem 7.3 that the function $Pbs = Pbs(Ns, ofr.Ys)$, defined through (7.2.11) in the NDT, is strictly monotone decreasing when $ofr.Ys > 0$. The absolute maximum of the function is 1 and the absolute minimum of the function is 0. There is only one solution for equation (7.3.1) for $adm.\ Pbs \in [0; 1]$, based on the Intermediate Value Theorem (Leduc, 2002).

Uniqueness of the solution: Let $Ns1 \neq Ns2$ be two different solutions of the equation (7.3.1) for $adm.Pbs \in [0; 1]$; therefore, $Erl\_b(Ns1, ofr.Ys) = adm.Pbs$ and $Erl\_b(Ns2, ofr.Ys) = adm.Pbs$ are simultaneously fulfilled. This contradicts Theorem 7.3. Hence, only one solution Ns satisfies equation (7.3.1) by determining administratively (and in advance) the QoS blocking probabilities *adm. Pbs* $\in [0; 1]$ and *adm. Pbr* $\in (0; 1]$.

7.4 Algorithm for Calculating the Values of the NDT Parameters:

1. Input administratively-determined values (7.1.1) of blocking probabilities *adm.Pbs* $\in [0; 1]$ and *adm. Pbr* $\in (0; 1]$, and M.
2. Input values for the known parameters (7.1.2).
3. Calculate the value of *Yab* based on the input parameters and equation (7.2.4).
4. Calculate the values of the unknown parameters (7.1.3) based on the resulting value of *Yab*, input data (7.1.1) - (7.1.2), and equations (7.2.6) – (7.2.10), (7.2.1), and (6.2.2.1).
5. Solve equation (7.3.1) numerically, according the number of switching lines Ns and based on the input parameters and the calculated values of the unknown parameters (7.1.3).

The algorithm can be used when *Pbr* $\in (0; 1]$. If *Pbr* = 0, the network loading is rather low and is an area for future research.

Therefore, we have shown that if *Pbr* $\neq$ 0 and *Pbs* $\neq$ (*S1* – *S2 Pbr*) / (*S3* – *S2 Pbr*), then the NDT can be solved using the proposed algorithm.

8. MODEL VALIDITY

The mathematical models presented herein are verified through mathematical analysis and confirmation tests of values of the main output variables. The models are validated using a detailed simulation model, a development of (Todorov, Poryazov 1985), and a qualitative behaviour comparison with measurements in real PSTN. The simulation model uses Poisson input flow of calls, and all holding times of the base virtual devices are random variables with usual distributions. From the theorem of Sevastianov (1957), we know that Erlang's loss formula is insensitive to the holding time distribution, and therefore our mathematical models are practically insensitive to holding time distributions in every base virtual device because we use Little's formula (which is insensitive to the holding time distribution). As a result, and despite inaccuracies in the assumptions (e.g., A-14) and mathematics (e.g., using Erlang's formula for a limited number of sources), we believe that the models are valid as a first approximation for telecommunication systems with homogeneous terminals and (virtual) circuit switching.

9. CONCLUSIONS

1. We present a detailed normalized conceptual model of an overall (virtual) circuit switching telecommunication system (e.g., the PSTN and GSM). The model is relatively close to real-life communication systems with homogeneous terminals.

2. We define and explain the terms "system tuple" and "base parameter set". They are used in the proposed classification of parameters, and to define different teletraffic tasks. This approach is useful for teletraffic metadata definition and constitution;

3. We formulate ten (A-1 − A-10) natural mathematical assumptions. A set of equations (all from (4.3.1) to (5.17)) of the Terminal Teletraffic Theory are presented and proven. They consider the main characteristics of A, B and all (AB) simultaneously-occupied terminals. These mathematical expressions use mean values of the parameters. They are exact, and are in force for every (virtual) channel switching system in stationary state under the assumptions made herein; they are also independent of:

    - the distribution of the input flow;

    - the distributions of every holding time;

    - the existence, or absence, of repeated attempts.

4. We formulate three additional natural assumptions (A-11 − A-13), and define and prove a mathematical expression for the probability of finding the B-terminal busy. This expression is exact and is independent of the distribution of the input flow and the distributions of every holding time, if there are no repeated attempts in the system.

5. We make assumption A-14 for the case of repeated attempts. It considerably simplifies mathematical considerations. A-14 is the only assumption in this paper causing systematic error. This relative error does not exceed 5% of the *Pbr* in a reasonable traffic load interval.

6. Analyses of numerical solutions of the described system of equations (Poryazov 2005a) allow quantitative presentation of the differences between values of the traffic characteristics of A, B and AB-terminals, and prediction of three qualitative phenomena in communication systems:

- The existence of a threshold ($thr.Fa > 0$) of the intensity of the input flow, so that in the interval $Fa \in [0, thr.Fa]$ busy terminals exist but there are no losses due to finding the B-terminal busy (Theorem 6.1.3.2);

- The difference between values of offered traffic in the cases with, and without, blocking (Poryazov 2005b);

- The existence of a local maximum of Pbs (Poryazov 2005b), and of the corresponding offered traffic at extremely high system load.

The last two phenomena must be experimentally confirmed through measurements in real systems, but in any case the ITU-T definition of offered traffic is not valid for these systems and must be reconsidered.

7. Phenomena with time and traffic characteristics (as described) are emerging at the network level. In this paper, we demonstrate a possible approach which extends the development of Network and Terminal Teletraffic Theory and helps determine many network parameters in present- and next-generation fixed and mobile networks.

8. We formulate the network dimensioning task (NDT) on the basis of preassigned values of the QoS parameters Pbr and Pbs;

9. We describe the conditions for existence and uniqueness of a solution of the NDT, and present an analytical solution for the NDT;

10. We propose an algorithm for calculating the values of eight unknown parameters (7.1.3) and the number of switching lines Ns;

11. The NDT results make network dimensioning, based on QoS requirements, simple;

12. The general equations, described in the paper, may reduce the volume of stored measurement data;

13. The equations may be used as a scientific basis for Internet-based Network and Terminal Teletraffic Engineering tools; (a preliminary version is at: http://cose.math.bas.bg/Teletraffic/);

14. The approach is directly applicable for every (virtual) circuit switching telecommunication system (e.g., GSM and PSTN), and may help considerably for traffic modelling in ISDN, BISDN and many other core and access networks. For packet switching systems like the Internet, the proposed approach may be used as a basis for comparison.

Acknowledgments

Many specialists have helped and are supporting research in this field, but without those listed below it would not continue: Network and Terminal Teletraffic Theory research in Bulgaria began in 1981, after P. M. Todorov (Telecommunication Research Institute - Sofia) pointed out its importance and lack of practical results; in 1983 V. B. Iversen (Danish Technical University) and M. A. Schneps-Schneppe (Riga Technical University, Latvia) confirmed this, and provided useful opinions; in 1984 A. D. Kharkevich, and other scientists from the Institute of Information Transmission Problems, Russian Academy of Science, began actively supporting the work; in 1985 B. V. Gnedenko (Moscow State University, Russia) officially endorsed our research direction; since 1997 N. Ince (Istanbul Technical University Foundation) has been encouraging efforts through EU Actions COST 256 and COST 285. This work was supported by the EU Action COST-285 "Modeling and Simulation Tools for Research in Emerging Multi-service Telecommunications".

REFERENCES

[1]   Burk P. J., 1956. The output of a queueing system. J. Op. Res. Soc. Amer. 4, 699-704, 1956.

[2]   Engset, T. O., 1918. The Probability Calculation to Determine the Number of Switches in Automatic Telephone Exchanges. English translation by Mr. Eliot Jensen, Telektronikk, juni 1991, pp 1-5, ISSN 0085-7130. (Thore Olaus Engset (1865-1943). "Die Wahrscheinlichkeitsrechnung zur Bestimmung der Wählerzahl in automatischen Fernsprechämtern", Elektrotechnische zeitschrift, 1918, Heft 31.)

[3]   ITU E.501. ITU-T Recommendation E.501: Estimation of traffic offered in the network. (Previously CCITT - Recommendation, revised 26. May 1997)

[4]   ITU E.600, ITU-T Recommendation E.600: Terms and Definitions of Traffic Engineering (Melbourne, 1988; revised at Helsinki, 1993).

[5]   ITU E.800. ITU-T Recommendation E.800: Terms and Definitions related to Quality of Service and Network Performance, including Dependability. (Helsinki, March 1-12, 1993, revised August 12, 1994).

[6]   Iversen V. B., 2003. Teletraffic Engineering Handbook. ITU-D SG 2/16 & ITC. Draft, December 2003, pp. 328. http://www.tele.dtu.dk/teletraffic/ (11.12.2004).

[7]   Iversen V. B., 2004. Teletraffic Engineering and Network Planning, Technical University of Denmark, pp.125

[8]   Johannesen Fr., 1908. "Busy" the Frequency of Reporting "busy" and the Cost Caused thereby. The Copenhagen Telephone Company, Copenhagen, October 1908, 4 pp.

[9]   Leduc S. A., 2002. Cracking the GRE Math – The Princeton Review

[10]   Little J. D. C., 1961. A Proof of the Queueing Formula L=λW. Operations Research, 9, 1961, 383-387.

[11]   Poryazov S. A, Saranova E. T., 2002. On the Minimal Traffic Measurements for Determining the Number of Used Terminals in Telecommunication Systems with Channel Switching. In: "Modeling And Simulation Environment for Satellite and Terrestrial Communication Networks - Proceedings of the European COST Telecommunications Symposium", edited by A. Nejat Ince. Kluwer Academic Publishers, 2002, pp. 135-144;

[12] Poryazov S. A., 1991. Determination of the Probability of Finding B-Subscriber Busy in Telephone Systems with Decentralized Control. Comptes Rendus de l'Academie Bulgare des Sciences – Sofia, 1991, Tome 44, No.3, pp. 37-39.

[13] Poryazov S. A., 1992. Determining of the Basic Traffic and Time Characteristics of Telephone Systems with Nonhomogeneous Subscribers. 4th International Seminar on Teletraffic Theory and Computer Modelling – Moscow October 20-25, 1992 – pp. 130-138.

[14] Poryazov S. A., 2001. On the Two Basic Structures of the Teletraffic Models. Conference "Telecom'2001" - Varna, Bulgaria, 10-12 October 2001 – pp. 435-450).

[15] Poryazov S. A., 2004. The B-Terminal Busy Probability Prediction. IJ Information Theories & Applications, Vol.11/2004, Number 4, pp. 409-415;

[16] Poryazov S. A., 2005a. Can Terminal Teletraffic Theory Be Liberated from the Main Illusions? In: Proceedings of the International Workshop "Distributed Computer and Communication Networks", Sofia, Bulgaria 24-25 April, 2005, Editors: V. Vishnevski and Hr. Daskalova, Technosphera publisher, Moscow, Russia, 2005, ISBN 5-94836-048-2, pp. 126-138.; COST-285 TD/285/05/04; COST 290 TD(05)009.

[17] Poryazov S. A., 2005b. What is Offered Traffic in a Real Telecommunication Network? COST 285 TD/285/05/05; 19th International Teletraffic Congress, Beijing, China, August 29-September 2, 2005, accepted paper No 32-104A.

[18] Poryazov, S. A., Bararova, M., 1999. A Model of Users' Traffic in a Telecommunication System with Blocking. International Conference "TELECOM'99" - Varna, October 26-28, 1999 – Vol. 2 – pp. 47 – 53; COST 256 Temporary document TD/256/99/26.

[19] Sevastianov B. A., 1957. An Ergodic Theorem for Marcovian Processes and it Application to Telephone Lines with Blocking. J. "Teoriya veroyatnostei i ee primeneniya", Vol. 2:1, 1957. (in Russian).

[20] Todorov P. M., Poryazov S. A., 1985. Basic Dependences Characterizing a Model of an Idealised (Standard) Subscriber Automatic Telephone Exchange. Proc. of the 11-th ITC, September 4-11, 1985 Kyoto, Japan, pp. 4.3B-3-1- 4.3B-3-8.

[21] Vere-Jones D., 1968. Some applications of probability generating functionals to the study of input-output streams. J. Roy. Statist. Soc., 1968, B30, No 2, 321-333.

[22] Ionin 1978. Ионин Г. Л. Исследование необратимых состояний системы массового обслуживания с повторными вызовами и конечным числом источников. В кн.: Информационные сети и их анализ. Москва, «Наука», 1978., с. 13-31. (in Russian). Ionin G L, 1978. "A study of non-reversible states in a system for mass service with secondary (repeating) calls and a finite number of sources". Book chapter from "Information networks and their analysis", Moscow, "Science", 1978 pp 13-31.

SUBJECT INDEX